D1321438

'*Burning Table Mountain* contributes a fresh and innovative exploration of more than a century of science, history and environmental policy around the emotional topic of fires at the wildland–urban interface. Simon Pooley has produced a book that is not only fascinating, well written and excellently researched, but one which will be regarded as definitive for many years to come. Covering the entire span of settler occupation of the Cape Peninsula but focusing on the twentieth century, by deftly weaving a multiplicity of sources and by contextualising fire policy, management and philosophy in the western Cape and beyond, his careful research fills in an immense gap in our understanding of the growth of knowledge about fire but also of South Africa's human history. Through the prism of Table Mountain, Pooley engages with more than a century of controversy among scientists, policy makers and the public about the role of fire in South Africa, particularly in the fynbos of the renowned Cape Floral Kingdom. Pooley's book documents the growth of environmental and ecological knowledge, and his work contributes to our scientific and historical understanding of this critically important driver of South Africa's vegetation.'

– Emeritus Professor Jane Carruthers, University of South Africa

'In this meticulously researched and lucidly written book, Simon Pooley exposes the reader to the myriad of contradictions and conflicts that arise when northern Europeans colonise a fire-prone ecosystem. Throw into the mix introduced invasive trees, a biota of exceptional diversity and a populace riddled with inequality, arrogance, ingenuity and passion – and what you have is the setting for a riveting tale. Pooley's cogent message is that the fire regime concept – currently couched in biophysical parlance – must embrace the messy world of humans, their values and their institutions. How else can fire be comprehended in the densely populated regions where it holds sway and will continue to do so, despite a plethora of well-intentioned attempts to suppress it? This book is a must-read for social and natural scientists that grapple with the human–wildland interface in the vast areas of the world's fire-prone ecosystems.'

– Professor Richard Cowling, Nelson Mandela Metropolitan University, South Africa

'South Africa is a fire-prone land. Simon Pooley has written an innovative and fascinating study of fire in the Cape Peninsula over the long term. This is the first sustained historical treatment of the subject, and it goes some way to answering a key question: have the fires that sear Cape Town become more dangerous to natural fynbos, people and property? In the process, the book tells us a great deal about environmental history, Cape landscapes, forestry, indigenous and exotic plants, South African science and growing ecological understanding. The peninsula is both beautiful and blessed in its biodiversity. But it is crowded by a rapidly expanding city and exotic plants. This is a rich and multi-faceted discussion of the processes, human and natural, that produce fire as well as the means by which it may be controlled.'

– Professor William Beinart, University of Oxford

'Ecologists have known for decades that fire is essential for the healthy functioning of fynbos ecosystems. However, fire is a complex and destructive force as well and therefore needs to be carefully managed. Simon Pooley's *Burning Table Mountain* is the most comprehensive account to date of the development of fire policy in the Cape Floristic Region and provides a detailed description of how fire practices have changed over the last 300 years. Using the Cape Peninsula as the focal area, for which most of the historical data are available, this book synthesises our knowledge of early burning practices undertaken by Khoisan and settler communities at the Cape. It then traces the development of fire research management in the wider Cape Floristic Region and how it is currently practised. The influence of several important characters and outcomes of key research initiatives are discussed in wonderful detail and woven into a storyline that provides a rich and interesting account of how science and management are inextricably linked. It concludes with a focus on the major fires that occurred at the end of the twentieth century on Table Mountain and warns of the potential impact of climate change on fire regimes in the Cape. *Burning Table Mountain* is essential reading for anyone interested in the complex wildland–urban interface where inappropriate fire management policies and practices have important repercussions for both people and the environment.'

– Professor M. Timm Hoffman, University of Cape Town

'Simon Pooley's *Burning Table Mountain* is a deeply researched but crisply written study of a flammable landscape and how people have understood, managed and perhaps mismanaged it. More than a local South African story, it shows how land-managers wrestled with issues that confront Mediterranean and semi-arid lands everywhere. An exciting addition to environmental history, history of science, South African history and the history of fire.'

– Professor J.R. McNeill, author of *Something New under the Sun* and *Mosquito Empires*

Palgrave Studies in World Environmental History

Editors:
Dr Vinita Damodaran, University of Sussex, UK
Prof Rohan D'Souza, Shiv Nadar University, India
Dr Sujit Sivasundaram, University of Cambridge, UK
Dr James Beattie, University of Waikato, Hamilton, New Zealand

The widespread perception of a global environmental crisis has stimulated the burgeoning interest in environmental studies. This has encouraged a wide range of scholars, including historians, to place the environment at the heart of their analytical and conceptual explorations. As a result, the understanding of the history of human interactions with all parts of the cultivated and non-cultivated surface of the earth and with living organisms and other physical phenomena is increasingly seen as an essential aspect both of historical scholarship and in adjacent fields, such as the history of science, anthropology, geography and sociology. Environmental history can be of considerable assistance in efforts to comprehend the traumatic environmental difficulties facing us today, while making us reconsider the bounds of possibility open to humans over time and space in their interaction with different environments.

This series explores these interactions in studies that together touch on all parts of the globe and all manner of environments including the built environment. Books in the series will come from a wide range of fields of scholarship, from the sciences, social sciences and humanities. The series particularly encourages interdisciplinary projects that emphasize historical engagement with science and other fields of study.

Titles include:

Simon Pooley
BURNING TABLE MOUNTAIN
An Environmental History of Fire on the Cape Peninsula

Forthcoming titles:

Vinita Damodaran, Anna Winterbottom and Alan Lester (*editors*)
THE ENGLISH EAST INDIA COMPANY AND THE NATURAL WORLD, 1600–1850

Richard Grove and George Adamson
EL NIÑO IN WORLD HISTORY, 3000 BCE–2000 CE

Palgrave Studies in World Environmental History
Series Standing Order ISBN 978-1-137-41537-0 (Hardback) 978-1-137-41538-7 (Paperback)
(*outside North America only*)

You can receive future titles in this series as they are published by placing a standing order. Please contact your bookseller or, in case of difficulty, write to us at the address below with your name and address, the title of the series and the ISBN quoted above.

Customer Services Department, Macmillan Distribution Ltd, Houndmills, Basingstoke, Hampshire RG21 6XS, England

Burning Table Mountain

An Environmental History of Fire on the Cape Peninsula

Simon Pooley

Junior Research Fellow, Imperial College London, UK

First published 2014 by
PALGRAVE MACMILLAN

Palgrave Macmillan in the UK is an imprint of Macmillan Publishers Limited, registered in England, company number 785998, of Houndmills, Basingstoke, Hampshire RG21 6XS.

Palgrave Macmillan in the US is a division of St Martin's Press LLC, 175 Fifth Avenue, New York, NY 10010.

Palgrave Macmillan is the global academic imprint of the above companies and has companies and representatives throughout the world.

Palgrave® and Macmillan® are registered trademarks in the United States, the United Kingdom, Europe and other countries.

ISBN 978–1–137–41543–1

This book is printed on paper suitable for recycling and made from fully managed and sustained forest sources. Logging, pulping and manufacturing processes are expected to conform to the environmental regulations of the country of origin.

A catalogue record for this book is available from the British Library.

Library of Congress Cataloging-in-Publication Data
Pooley, Simon Preston, 1970–
    Burning Table Mountain : an environmental history of fire on the Cape Peninsula / Simon Pooley.
        pages   cm
    Summary: "Cape Town's iconic Table Mountain and the surrounding peninsula has been a crucible for attempts to integrate the social and ecological dimensions of wild fire. This environmental history of humans and wildfire outlines these interactions from the practices of Khoikhoi herders to the conflagrations of January 2000. The region's unique, famously diverse fynbos vegetation has been transformed since European colonial settlement, through urbanisation and biological modifications, both intentional (forestry) and unintentional (biological invasions). In all the diverse visions people have formed for Table Mountain, aesthetic and utilitarian, fire has been regarded as a central problem. This book shows how scientific understandings of fire in fynbos developed slowly in the face of strong prejudices. Human impacts were intensified in the twentieth century, which provides the temporal focus for the book. The disjunctures between popular perception, expert knowledge, policy and management are explored, and the book supplements existing short-term scientific data with proxies on fire incidence trends recovered from historical records" — Provided by publisher.
    ISBN 978–1–137–41543–1 (hardback)
    1. Wildfires—South Africa—Cape Peninsula—History.   2. Cape Peninsula (South Africa)—Environmental conditions.   3. Cape Peninsula (South Africa)—History.   4. Table Mountain (Western Cape, South Africa)—Environmental conditions.   5. Table Mountain (Western Cape, South Africa)—History.   6. Wildfires—Social aspects—South Africa—Cape Peninsula—History.   7. Human ecology—South Africa—Cape Peninsula—History.   8. Cape Peninsula (South Africa)—Social conditions.   I. Title.
    SD421.34.S6P66 2014
    363.37096873—dc23   2014022927

Contents

List of Illustrations vii

Preface ix

Acknowledgements x

Introduction 1

Part I   Fire at the Cape from Prehistory to 1900

1   Fire at the Cape: From Prehistory to 1795 15

2   Fire at the Cape: British Colonial Rule, 1795–1900 29

Part II   Fynbos, Fire Research and Management, c.1900–99

3   Science, Management and Fire in Fynbos: 1900–45 47

4   Science, Management and Fire in Fynbos: 1945–99 80

Part III   Fire on the Cape Peninsula, 1900–2000

5   Fire Geography and Urbanisation on the Cape Peninsula 117

6   Conserving Table Mountain 135

7   Afforestation, Plant Invasions and Fire 162

8   Socio-Economic Causes of Fire: Population, Utilisation and Recreation 184

9   Fire on the Cape Peninsula, 1900–2000 197

Conclusion 230

Appendix 1: Cape Peninsula Vegetation, Climate, Weather and Fire 236

Appendix 2: Fire Causes   239

Notes   240

Selected References   276

Index of People   291

Index of Places   295

Index of Plants and Animals Species, Genera   298

General Index   301

Illustrations

Maps

1.1 Western Cape, South Africa   17
5.1 Indigenous vegetation of the Cape Peninsula, past and
     present   122
6.1 The Northern Cape Peninsula, c.1908   140
6.2 The Peninsula showing Table Mountain National Park
     (shaded)   160

Photos

3.1 Rudolf Marloth   49
3.2 John William Bews   50
3.3 Forestry Department Conference, 1931   62
3.4 Jonkershoek Forest Reserve, July 1970   72
4.1 Foresters at Jonkershoek, 1971   92
4.2 Brian van Wilgen monitoring vegetation on a firebreak,
     Jonkershoek, c.1980   102
6.1 Jarman Memorial on Devil's Peak, with felled tree, and
     King's Blockhouse behind   147
6.2 Rhodes' Memorial, with stone pines   148
6.3 The iconic 'Bokkie' anti-wildfire awareness poster   155
7.1 A hybrid landscape at Constantia   183
9.1 City Council plantation and firebreaks on Lion's Head,
     photographed from east of the cableway on
     Table Mountain, c.1936   203
9.2 The Christmas Day 1935 fire on Devil's Peak   206
9.3 The December 1986 fire above Cape Town
     city bowl   222
9.4 Firefighters struggle to save the *kramat* on Lion's Head,
     23 March 1995   225

Figures

4.1 Total number of fires on state forestry lands per annum,
     1915/16–1989/90   85
4.2 Direct costs of fire protection, 1955/56–1987/88   112

7.1   Trends in state afforested area, Cape Peninsula, 1905–83   166

8.1   Growth in population (Cape Town Municipality) and
      incidence of fires in vegetation attended by the
      municipal fire brigade, 1914–1992/93   186

9.1   Comparison of distributions of monthly rainfall
      (average for 1899–1999, Cape Town Fire Station) and
      monthly vegetation fire incidence (fires attended by the
      fire brigade 1931–70), for the Cape Peninsula   198

9.2   Comparison of calls to fires, total actual fires in all
      locations (excluding bush fires), and bush fires, from
      Fire Brigade records, 1901–53   202

9.3   No. of fires in vegetation the Cape Town Fire Brigade
      attended, 1901–39   204

9.4   Fires in vegetation attended by the Cape Town Fire
      Brigade, 1940–60   212

9.5   Fires in vegetation attended by the Cape Town Fire
      Brigade, 1960–92/93   217

A1.1 Annual rainfall (mm) for three Cape Peninsula stations
      and the number of bush fires 1939–62   237

A1.2 Annual rainfall (mm) for Cape Town, 1899–1999, with
      band representing one std deviation above and below
      the 100-year average (590mm)   237

Preface

I have two generations of family members who have been engaged with the effects of fire on the fynbos of the Cape Peninsula. In the December 1954 edition of *Veldtrust*, my grandfather John Bond wrote in support of a Table Mountain Trust, warning that 'Australian wattles are creeping up from the sandier slopes below, and obliterating scores of species of rare and beautiful flowers.... Finally, larger funds are needed for preventing the 100-odd fires that ravage the mountain every year' (Vol.25: 12, pp.21, 27, 31). My uncle William Bond, until recently Harry Bolus Professor of Botany at the University of Cape Town, is an expert on the effects of fire on fynbos who has done much to help us understand that fynbos needs fire within certain ecological parameters, and what those parameters might be. My contribution is that of an environmental historian, and I am interested in the ways in which thinking about fire on Table Mountain has (and hasn't) changed – for instance from my grandfather's to my uncle's time – sometimes underpinned by almost identical values about the need to preserve the indigenous flora and fauna and the Peninsula's majestic landscapes.

The histories of scientific research and institutions are a central part of this history, and I have attempted to spare the reader as much technical detail as possible without losing the essential lineaments of the story as I see it. To encompass the social and ecological dimensions of this history, I have inevitably ranged more widely than my expertise extends, and no doubt this will reveal some of the shortfalls – but hopefully also some of the strengths – of taking an interdisciplinary approach. I have supplemented my gaps as best I can with wide reading and extensive consultations, but all errors are my own responsibility.

Every source is documented in the endnotes to the chapters, and the Selected References includes all the books, journal papers, grey literature and archival sources referred to. Individual newspaper stories, maps and most websites are only listed in the endnotes. Please refer to the sections on state forestry reports and municipal sources for an overview and explanation of these sources and some detail on institutional structures over time. For place names on the Cape Peninsula, refer to Maps 6.1 and, particularly, 6.2. To visualise the extent of the study area tackled in Part III, see Map 5.1.

Acknowledgements

Without the support of my family – emotional, intellectual and material – this research and this book would not have been possible. Above all I am grateful for the patience, advice and support of Susan Pooley, and Alexander and Lara. John and Rosemary Lavers provided support, in every way, throughout. My parents Tony and Elsa instilled in me my fascination with the natural world, and people, shared with my brothers Justin and Thomas. William Bond inspired me with his deep knowledge of fynbos and fire ecology, and he and Winifred and family have offered unfailing hospitality on my field visits to Cape Town.

William Beinart shared his deep knowledge and field experience of South African environmental history. Saul Dubow and Vinita Damodaran have been very supportive of my research. I appreciate conversations with Karen Brown, Richard McKay, Tamson Pietsch, Brett Bennett and Thomas Pooley about research, writing and field work. My thanks to Fred Kruger, Brett Bennett and Lance van Sittert for sharing data and sources.

Many fire researchers and managers have been generous with their time and knowledge, particularly William Bond, Winston and Lynn Trollope, Brian van Wilgen and Fred Kruger. I greatly benefited from interviews with Val Charlton, Richard Cowling, Neels de Ronde, Zane Erasmus, Greg Forsythe, Timm Hoffman, Brian Huntley, David Le Maitre, Guy Preston, Philip Prins, David Richardson, John Rourke and Armin Seydack.

My thanks to John Yeld for comments on several chapters. Thanks also to Peter Slingsby (Slingsby Maps); Patricia Holmes, Amalia Pugnalin and Nasiphi Sityebi (Environmental Resource Management Department, City of Cape Town); Amanda Willet (Cape Town's Fire and Rescue Service); Deborah Jean Winterton and Zishan Ebrahim (SANParks); Greg Forsythe (CSIR); Kayleigh Roos (UCT Libraries); and Martine Barker (*Independent Newspapers*). The Kew Gardens Library and Archive (particularly Anne Marshall and Kiri Ross-Jones); the Rhodes House and (now defunct) Plant Sciences libraries in Oxford; and the University of Cape Town's Government Documents and African Studies libraries were particularly helpful. Thanks to Les Powrie (SANBI), and my friend Carolyn Schnell for help with Map 1.1.

I am grateful to the Trustees of the Royal Botanic Gardens, Kew, and the Bodleian Library, Oxford, for permission to quote from sources in their collections, and to A.A. Balkema Publishers and the University of KwaZulu-Natal Press for allowing me to reproduce photographs from their titles.

My thanks to the Council for Scientific and Industrial Research (CSIR), South African National Biodiversity Institute (SANBI), City of Cape Town, University of Cape Town libraries, *Independent Newspapers* and SANParks for allowing me to reproduce images from their collections. I am grateful to the Arts and Humanities Research Council of the UK (AHRC), the Beit Fund and St Antony's College, Oxford, for funding the early phases of my research on this project.

Finally, thanks to Holly Tyler and Jenny McCall at Palgrave Macmillan and Arvinth Ranganathan and the team at Integra.

Introduction

The spark

In January 2000, two wildfires swept through the Cape Peninsula's UNESCO World Heritage Site for Nature, the Table Mountain National Park, burning down houses and destroying property on the wildland–urban interface (WUI) of South Africa's parliamentary capital. There were more than 120 fires in the region on that one 'fire-storm Sunday'. These fires made a big impact regionally, nationally and (briefly) internationally, assisted by media images of flames and smoke racing over the Peninsula's iconic Table Mountain chain and threatening homes on the slopes below. A book was published on 'The Great Fire of January 2000', with all proceeds going to the 'Santam/Cape Argus Ukuvuka: Operation Firestop Campaign', an initiative supported by local communities, conservation organisations, non-governmental organisations (NGOs), local authorities and the private sector. The Minister of Water Affairs and Forestry, African National Congress (ANC) struggle hero Ronnie Kasrils, commissioned an inquiry which was subtitled 'towards improved veld fire management in South Africa'. Following a decade of political turmoil and environmental management neglect, the fires were seen as an environmental wake-up call not just for the Peninsula, but for the country as a whole.[1]

Walking the sites of the 2000 fires, high on the Steenberg Plateau amid the charred sugarbushes, I tried to make sense of the outbreak of overheated prose burning up so many column inches in the aftermath of the fires. How did they know that the fires of the 1990s were unusually frequent or extensive when they produced no evidence from earlier decades? If the fires were getting worse, why was this happening, and why hadn't anyone done something about it? Was it a natural phenomenon, or a human one, or both: could you even separate the

1

two? Why did everyone seem so hung up on the idea that the extent and intensity of the fires were caused by the invasion of the indigenous flora of the Peninsula by exotic or 'alien' species? Why did so many observers assume the fires had 'destroyed' the fynbos, when I had thought it was well known that fynbos needs fire to regenerate?

The dissonance I experienced over the following months with the otherwise admirable outpouring of expert and civil concern over the state of the Peninsula's natural environment sprang from two sources. First, I had walked the Peninsula's mountains and the Smitswinkel Flats near Cape Point with my uncle William Bond, a botanist who had worked on fire in fynbos for a quarter of a century. I had seen the bush regenerating after fire in the nature reserves I grew up in, and marvelled anew at the seedlings and sproutings of the fynbos after fires. I also knew William felt he was waging a battle within academic ecology to have fire recognised as a major ecological driver of vegetation distribution and diversity. I wanted to unravel the reasons for this apparent scientific blind spot, as well as the glaring disconnect between local expert thinking and public opinion about the effects of fire on fynbos.

Second, my academic mentor at the time was the poet and critic Stephen Watson, the Peninsula's foremost poet of place and landscape. Stephen hymned the now apparently detestable pine trees that we were being told 'infested' the slopes of Table Mountain. In his poems, the stone pines are as integral to the physical identity of Cape Town as the stone road bisecting the Bo-Kaap. Cleansing rains fall even-handedly on switchyards, weeds, gravel and neglected parks, on cloud-fed fynbos and stone pines. Amidst the enthusiastic clamour of heightened nationalist rhetoric in our young democracy, I was taken aback to observe an ecological intolerance tainting reasoned arguments about the ill effects of invasive species. The idea that the Peninsula could be cleansed of its extra-African past was disturbing, and seemed ecologically naive. After all, the Cape's 'natural' landscapes had been extensively transformed over a period of nearly 350 years, and many of its introduced plants had been endowed with deep cultural significance. Much of the flora shown on the vegetation maps had long been buried beneath urban sprawl and infrastructure, parks, gardens, orchards and plantations. I knew what kind of an uproar would greet any suggestion that we should rip out the grapevines or the camphors of Constantia or the shady pines of Cecilia. Could we, or should we, really try to create airtight 'natural' spaces on the Peninsula? What about those precious fynbos species that cling on in parks, commons, along canalised rivers and racecourses? I knew that in order to answer all my questions I would have to descend into the tangle below.[2]

I didn't yet know that several years later, wrestling with this Gordian knot of ecological, social and cultural issues would lead me to environmental history. I was fortunate to have my doctorate on the history of wildfire in South Africa in the twentieth century supervised by the Africanist and environmental historian William Beinart in Oxford. I interviewed many of the key individuals who have influenced fire research and management in the country and began to appreciate the many linkages (and crossed wires) between different land management traditions developed in the country's inflammable (easily set on fire) vegetation types. This is the journey which led me to writing this history.

The clay

This book is a history of the tangled relations of humans and fire on the Cape Peninsula from prehistory until the fires of January 2000. The focus is on the twentieth century, the period in which humans came to transform the Peninsula's fire regimes. Taking fire as my theme both illuminates new dimensions of the history of the region, so often dominated by its political history, and leaves other dimensions in darkness. For instance, as the son of conservationists I had always taken a dim view of foresters, planting up the shores of my childhood haunts on the Eastern Shores of Lake St Lucia in KwaZulu-Natal, South Africa, with pine trees. My historical research revealed to me the pioneering role that state foresters had played in researching and conserving many of the country's indigenous forests and catchment areas. I was struck by the foresters' sense of duty and service, of moral obligation to protect the country's natural resources understood in terms of ecosystem services (to use a more recent term), but also as a spiritual and social resource. This was something they articulated as a patriotic duty, but also as intrinsic to their identity as foresters, influenced by a wider sense of mission shared with members of the British Empire (later Commonwealth) forestry network.

As was the case in Australia, India and in other British colonial territories, the implementation of these admirable ideals could have disastrous impacts on local peoples. What I don't investigate is the extent to which the foresters' approach to management excluded rural Africans and also poor white and coloured (mixed-race) woodcutters from the country's forests and catchments. Many who had subsisted on the resources offered by these areas found themselves criminalised for gathering wood, honey, hunting game, burning grass to create fresh grazing and other activities deemed harmful to the ecological health and integrity of these now nationalised resources. Flower pickers and wood collectors on Table Mountain, and hunters and honey collectors in the forests of the Eastern Cape and KwaZulu-Natal, were prosecuted for harming the indigenous flora through lighting fires. The forests and mountain pastures were being protected for 'the people': just not these particular local people. This was, and remains, a central challenge for conservation: how to protect natural resources from destructive subsistence practises and commercial development, while recognising human rights and encouraging and incentivising sustainable use.

Most of this book focuses on revealing how European settlers and their descendants' encounters with the natural environment and indigenous peoples at the Cape shaped their thinking on fire and how to manage it – rather than the injustices of the policies of governance they evolved as a consequence, or on the relationships between experts and those living off the land. There are some excellent studies of precisely these latter relationships and their consequences in South Africa, for instance Jacob Tropp's work on forestry in the former Transkei, or William Beinart's work on rural rebellions and agricultural and soil conservation expertise. An area I do briefly engage with is past and present subsistence utilisation of the natural resources of the Cape Peninsula. My silences on these kinds of issues reflect the limitations of one already complex project, rather than a failure to appreciate their cultural and environmental significance.[3]

My study region, the southwestern Cape, is untypical of South Africa in that by 1800 the precolonial inhabitants had been conquered and absorbed into settler society. Their culture was smashed and even the deluded fantasies of the architects of Apartheid could not fashion ethnic 'homelands' here to herd non-Europeans into. Opposition to attempts to control the use of fire thus came principally from settler farmers burning off the indigenous grasses or stubble from croplands for agricultural purposes.

While arsonists have been active throughout the past three centuries in the region, fire has (perhaps surprisingly) seldom been used as an offensive weapon. It is possible that Khoikhoi people burned down a few settlers' homes in the first conflicts with Europeans in the late 1600s, and there is at least one confirmed case of rebellious slaves deliberately igniting a major fire in Cape Town. The only sustained offensive use of fire in the region was perpetrated by the British during the South African (or Anglo-Boer) War of 1899–1902. Lord Roberts resorted to farm burning to try to stamp out resistance from Boer guerrillas, but this was largely implemented in the Boer republics of the Orange Free State and the Transvaal.[4]

In the postcolonial period, which in South Africa included much of the Apartheid period (1948–94), there has been a tendency among historians and social scientists to provide a somewhat two-dimensional portrait of state expertise and employees as instruments of oppression. Africanists took the part of the peasants, celebrating indigenous knowledge and rural development initiatives. In the 1970s and 1980s, many criticised 'colonial scientists' for being parachuted into unfamiliar contexts, ignoring local knowledge and social and ecological circumstances and recommending unsuitable management solutions based on inadequate research before skipping on to new pastures. More recently, James Scott's influential book *Seeing Like a State* (1998) provided a framework for criticising the limitations and failures of the large-scale modernist planning so popular in Africa following World War II. Here experts like state foresters and agricultural and nature conservation planners are characterised as the 'hedgehogs', who know 'only one big thing', trying to impose their simplified models of the world onto complex reality. Rural people and naturalists are the 'foxes', who know 'a great many things'.[5]

There is some justice in these characterisations of expertise. However, since the 1990s, there has been a turn towards taking a more nuanced view of the lives, expectations and motivations of the motley crew who worked for colonial departments and agencies across the European colonial territories and their successors. A body of work on the history of veterinary science in South Africa has been notable in this regard. For one thing, the country developed its own institutions and specialisms within an African context, and both immigrant and local-born researchers and managers had long careers in situ. This enabled them to build up a detailed theoretical and field knowledge of local conditions. Among ecologists, mobility between different ecosystems in the region was often an advantage because it generated different perspectives on problems previously defined within observed local conditions. This book is a contribution to this impetus towards studying the origins and generation of scientific ideas and knowledge in colonial and postcolonial territories, rather than concentrating only on the political and social implications of their implementation. Considering the scale and long-term impacts of the scientific thinking, policies and management practices of experts and managers, and the institutions that employed them, on physical environments and land use practices, it seems unwise to neglect or caricature their histories or their legacies.[6]

As Saul Dubow has shown, the colonial authorities at the Cape were never merely exponents of the views and authority of the colonial metropole.[7] The life and career trajectories of individuals recounted in this book should make it clear that those responsible for influencing official (later scientific) thinking on the effects of fire, and for shaping fire policy and official management practices in the region, came from a wide variety of geographical, intellectual and professional backgrounds. Some were local-born, some were immigrants or visitors from European countries, and others had previously served in the East Indies or in India. Some acknowledged experts on the local flora were largely self-educated, while others were professionally trained in Europe, the United States and later South Africa. They were connected to well-established scientific networks, notably those centred at Kew, Uppsala and Leiden. They variously drew on personal experiences and written expertise – from northern and Mediterranean Europe, the United States, Caribbean and Indian Ocean islands, Australia, India and the East Indies, as well as other African territories – in attempting to understand local ecosystems and formulate management approaches that would work at the Cape. They were trying both to manage the indigenous vegetation of the region (primarily as a natural resource) and to transform the region through the introduction of species from Australia, Europe, Asia and the Americas. Foresters in particular were aware they were managing (and creating) hybrid landscapes, rather than trying to protect 'pure' indigenous ecosystems (a latter day term) or manage northern hemisphere forests.

History of science offers a framework for unravelling the theories, ideas and networks of influence that shaped fire policy and management at the Cape. Fortunately, the history of ecological and botanical science and the applied sciences of agriculture (broadly defined) and forestry have been relatively well served by historians of the region. They have focused, as Roy McLeod recommended, on understanding experts as scientists in the colonies and not only as colonial scientists. This includes seeing them as part of what Wade-Chambers and Gillespie have referred to as the 'polycentric communications network' that comprises modern science, characterised by reciprocal information flows. This book unravels the ways in which organisms, people, expertise and ideas circulated between a series of key individuals, institutions, ecosystems, regions and places.[8]

I have aimed to avoid a narrowly technical approach focused only on experts and research institutions, or which privileges scientific papers and policy documents over actual management practices (and failures to implement research and policy). I have tried to show when and why good research and excellent policies have not been implemented. This is interesting in part because much of the research I survey, at least from the 1930s, was applied science, intended for use by humans and with humans (and not some disembodied 'natural world') in mind. In any place, at any time, human management of fire – and the physical environment in general – is profoundly enmeshed in long-term cultural, socio-economic and political interrelationships, as well as being a product of the co-evolution of humans and particular ecological systems.

It is glib to suggest that if we understand the mistakes of the past we will not repeat them in the future. We should also not presume that all problems are solvable – critiques of scientific experts often assume that they are. As is becoming increasingly obvious, most environmental problems are bound up in complex and dynamic social-ecological systems which are difficult to unravel and understand, and negotiating and implementing solutions can be even more challenging. Untangling 'natural' from anthropogenic fire is difficult, often impossible, and the best scientific recommendations may fail in the face of inadequate resources and social opposition. However, at the very least a historical perspective on current social-ecological challenges can reveal why we are framing these in particular ways. It offers a long-term perspective on the limitations of current priorities and interpretations. Historical analysis recovers alternative ways of understanding and tackling such challenges and some cautionary tales of unexpected outcomes. It provides a more comprehensive account of the many factors that influence our environmental interventions. When it comes to fire and other natural disasters, the boom and bust cycles of controversy and forgetting are certainly worthy of consideration when attempting any long-term environmental planning. Fire is not a 'problem' to be 'solved'; in many ecosystems it is a naturally recurring driver of biodiversity which we need to learn to live with.

The fire

Wildfire is a good subject for a historical study of how humans have understood and tried to manage their environments. It provides a salutary example of how northern-hemisphere scientific thinking has been misapplied to the global south, and more specifically, theories developed in temperate regions have been misapplied to fire-prone Mediterranean climate regions. Deep and long-running influences on ecological thinking developed in the temperate north even today marginalise the role of fire as a central driver of how many ecosystems function. This has been recognised by some environmental historians and social scientists who have focused on the impacts of colonial research and management (including their postcolonial legacies) on local peoples and environments.[9]

Within the natural sciences, researchers working on the Mediterranean climate regions have been at the forefront of attempts to address the 'fire-blindness' of mainstream ecology and biology. In 1996, a quarter of a century after the Mediterranean Ecosystems Conferences (MEDECOS) began comparative studies of vegetation in these regions including the Cape, with fire noted as a key driver of ecosystem function, fynbos ecologists William Bond and Brian van Wilgen could still grumble that although 'in the part of the world where we live, ecology cannot be studied without reference to fire...when we turn to the general ecological literature for theory or models relevant to these systems, we find them disturbingly mute'. Much fire ecology had been developed by foresters and agricultural researchers as applied science, in isolation from mainstream ecology and biology, with unfortunate consequences for both. The authors warned that ignorance of fire ecology among biologists who focused on biodiversity conservation had 'resulted in serious misjudgements in the conservation or management of fire-prone ecosystems' where fire was regarded as 'an agent destroying the "balance of nature" rather than a process that helps preserve much of the world's diversity'.[10]

In their 2012 book *Fire in Mediterranean Ecosystems*, Jon Keeley and co-authors could still lament that 'fire is a global ecosystem process...whose role in shaping the distribution of fauna and flora is widely underappreciated'. They note that climate and geology are still considered to be the key factors shaping the assembly and distribution of ecosystems. The equally, or sometimes more, important influence of fire in some ecosystems has been neglected. Plants have evolved to cope with the interactions of all three of these environmental factors.[11]

As a few historians and more recently fire ecologists have therefore recognised, a key ingredient for understanding global environmental change is the influence of anthropogenic fire on ecological systems. We humans have a monopoly over fire use, and fire has played a central role in our relations with the physical world, from cooking to farming, baking clay to clearing land, to the use of fossil fuels in steam and internal combustion engines and numerous other technological applications. This is an inherently social-ecological research area. As Bowman et al. (2011) observe, histories of the evolving relationships and spatial patterns of human landscape burning are vital for comprehending the historical ranges of variability in burning which determine the survival of numerous plants and animals and ecosystems. This requires a multiple disciplinary approach which integrates biophysical with cultural dimensions of 'wild' fire.[12]

The concept of 'natural fire regimes' raises interesting questions. For some ecologists and managers, these are the fire regimes which existed before humans arrived in a region. (This doesn't work well for Africa, where humans evolved.) However, some approaches, including the United States' Landscape Fire and Resource Management Planning Tools (LANDFIRE), include all prehistoric burning when describing 'natural fire regimes'. Thus LANDFIRE includes Native American burning within 'natural' fire regimes. Bowman and co-authors are surely correct to argue that today, the idea 'of "restoring natural fire regimes" without anthropogenic influence is neither possible nor useful'.[13]

From the evidence available, it remains difficult to separate 'natural' from 'anthropogenic' burning – to disaggregate and study human influences on fire regimes. For the Cape Peninsula, as in several other inflammable regions, we lack the quantitative evidence from tree rings and charcoal that would enable us to assemble a long-term picture of burning in the Holocene. Remote sensing information only goes back to the 1970s, and so we are reliant primarily on documentary evidence. In this book I cannot offer a definitive quantitative fire history of the region. Rather, I utilise the available historical records to reconstruct a history of how fire has been understood and managed in the region, and use the available quantitative evidence to record major fire events and give a tentative picture of trends in fire incidence on the Cape Peninsula in the twentieth century.

The kiln

This book is predicated on the fact that the fire history of Table Mountain and the Cape Peninsula is of more than autobiographical or regional interest. It is of intrinsic but also of comparative international interest for three main reasons. First, the exceptionally floristically biodiverse Fynbos Biome is unique: it comprises one of the world's six floristic regions, confined to the southernmost region of South Africa. Second, the Peninsula offers a compact example of the development of what has become known in fire studies as a WUI area, with the development of a modern city with its suburbs, industries and infrastructure in the context of a striking physical landscape and biologically unique fire-adapted flora of considerable scientific, economic and cultural importance. Third, the WUI is a fascinating liminal zone, which often falls between the disciplines of urban or rural environmental history, or is interpreted from the perspective of one or the other side of its permeable borders. Fire management in such areas has been baffled by the conflicting or simply incongruent concerns and priorities of the residents, owners and authorities responsible for their management. Ecologically, they contain areas of transformed vegetation which a small but growing band of ecologists have begun to think of as 'novel ecosystems', trying to avoid the pejorative notions of 'invaded' or 'degraded' formerly 'natural' ecosystems. Steve Pyne is surely correct that the WUI is better understood as a 'scrambling of landscape genres' created by the disintegration of long-established land management traditions and the reclamation of these landscapes by tourists and exurbanites.[14]

The Cape Floristic Region is of great comparative interest because it is one of the Earth's Mediterranean-type climate regions. These regions are spatially the most degraded (93 per cent) on Earth in terms of having fire regimes falling outside of natural rates of variation. (Fire regimes are characteristic regional patterns of the distribution in space and time, frequency, size and intensity of fire events.) This is unsurprising considering the attractiveness of these regions for human development, together with their ecological fire-dependence. In brief, these systems cover the coastal plains and mountains of southwestern and southern South Africa, southern Australia, most of California and small parts of Oregon in the United States and Baja California in Mexico, the middle third of Chile and a large area around the Mediterranean Basin. Although together these regions cover only 2.25 per cent of the earth's land surface, they support 15–20 per cent of all named vascular plant species (land plants with vascular tissues which transport fluids and nutrients internally). They also support 12 of the 100 largest urban agglomerations on earth and attract massive seasonal influxes of tourists over their long summers – peak fire season.[15]

The vessel

This book has three parts. Part I comprises two chapters giving an account of the long-term history of human relations with fire on the Cape Peninsula, from prehistory to c.1795, and from then to 1900. Part II comprises two chapters investigating the history of research, policy and management of fire in fynbos in South Africa, from 1900 to 1945, and from 1945 to 2000. Part III comprises five chapters, each of the first four focusing on a key theme in the history of fire on the Cape Peninsula in the twentieth century.

Chapter 5 explores the biophysical template of the Peninsula and describes how Cape Town's urban development in the twentieth century altered the fire environment. Chapter 6 explores cultural dimensions of residents' relationships with the landscapes of the Peninsula and the local history of nature conservation. Chapter 7 narrates the history of the afforestation of the Peninsula with introduced trees and shrubs and discusses the social and ecological impacts of the ensuing biological invasions. Chapter 8 focuses on the contested influence of population growth and socio-economic impacts on fire incidence. These thematic discussions inform Chapter 9, which provides an integrated history of fire incidence and management on the Cape Peninsula in the twentieth century. The Conclusion reviews the main themes and findings of the book.

Voltaire is said to have remarked that: 'The instruction we find in books is like fire. We fetch it from our neighbours, kindle it at home, communicate it to others, and it becomes the property of all.' My hope is that this book sparks further questions, critiques and investigations of its motivating themes and questions.

Part I

Fire at the Cape from Prehistory to 1900

1
Fire at the Cape: From Prehistory to 1795

Fire ... is very common in the dry season, when the Hottentots usually set fire to the dry herbage and grass everywhere.... [I]t is very difficult to keep the fire from the ripe grain or to extinguish it.

Journal of Jan van Riebeeck, August 1659[1]

The ground is, indeed, by [burning] stripped quite bare; but merely in order that it may shortly afterwards appear in a much more beautiful dress, being, in this case, decked with many kinds of annual grasses, herbs, and superb lilies.

Anders Sparrman, mid-1770s[2]

Night comes in; and the city, the road, and the whole neighbourhood, enjoy a spectacle so much the more magnificent, as they are not ignorant of the cause of it [bush fire] ... for the extent and height of this conflagration give [Table] mountain a more dreadful appearance than the lava does to Vesuvius.

François le Vaillant, Cape Town, early 1780s[3]

Hominids have inhabited what is now South Africa for around three million years, and humans have occupied the southwestern Cape for nearly half of that time. This region falls within the fynbos biome which occupies the mountains and coastal strip of southern South Africa, within the winter- and 'all-year-round' rainfall regions. (A biome is a unit defined by climate, major disturbances like fire and corresponding life-form patterns.) The fynbos biome is actually made up of three main vegetation types: fynbos, renosterveld and strandveld (and – relatively – small patches of forest). Because of its fragmented nature, Takhtajan's idea of a Cape Floristic Region is often used to enable researchers and managers to think coherently about conserving and managing the region.

The plants of the three major vegetation-types of the Cape Floristic Region are adapted to local fire regimes, meaning patterns of burning. The fire historian Steve Pyne compares fires and fire regimes to storms and climate: within a climatic region, different kinds of storms occur in a seasonal and annual pattern recurring over time. Fire regimes are an amalgam of the following aspects of fires in landscapes over time: fire spread patterns, fire intensity, the size and distribution of fire patches, fire frequency and seasonality. Major shifts in fire regimes (e.g. more frequent, or more intense, fires) can kill these plants (they are not adapted to fire per se). Weather events and climatic variations, or changes in human burning practices (including fire exclusion), can change the structure and species composition of plant communities. For instance, wet winters followed by long, hot summers encourages vegetation growth and then creates ideal conditions for fires to spread – lightning or falling rocks providing natural ignition sources. Frequent fires encouraged by climatic cycles and good fire weather, or human veld (bush) burning, favour grasses over some of the larger shrubs in renosterveld and fynbos.[4]

The first human explorers of the region were hunter-gatherers, using fire for warmth, cooking and hunting. They encountered numerous antelope – including eland, mountain zebra, quagga (extinct by 1883), red hartebeest and grey rhebuck – as well as large browsers like rhinoceros. The grazing animals and their predators prospered on the more nutritious renosterveld rather than the nutrient-poor fynbos, but within these broad constraints, veld burning would have encouraged grassy species more palatable to grazers. Based on observations of hunter-gatherers in the wider region dating back to the seventeenth century CE, it seems likely that the region's first human inhabitants burned away old vegetation to attract game to the resulting young herbage. Deacon suggests that fire-stick farming has been practised in the southwestern Cape for at least 100,000 years. Here, hunter-gatherers burned seasonally to find or maintain populations of fire-adapted food plants including carbohydrate-rich geophytes like watsonias and gladioli (which propagate from underground corms or bulbs). Evidence from rock shelters indicates that hunter-gatherers moved seasonally between the coast, which they occupied during the wetter winters, and the mountains with their dependable water sources during spring and summer.

The arrival of Khoikhoi pastoralists around 2000 years ago changed this pattern of life. By 1800 years ago pastoralists and their herds were forcing hunter-gatherers, and probably wild animals, into the more

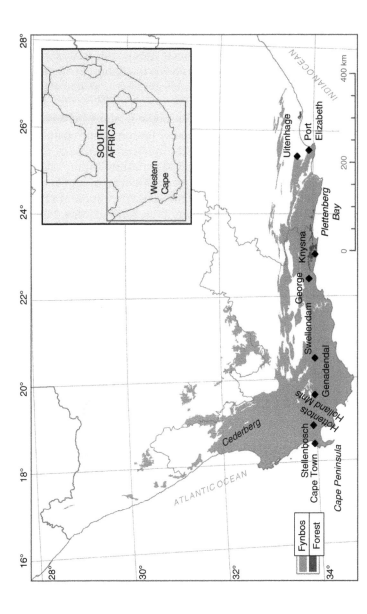

Map 1.1 Western Cape, South Africa, showing fynbos biome and indigenous forests

Source: Base map from M.C. Rutherford, L. Mucina, L.W. Powrie, 2006, 'Biomes and Bioregions of Southern Africa' in L. Mucina, M.C. Rutherford (eds), The vegetation of South Africa, Lesotho and Swaziland, Strelitzia 19 (Pretoria: South African National Biodiversity Institute), pp. 30–51.

mountainous and marginal environments of the fynbos biome. These hunter-gatherers introduced more frequent burning (than caused by lightning or falling rocks) to the mountainous areas. They may have remained there, with their fires, for most or all of the year. On the plains, the Khoikhoi burned the vegetation seasonally to provide fresh grazing for their sheep and cattle. Sheep were present in significant numbers from around 1600 years ago, with cattle appearing in the archaeological record around 1300 years ago.[5]

Smith has reconstructed the Khoikhoi peoples' seasonal movements of livestock from one grazing area to another (transhumance patterns) from historical records dating back to the mid-seventeenth century. The Goringhaiqua (called 'Caepmen' by seventeenth-century Dutch settlers) and Gorachoqua (called the 'Tobacco Thieves' by the Dutch) shared the southern pastures of the Western Cape. They appear to have used the pastures around the Cape Peninsula in early summer, before firing the vegetation and moving east onto the coastal renosterveld towards the Hottentots Holland Mountains, and thence anti-clockwise to return the coast along the course of the Diep River. This seasonal movement allowed the pasture to recover from heavy grazing and pre-vented the trace-element deficiency and related livestock illnesses that would result from year-round use of nutrient-poor vegetation types.[6]

Not much is known about the vegetation before the Khoikhoi arrived, but there is little evidence that the 1500 years of Khoikhoi pastoralism before Europeans arrived degraded the environment of the region. This lack of evidence is partly because the Khoikhoi concentrated their graz-ing and burning on the more nutritious lowland renosterveld, more than 90 per cent of which is now under crops. Signs of land degrada-tion in colonial times were first noted in the late 1720s, and a shift to a shrubbier state in the Renosterveld was commented on by Sparrman in the 1770s, associated with the effects of overgrazing and selective grazing on European settler farms. There were about 50,000 Khoikhoi in the region at the time the Dutch arrived in 1652, owning hundreds of thousands of sheep and cattle. It is possible that their grazing herds and burning of the veld maintained the renosterveld in a grassier rather than a shrubbier state.[7]

VOC (Dutch East India Company) rule, c.1652–1795

European settlement and environmental impacts

In April 1652, a 35-year-old Dutch East India Company (Vereenigde Oost-Indische Compagnie, or VOC) merchant, Jan van Riebeeck, sailed into Table Bay with his young family and a motley assortment of VOC hirelings to set up a refreshment station for the fleets sailing between Amsterdam and Batavia in the Far East. The journey from Europe to Batavia took five-and-a-half to seven months and scurvy was a major problem. Van Riebeeck had been disciplined and dismissed for private trading on a company ship and volunteered to lead the Cape mission in order to redeem his reputation (he was to be stuck at the Cape until 1662).[8]

The Dutch established a small settlement and fort on the small renosterveld-clad coastal plain between Table Bay and the sheer cliffs of Table Mountain. They found clans of Khoikhoi pastoralists (whom Europeans had named Hottentots, which is probably a rendering of a Khoi greeting) making use of the Cape Peninsula for seasonal grazing, and *strandlopers* (beach walkers) who lived off gathering and scavenging along the coastline. On expeditions inland and further north, they came across small bands of hunter-gatherers they called bushmen.[9]

There were tensions over access to grazing on the Peninsula from early 1655. Once the Dutch established their own herd, and in 1657 granted nine soldiers the right to set up as independent farmers ('free burghers') on land the peninsular Khoikhoi were accustome 6 to graze their herds on, conflict was inevitable. VOC officials and the free burghers feared the tens of thousands of livestock the Khoikhoi brought would damage their crops and exhaust the Peninsula's grazing resources. Open conflict broke out in May 1659, but a year later peace had been restored, the Khoikhoi retaining the livestock they had captured, but fatefully acknowledging the VOC's sovereignty over the lands settled by the free burghers.[10]

By 1700 many Western Cape Khoikhoi depended on the colony for their livelihoods, working for free burghers or the company in Cape Town and new settlements like Stellenbosch (settled in 1679) and Drakenstein (1687), or on settler farms. As a result of political fragmentation, settler expansion and a series of smallpox epidemics, by 1800 European settlers dominated all the regions of the southwestern Cape suitable for agriculture or pastoral farming. In terms of their original plan to set up a refreshment station at the Cape, the Dutch failed to establish intensive mixed agriculture. Economic conditions, largely determined by the VOC's low salaries and strict price control, meant that the free burghers at the Cape were driven to reduce their investment of labour and capital to a minimum. They focused on crops that gave a fast return, and many turned to purely livestock farming.[11]

The impact of Dutch settlement on the indigenous vegetation, and fire regimes, was threefold. First, sedentary livestock farming and regular, uniform veld burning resulted in overgrazing and selective grazing and, it has been argued, a transformation of grassy renosterveld to a shrubland state of this vegetation type. Second, the extermination of large indigenous browsers like black rhino and eland also contributed to a shift to a more shrubby vegetation state. In 1655, Van Riebeeck's men killed a rhinoceros stuck in a salt pan near Table Bay. A century and a half later, the leopards which had eaten Van Riebeeck's geese, the lions which had attacked his oxen and the hippos after which Zeekoevlei ('Hippo Lake') on the Peninsula was named, were long gone, along with all the other megafauna. By 1700 there was no game within 200km of Cape Town, hunters having wiped out the animals for food or sport. Third, the conversion, particularly of the more fertile renosterveld areas, to cereal crops and vineyards replaced the indigenous vegetation and disrupted natural fire regimes. Both wine and wheat production were well established by 1700.[12]

Fire policies and management under VOC rule

Dutch officialdom was opposed to veld burning from the outset. In December 1652 their daily journal noted that 'to-day the Saldanhars set fire to the grass, and as the fire came rather close to our pasture grounds we requested them not to come so near with their fire'. During the conflicts of 1658 and 1659 they feared the Khoikhoi might use fire as a weapon to burn down buildings or set ripe crops alight. In 1659, Van Riebeeck planted a strip of sweet potatoes '2 roods wide outside his grain-fields and vineyard' to

> protect the ripe corn from fire, which is very common in the dry season, when the Hottentots usually set fire to the dry herbage and grass everywhere. Without a crop of low growing vegetables or an open belt in between, it is very difficult to keep the fire from the ripe grain or to extinguish it.

In fact there are no records of runaway grass fires started by Khoikhoi seriously damaging the settlement or its fields and pastures, and the first arson fire is recorded in 1715, when Khoikhoi torched the house of one Pieter Rossouw.[13]

The authorities experienced considerably more problems over burning from their employees, the free burghers and imported slaves. The structures of the settlement were mostly thatched and there were numerous fires in the early years, most caused by smokers and by sparks flying out of chimneys.[14] In February 1691 Commander (later Governor) Simon van der Stel instigated a resolution aimed at fire prevention, and in April a fire brigade was established.[15] Between 1691 and the end of Dutch colonial control in 1795, there are 14 subsequent resolutions of council pertaining directly to firefighting and the fire brigade.[16] This chapter focuses on fires in vegetation rather than structures, though as we shall see, in Cape Town fires frequently crossed the so-called wildland–urban interface. Living in a town of mostly reed-thatched houses bordering on a mountain covered in inflammable vegetation was a constant worry for the VOC authorities. This was exacerbated by the lack of control they felt they exercised over the mountains' unregulated open spaces.

On the night of 11 March 1736, a group of fugitive slaves set Cape Town on fire with the (alleged) aim of robbing and murdering in the ensuing confusion. When a strong southeasterly wind sprang up, they started a fire at the southeastern corner of the town. Only desperate firefighting managed to confine it to burning five houses to the ground, including the house of Lieutenant Siegfried Allemann, commander of the colony's military forces. Otto Mentzel (1710–1801), a Prussian who lived at the Cape from 1733 to 1741, described the terrible punishments exacted on the incendiaries. Three of the captured perpetrators managed to cut their own throats before punishment – with a razor tossed into their cell by an accomplice. Five were impaled, four broken on the wheel, four hanged and two women were strangled slowly while a burning bundle of reeds was waved in their faces. The Council also assembled a commando to rid the mountain of vagabonds, 'be it Europeans or blacks', who might have willingly or otherwise started such a fire. The commando were authorised to use hand grenades to drive 'vagabonds' out of holes and cracks in the mountain. This was the beginning of a long tradition of blaming mountain fires on vagrants, though fighting fire with fire was perhaps never again so brutally implemented.[17]

Mentzel described a 'massive fire station' on the corner of the square between the castle and the town, and was of the opinion that the firefighting 'organisation is so perfect that many European cities might profitably take a lesson from the Cape'. All slaves, whether privately or Company-owned, were obliged to assist in firefighting. Slaves carried the hoses, worked the pressure engine and fetched water. Soldiers did not assist with firefighting, retreating rather to the castle where they locked the gates and stood by in readiness to cope with any rioting by slaves.[18] This provides a vivid picture of the tenuous nature of colonial control as perceived by the Dutch authorities.

Mentzel also provides one of the first descriptions of a major bush fire on Table Mountain. Referring to Peter Kolb's book about the Cape (of 1719), he confirms that he too saw Kolbe's 'carbuncle' and 'crowned serpent' on Table Mountain at night. He explains that these are descriptions of bush fires, which he attributes to the spontaneous combustion of dry bushes by solar rays concentrated on Table Mountain through reflection off Devil's Peak and Lion's Head. The burning bushes resemble Kolbe's 'carbuncle', and where a fire spreads upwards from bush to bush, a 'crowned serpent' appears. Mentzel writes:

> Upon one occasion during my stay at the Cape the whole of Table Mountain was illuminated in this fashion; it happened after several days of torrid heat without a breath of wind, and formed a magnificent spectacle at midnight. A singular feature about these bushes is that though they are of soft wood and appear to be little more than an inch in diameter they burn very slowly. In my opinion they either contain a fire-resisting substance like phlogiston or the slowness of the combustion is accounted for by the ignition taking place at the top, the tips of the branches catch fire, which then pushes its way down the stem to the roots. Each bush usually contains some moisture from the evening dews, and this hampers the spread of the flames. Even if the fire spreads to the roots the bushes spring up anew the following year.[19]

In 1742, making fires on the beaches at night (usually from old kelp), or in the nearby veld, was prohibited to prevent ships coming to grief in Table Bay through mistaking these fires for the watch fire on Robben Island. Other resolutions of council condemned torch-carrying drunks and smokers and forbade shooting in the streets at night, particularly on New Years' Night.[20]

Lessons from the Khoikhoi

From the outset the Dutch had noted the Khoikhoi's practice of firing the veld before moving on to fresh grazing grounds. In the 1650s, the sight of smoke rising over the flat plains to the north had usually been a welcome one because it meant Khoikhoi were approaching the Peninsula, heralding an opportunity to trade copper and tobacco for livestock. When some free burghers took up livestock farming, letting their cattle and sheep range widely on the veld, they copied Khoikhoi practices including veld burning. Following the collapse of Khoikhoi society, many Khoikhoi skilled in livestock handling found employment on Dutch settler farms, and it seems likely these men continued their burning practices.

The first *placaats* (resolutions of the Council of Policy of the Cape of Good Hope) forbidding grass burning around the settlement in the dry season, were promulgated in November 1658 and December 1661.[21] Johan Vogel, who visited the Cape in 1679, noted that

> when [the grass] is too old and tough to be any more eaten by the beasts, it is set on fire by the inhabitants.... But in order that the fire may go no further than the inhabitants wish, they dig out a trench, at which the fire decreases and dies out when it reaches it.... The ashes of such burnt grass manure the land where the fire was, and make it so fertile that, when light rains fall, in a short time new or young grass grows up, into which the animals are driven to graze.[22]

Peter Kolb (variously Kolbe, Kolben), a German astronomer and mathematician who lived at the Cape from 1705 to 1713, noted that the 'Hottentots' burned grass that was old and rank to renew the pasture, and was of the opinion that '[i]n this the Hottentots are imitated by the Europeans at the Cape, with this difference only, that the Europeans make ditches round the grass they would burn, to stop the course of the fire, whereas the Hottentots give themselves no such trouble'.[23]

Settler farmers also burned off stubble from old fields to prepare the ground for new crops. An almanac (written c.1708–1711) instructs that the correct times to burn are in the dry summer season, commencing when 'the southeaster... [established] its sole supremacy' in November, with the best month being February as 'the roots of the grass burn, and the great heat dries up the rest'. Any remaining unburned areas should be burned in March.[24]

A series of *placaats* forbidding unauthorised veld burning, including the threat of hanging for a second offence (*placaat* 215 of 19 February 1687), suggests that veld burning was common throughout the latter half of the seventeenth century. The 1687 *placaat* followed an episode of careless veld burning which destroyed 400 muids (or wagonloads) of gathered grain near Paarl. It specified a wide range of lands which could not be burned without permission, including forests, bushes, grazing land, stubble, ploughed and unploughed fields.

In January 1741 Governor Hendrik Swellengrebel passed a detailed resolution forbidding unauthorised veld burning, grumbling that the practice persisted despite his resolution on veld burning of 25 October 1740. The 1741 resolution notes the grave risks posed to company property, buildings, crops and timber forests by reckless burning and forbids all unauthorised burning by burghers, Khoikhoi and slaves alike, of all lands as specified in the 1687 *placaat*. Once burning was authorised, prior warning had to be given to all neighbours beforehand, and burning could only be done in calm weather and in places where there was no danger of destructive accidents. A first offence was punishable by a whipping, and a second by strangulation with a rope 'without mercy' on the gallows unto death, in addition to paying damages as determined by the judge. Masters were held accountable for the actions of their dependants and slaves, and were liable for damages. They were instructed to order their herders to take extreme care when smoking and otherwise using fire while tending their flocks.[25]

As settler farmers moved from cattle to sheep farming and then to cultivating wheat and vineyards on the more fertile areas of the region, the practice arose of grazing livestock on the mountains in summer and grazing the harvested fields or trekking into the Karoo in winter.[26] This may have reduced late summer burning for pasture on the lowlands. On the Cape Peninsula, the first experiments in growing cereal crops did not translate into extensive cultivation, which was developed farther afield on better soils. Fynbos on the lower eastern slopes of Table Mountain, which had previously been vulnerable to mountain fires, was replaced with vineyards and orchards. However, large numbers of livestock were kept on the Peninsula and the Cape Flats. They were used for transport (locally to serve the harbour at Cape Town and later Simon's Town, and for trade with the surrounding region notably in foodstuffs, firewood and timber) and for food. Dairy herds were kept on the Peninsula well into the twentieth century. Where livestock was kept, the vegetation was burned to improve grazing.[27]

Scientific explorers and collectors, c.1772–84

The 1770s and 1780s were an exciting period in the exploration of the natural history of the Cape. Two of the most observant visitors had studied under the great Swedish botanist Carl Linnaeus at Uppsala University, namely Carl Thunberg (1743–1828) and Anders Sparrman (1748–1820). François le Vaillant (1753–1824), a Frenchman born in Surinam, explored the Cape from 1781 to 1784.[28]

Thunberg, often described as 'the father of Cape botany', was based in Cape Town from April 1772 to March 1774 while he learned Dutch to enable him to visit Japan (only the Dutch had access). He collected plants on the Cape Peninsula and undertook three long journeys into the surrounding region. While travelling eastward in the vicinity of Swellendam, he wrote,

> in many places I observed the land to have been set on fire for the purpose of clearing it .... Divers plains here, produce a very high sort of grass, which being of too coarse a nature, and unfit food for cattle, is not consumed, and thus prevents fresh verdure from shooting up; not to mention that it harbours a great number of serpents and beasts of prey. Such a piece of land as this, therefore, is set on fire, to the end that new grass may spring up from the roots. Now if any of these places were overgrown with bushes, these latter were burned quite black, and left standing in this sooty condition for a great length of time afterwards, to my great vexation, as well as that of other travellers, who were obliged to pass through them.[29]

Thunberg did not pass judgement on the practice of veld burning, other than complain of the effects on his wardrobe. He noted the advantages to farmers of burning, including replenishing pastures and pest control.

Sparrman arrived in Cape Town on 13 April 1772, where he spent a few days botanising with Thunberg before taking up a post as a tutor in Simon's Town. Six months later he left to accompany Captain James Cook's eastward circumnavigation of the globe, returning on 21 March 1775. He undertook his major Cape journey from July 1775 to 15 April 1776, covering an estimated 2100km. His two-volume account of his travels is renowned for its good judgement on agricultural and environmental matters. It is also one of the earliest sources of the narrative of environmental degradation, particularly through bad farming practices, that was to become prevalent in the second half of the nineteenth century.

Sparrman relates that farmers complained about the spread of the shrub renosterbush ('rhinoceros bush', *Elytropappus rhinocerotis*, characteristic of the renosterveld and after which the vegetation type is named). One farmer tried to destroy the renosterbushes on his farm through burning, with the result that 'they grew up again more vigorous than ever'. Sparrman compared the impact of a swarm of locusts to the effects of fire, fire being used by colonists and 'Hottentots' for clearing fields of weeds. In November 1775 he wrote:

> The ground is, indeed, by this means, in both cases, stripped quite bare; but merely in order that it may shortly afterwards appear in a much more beautiful dress, being, in this case, decked with many

kinds of annual grasses, herbs, and superb lilies, which had been choked up before by shrubs and perennial plants. These last, moreover, which...were hard, dry, withered, and half dead...and unfit for fodder, have now an opportunity of springing up again, so as to produce with their young shoots and leaves, pastures adorned with a delightful verdure for the use of cattle and game.

Like Thunberg, Sparrman was not critical of burning per se, commending the positive effects both for the natural vegetation and grazing for livestock and wild animals. However, while returning to the Cape in March 1776, he and his companions were nearly caught up in a runaway bushfire ignited by a farmer to rid his lands of desiccated grass and bushes, and he criticised unsafe burning practices. Sparrman noted that veld burning destroyed 'a number of serpents, lizards, scorpions, and several other kinds of insects, together with young birds...in their habitations'.[30]

François le Vaillant arrived in Cape Town in April 1781, and spent three years amassing a natural history collection. His two works describing his South African travels were published in the 1790s and became best-selling travel books in the period.[31] Le Vaillant has left us one of the most vivid descriptions of a bush fire on the Cape Peninsula. He also describes the causes of some of these fires, writing that slaves made fires wherever they worked, for lighting their pipes, and to warm or cook their food. Slaves were sent to collect firewood for their masters, and some sought it on the back slopes of Table Mountain (the wooded eastern slopes above modern-day Rondebosch and Newlands). Having completed their labours in the evenings, they sometimes forgot to extinguish these fires, which spread to the dry grass and roots in the area. Le Vaillant writes:

The flames then spread rapidly on all sides and soon reach deep valleys, where all the wood, both dry and green, without exception, blazes forth with surprising fury; having the appearance of so many small volcanoes, connected together by strings of fire. The flames rise in clouds of different shades, according as the caverns have a greater or less depth; night comes in; and the city, the road, and the whole neighbourhood, enjoy a spectacle so much the more magnificent, as they are not ignorant of the cause of it; and the people are therefore quite free from that great terror which such a phenomenon would otherwise occasion: for the extent and height of this conflagration give the mountain a more dreadful appearance than the lava does to

Vesuvius, when it burst forth with the utmost fury. I never but once saw this majestic illumination; and I can say that it afforded me very great pleasure.

Following Kolb and Mentzel, this early literary description of a bush fire celebrates the natural spectacle and expresses the combination of aesthetic pleasure and dread which characterises descriptions of big bush fires on the Peninsula's mountains well into the first half of the twentieth century.[32]

Conclusion

Humans have been a part of the fire regimes of the Western Cape region for around 1.5 million years. Although burning of the grassy lowlands by Khoikhoi herders probably became more frequent and widespread from c.1600 years ago, as the size of their herds increased, their transhumant lifestyle prevented sustained overutilisation of any area. Ecologists do not believe that precolonial societies degraded the environment of the region, though heavy annual grazing, and burning, at the Cape Peninsula must have affected the structure and composition of the vegetation.[33]

In the first century of Dutch settlement, the major changes to the region's fire regimes were caused by the establishment of sedentary farming, the extermination of large browsers and the transformation of the veld into croplands and vineyards. By the late seventeenth century the subjugation of the Khoikhoi had ended their annual visits to the Peninsula *en masse*. This, together with the development of the settlement and farming – with attendant attempts to regulate fire use – probably reduced burning below the northern and eastern slopes of Table Mountain. However, we know from contemporary descriptions that large mountain fires higher up on the eastern and northern slopes of Table Mountain were a seasonal commonplace throughout the period of Dutch rule. These fires were attributed to the carelessness of slaves and to vagrants.

Lack of labour, tight controls by the VOC and infertile soils meant many of the first European farmers at the Cape turned to livestock farming, learning to burn the veld to replenish grazing from the Khoikhoi. However, they tended to settle near permanent water sources and localised annual burning and grazing had a heavier environmental impact than Khoikhoi transhumance patterns. The first signs of degradation were noted in the late 1720s, and in the 1770s Sparrman noted that burning to get rid of unpalatable renosterbos shrubs was having the opposite effect.

By the 1680s, livestock farming was being displaced by cereal crops and vineyards in some more fertile inland locales. Transhumance patterns were established where livestock were grazed on the mountains in summer, and in winter were allowed to graze the harvested fields or trekked into the Karoo. This may have reduced late summer burning of indigenous vegetation in the lowlands, but increased burning on the mountains.

The Dutch authorities passed a string of Resolutions of Council forbidding grass burning, beginning in 1658 and repeated at intervals into the 1790s. Clearly farmers and others continued to burn. Draconian proscriptions were passed in 1687, and again in 1740 and 1741, suggesting that enforcement was difficult and intermittent. Absolute prohibition was not the objective; rather, attempts were made to regulate burning practices.

The Dutch authorities feared fires because the grazing and crops on their fixed properties were vulnerable, as were their reed-roofed buildings. The first legislation to prevent structural fires in the settlement was issued in 1659, and reeds were replaced with tiles on VOC company buildings from 1660. The first firefighting authority was instituted in 1691. However, in the 1730s most burghers still lived in brick houses with thatched roofs. A large fire in 1736 persuaded the more well-heeled inhabitants of Cape Town to adopt Italian-style flat-tiled, plastered roofs sealed with fish, seal or whale oil. Finally in 1798 Cape Town's 'great fire' of 23 November resulted in regulations forbidding the construction of new thatch roofs in the city. Thus the threat of fire played a formative role in transforming the materials, organisation and architecture of the town.[34]

2
Fire at the Cape: British Colonial Rule, 1795–1900

And having made an extensive tour through forests and forest lands of the Colony [in 1864], on my return to Capetown I submitted to the Government a memoir on the conservation and extension of forests as a means of counteracting disastrous consequences following the destruction of bush and herbage by fire.

John Croumbie Brown, 1887[1]

The picturesque slopes of the mountains of the Cape Peninsula are disfigured at frequent intervals along the entire sea frontage, from Cape Point, via Sea Point, to Hout's Bay [*sic*], by the brown patches left by this season's fires.

Cape Times, 27 February 1893[2]

Great misapprehension prevails on the subject [of fire], caused mainly by looking at it from the point of view of an inhabitant of Northern Europe. The veldt fire here is not an incendiary disaster but a natural process, that usually is only dangerous when ignorantly interrupted or for some reason or other, too long deferred.

David Ernest Hutchins, 1893[3]

On 11 June 1795 a British fleet sailed into False Bay to capture the Cape Peninsula, this 'Gibraltar of the Indian Ocean', to keep it out of the hands of revolutionary France. The Dutch capitulated after a brief, token defence, and the British stayed until 1803, when they ceded the colony to the Dutch Batavian government under the Treaty of Amiens. Following renewed hostilities with France in 1806, the British returned, and following a brief battle on Bloubergstrand on the shores of Table Bay, reoccupied the Cape. This time they remained for more than a century.[4]

From 1795, the Cape was integrated into the British Imperial trade network, which resulted in major environmental impacts. Wars in Europe and the United States brought ships and soldiers to the Cape, and wheat and wine production soared. Forests at Plettenberg Bay on the east coast were reserved for the use of the Royal Navy in 1811. The Navy Board's subsequent judgement that these forests were unsuited to their needs provoked a wider debate over ownership and management of the colony's forests and undeveloped veld. Before investigating the terms and outcomes of this debate, however, a brief account of the region and local burning practices at the outset of the 1800s is in order.[5]

Scientific travellers: Lichtenstein and Burchell

The best-known scientific travellers to visit the Cape in the early nineteenth century were the German physician and naturalist (Martin) Heinrich Lichtenstein and the English naturalist and traveller William Burchell. Lichtenstein (1780–1857) arrived in the Cape on 23 December 1802, as family doctor and tutor to the children of the last Dutch governor, J.W. Janssens. He made three long journeys between 1803 and 1805, returning to Europe with Janssens after the British took over in early 1806. Travelling through the Outeniqua region of the southern Cape, Lichtenstein wrote:

> The soil is poor, and is of that description that belongs to sour-fields... [I]t is common to burn the lands every year, by which means they are manured, and the foundation laid for a wholesome vegetation. But this must be done with great caution, lest the fire should spread too far and catch the bushes, by which means it might be communicated to the forests, when incalculable mischief would ensue.

Lichtenstein observed that the region's poor soils are 'manured' through burning, and didn't condemn veld burning per se. He did worry, prophetically, that runaway grass fires were a threat to the region's forests.[6]

William Burchell (1781–1863), who had served a botanical apprenticeship at Kew, arrived in Cape Town in November 1810. He undertook an epic journey into the interior between June 1811 and April 1815, and the two volumes he published about his travels made his reputation. He is commemorated in the scientific names of many plants, animals and birds. On 23 January 1811, Burchell described a bush fire above Cape Town:

This night the Devil's Mountain presented a curious sight. At about two-thirds of its height, it was encircled by an irregular line of fire, which continued slowly advancing towards the summit, varying in direction and in brightness. It had been constantly burning during the day-time; but was not visible till darkness came on; and, having expended itself, went out before the next morning.[7]

While amused by 'the progress of this line', Burchell worried that the fire might spread, 'consuming all the vegetation in the way, [and so] disappoint my expectation of seeing the multitude of curious and beautiful plants which were said to grow on this celebrated mountain'. Like Le Vaillant, he reported that the carelessness of slaves resulted 'almost every year, [in] a conflagration of this kind, which not only destroys the beauty and verdant clothing of the mountain, but, in time, renders firewood more scarce than ever'.[8]

Burchell was informed that farmers burned old dry grass and bushes seasonally, to make way for 'new herbage' for their livestock.[9] Once embarked on his own travels, some 120km (75 miles) east of Cape Town in the Riviersonderend Mountains near Genadendal, Burchell observed altogether more judgementally that

the face of the country was shrubby; but not a single tree of any magnitude was to be seen … firewood … is rendered [scarce] by the wasteful and destructive practice of annually setting fire to the old withered grass, as the means of clearing the pastures. The flames … destroy every shrub and plant in their way, and pervade the whole farm, unless stopped by a river, or a beaten road.[10]

Of the 'Dunkerhoek mountains' in the nearby Baviaanskloof chain, Burchell wrote that 'the devastation occasioned by this custom was very striking, in the fatal havoc it had lately made amongst the finest plants of proteas, heaths, and the richest variety of shrubs we had passed in our ride'. Like so many keen observers of natural history to come, Burchell assumed that fires destroy the fynbos: 'Every thing was nearly consumed or destroyed, and the black, charcoal-like state of this shrubbery presented a singular, but melancholy appearance.'[11] He also attributed the treeless state of fynbos to humans. On coming face-to-face with burned fynbos, then, Burchell the naturalist condemned the farming practice he had recorded with such equanimity on first arriving at the Cape.

Fulminations of the colonial botanists, 1858–66

Carl Wilhelm Ludwig Pappe (1803–62) studied medicine at Leipzig University and wrote his dissertation on the local flora. He arrived at Cape Town on 4 January 1831, where he opened a medical practice. He also immediately began to collect plants on the Cape Peninsula and to publish on the medical and other uses of indigenous plants. By the 1840s, Pappe had established himself as 'the Cape botanist', a non-stipendiary role which allowed him some influence with the colonial authorities. He was active in the debates following the Navy's rejection of the Plettenberg Bay forests, and particularly in the public debate over the driftsands and generally poor state of pastures and scrublands on the Cape Flats. The authorities were pressurised into passing an Ordinance for the 'better preservation' of the 'Cape downs' in 1846.[12]

From 1846, Pappe corresponded with Sir William Hooker, Director of Kew. He complained about the poor state of the indigenous vegetation of the region (which he knew Hooker was greatly interested in) and the lack of scientific botany at the Cape, motivating for the appointment of a Colonial Botanist. When this post was created in 1858, Pappe was appointed, serving until his death in 1862. As an adjunct to the post, Pappe also became South Africa's first (unsalaried) Professor of Botany.[13]

Grove has noted Pappe's importance for the formalisation and spread of a 'gospel' of conservation at the Cape. In his extensive travels undertaken to research his *Silva Capensis* (1854), a pamphlet on the trees and shrubs of the region, Pappe was particularly dismayed by the destructive effects of veld fires on woodland, especially on Table Mountain and in the Cederberg Mountains to the north. In *Silva Capensis* he fulminated that:

When, twenty-four years ago, I arrived in the Colony, *Protea coccinea*, R. Br. [Red Bearded Protea or red sugarbush, *Protea grandiceps*], one of the handsomest of the Proteacean tribe, adorned the slopes of the Devil's-head Mountain. Since that period this beautiful shrub has gradually disappeared, and seems now to have been altogether annihilated. It is well known besides that wooded ravines and forests increase the moisture, and produce springs and running streams, the scarcity of which is but too often felt in this dry country. That the legislature of our day has not yet enacted sound and stringent laws against the repetition of an outrage of such magnitude, is a matter of deep regret and just astonishment.[14]

As Colonial Botanist, Pappe's concerns carried some authority, particularly as he couched them in terms of economic losses to the Colony. He had two important allies in the Colonial Secretary the Hon. Rawson W. Rawson (whose previous posts were Canada then Mauritius), with whom he had co-authored a book on ferns, and Sir William Hooker. Fortuitously for the nascent conservation lobby at the Cape, a Cape Legislative Assembly was formed in 1854, allowing settlers to influence local legislation. In 1854 the Assembly appointed a commission to look into the management of Cape forests, and in 1856 conservancies were set up to administer the indigenous forests around George. With the backing of his influential allies, Pappe steered through the Forest and Herbage and Preservation Act no.18 of 1859, which according to Grove is 'the most comprehensive conservation legislation passed in British colonies during the nineteenth century'. This Act stipulated sentences of a fine not exceeding £100 or imprisonment with or without hard labour for up to six months for the wilful or grossly negligent ignition of all vegetation on the property of others.[15]

Pappe drew on ideas being successfully deployed by East India Company surgeon-botanists to justify the conservation of indigenous vegetation in the Madras and Bombay Presidencies. They argued that the destruction of forests was directly contributing to drought. In India, government figures were spurred to action by the fear that deforestation would result in droughts which would cause famine. In the Cape, there had been a serious drought in 1847 which had caused great hardship. The 1859 Act was the first legislation on veld burning aimed at protecting forests and the veld, rather than simply at disaster prevention.[16]

Pappe died in 1862 and was succeeded as Colonial Botanist at the Cape by a Scot, Reverend John Croumbie Brown (1808–95). Brown had worked for the London Missionary Society at the Cape from 1844 to 1948. He then returned to Aberdeen where he studied botany, taking up a position as a university lecturer there in 1853. In 1862 Brown was contacted about the position of Colonial Botanist at the Cape, which he took up on 15 April 1863.[17]

Brown's first action was to visit the forest districts of the Cape (mostly Afrotemperate forest around George and Knysna), and on his return to Cape Town he submitted a report 'On the conservation and extension of the forests as a means of counteracting disastrous consequences following the destruction of bush and herbage by fire'. Pappe's admirable 1859 Act was, he thought, already 'a dead letter'. Brown believed, partly because he was told so by local inhabitants, that in the recent past the region had been well-watered and more forested than he found it. Brown was influenced in his views on desiccation by Pappe's reports and his *Silva Capensis*, and by his own experiences of the effects of the serious droughts of 1847. Initially, he found a ready audience for his views because there had been a very serious drought in the Cape in 1862.[18]

The most disastrous consequence of veld burning, argued Brown, was the resulting desiccation. Removal of the vegetation resulted in hotter soil temperatures, possibly reducing precipitation and more certainly increasing evaporation and reducing the soil's capacity to retain moisture. The Scot combined scientific argument with missionary zeal and Victorian moral disapproval. He drew on examples of deforestation and desiccation attributed to fire observed among the Tswana, north of the Cape Colony, and in Barbados and Jamaica. Brown was well aware, and professed himself saddened, that veld burning was not just an African and Boer practice but also one applied by British settlers. He gave numerous lectures as Colonial Botanist and corresponded with local farmers on the evils of veld burning, frequently resorting to warnings from the scriptures to 'cease to do evil; learn to do well'.[19]

Brown also believed veld burning had resulted in 'extensive districts, once covered with grass, [being] covered with the useless Rhinoster bush (*Elytropappus rhinocerotis*, Less.)'. He lamented the destruction of indigenous species, describing mountain slopes near Wynberg (now a southern suburb of Cape Town) covered with silver trees (*Leucadendron argenteum*), yellow pincushion (*Leucospermum conocarpodendron*) and other fynbos shrubs, which had been 'burned and blackened, killed by fire'. In 1864, Brown gave evidence to an Assembly Select Committee on soil erosion and drought. He made five key recommendations to improve existing environmental legislation, and the first of these advocated banning veld burning.[20]

The post of Colonial Botanist was abolished during a wave of budget cuts in 1866, but Brown's ideas continued to be influential in expert circles after he left the Cape. He wrote up his views in a series of books which offered the first comprehensive syntheses of the available material on environmental issues in the region.[21]

The great fires of the 1860s spark a revolution in forestry

Brown's fears about fires and forests received, it seemed, ample justification when a fire broke out in the Van Staden's River valley on

13 December 1865, burning 24km (15 miles) of forest and creating a plume of cloud that spread as far east as Adelaide in the Eastern Cape. Worse, from the point of view of the Conservator of Crown Forests in the district, Captain Christopher Harison, was another fire that broke out soon after in the Tsitsikamma (east of Knysna), damaging a more extensive area of government plantation and also private forest. Ironically, the new legislation intended to curb the ill effects of fire had some disastrous unintended consequences. Harison acknowledged that 'science is opposed to veld fires, and rightly so'. However, because farmers burned the lands around the forests and plantations, it remained necessary for foresters to burn 'in self defence' to keep their runaway fires out of forests and plantations. In the past Harison had burned regularly and safely in winter to achieve this end, but the Forest Act of 1859 had forbidden it – the result being 'the late disastrous fire'.[22]

Worse was to come, when Lichtenstein's prediction of the 'incalculable mischief' that grass fires might cause came to pass in 'The Great Fire of 1869'. Following a year of heavy rainfall (good for vegetation growth) in 1868, in January 1869 a severe drought set in. Table Mountain experienced a severe fire in early February. By 9 February the vegetation was very dry and a light northeasterly Berg (off-the-mountain) wind did nothing to ease the stifling heat across the Southern Cape. A number of small fires broke out, as it was customary to burn the veld with the Berg wind in the late summer. Then the northeaster suddenly stiffened and fanned several of these into uncontrollable blazes. By the time the fires had burned themselves out, hundreds of farms and thousands of livestock had been lost and 31 people had been burned to death. Fires had torched a strip of country 400 miles (644km) long and 15–150 miles (24–240km) wide extending from the district of Swellendam in the west through to Uitenhage in the east. It was an unprecedented disaster.[23] These big forest fires had a galvanising effect on forest management and a parliamentary commission was appointed to examine how best to protect the Colony's forests. They also added to the impetus for afforestation with introduced timber trees.

In 1876 the 'old regime' of disorganised and ineffective forest management was supplanted with the establishment of a Department of Forests and Plantations, run by a professional forester. Joseph Storr Lister served as Superintendent of Plantations at Cape Town from 1876 to 1888 and later as Chief Conservator of Forests at Cape Town (1905–10) and the first Chief Conservator of Forests for the Union of South Africa (1910–13). His father worked as an inspector for the Cape Road Board, overseeing tree-planting and drift-sand reclamation along hard roads across the Cape Flats. Lister's sister married an Indian Army Staff Officer and in 1869 he sailed to India, studying Hindustani and surveying in Hooghly, before being appointed Assistant Forest Officer at Chunga Munga plantation near Lahore. He was sent to England after contracting typhoid, from where he applied for the post of Superintendent of Drift Sands Plantations on the Cape Flats.

Following in his father's footsteps, Lister oversaw the planting of introduced trees [Port Jackson wattle (*Acacia saligna*) and various eucalypts and pines] around the Cape Peninsula and had great success using Cape Town street litter to stabilise driftsands on the Cape Flats. Indigenous *Mesembryanthemum* (low-growing succulents with attractive daisy-like flowers) were used successfully too. Most importantly he established the Colony's first extensive plantations at Worcester from 1876. The town's Mayor had asked him to help plant up low-lying swampy ground which was believed to be causing illness in the town. Lister had heard about the use of eucalypts to improve the fever-stricken Campagna Swamps near Rome, and in a few years' time his Worcester plantation comprising mostly *Eucalyptus globulus* (blue gum) proved itself a great financial success into the bargain.[24]

The revolution in forestry management was completed with the appointment of a French professional forester, Count Médéric de Vasselot de Regné (1837–1919), as Superintendent of Woods and Forests in 1880. He had been trained at the French National School of Forestry in Nancy. Lister was pleased with his new chief, later recalling de Regné's arrival on the mail steamer, arrayed in full French forest officer's uniform complete with a sword. 'The Count', as he was known, was a short, dark, stocky man, whom Lister's daughter recalled as a 'vivacious and amusing talker'. Despite his limited command of English, with the assistance of his talented protégé Henri Georges Fourcade, of David Ernest Hutchins who had trained at Nancy and was seconded to the Cape from India in 1883, and Lister's friend A.W. Heywood (who translated the Count's French into English), the Count soon overhauled the management of the Colony's forests and plantations. Tokai, Cecilia and Uitvlugt plantations were established on the Cape Peninsula in 1884. The nursery was moved from Uitvlugt to a more suitable site at Tokai, and in 1886 an arboretum was established at Tokai. Croumbie Brown notes approvingly that annual gross revenues from forestry in the Colony improved from around £250 in the 1860s to £7,680, just two years after the Count had taken control.[25]

The Count was well versed in then contemporary European scientific thinking about the environmental influences of forests, and in addition to setting out to make the Colony self-sufficient in its timber needs, he aimed to 'regulate the water-courses, and render drought...less frequent'. French scientists, he maintained, had proved that forested areas are cooler and moister, more rain falls in such areas, evaporation is reduced and fast runoff and thus erosion are prevented by the trees and the humus that forms beneath them, which also absorbs more moisture than bare soil. He gives the examples of Mauritius and St Helena islands, where forest cutting resulted in reduced stream flow and droughts.[26] In the context of the Cape, the Count argued that summer drought at the Peninsula existed despite the clouds driven over by the southeasters because the 'bare surface of [Table Mountain], baked by the scorching sun [offered] no attraction to them'. If the mountain was afforested, the clouds would be attracted by the infinitely cooler foliage and 'would frequently be condensed into copious rains'. Locals told him that 20 years ago the mountain had been covered with trees and bushes and rains had been frequent. The plantations planned for the 'bleak plateau' atop Table Mountain in 1884 and begun the following year were conceived with the water supply of Cape Town in mind.[27]

The Count toured the Colony's forests in 1881, accompanied by Captain Christopher Harison. In the Division of George, he noted, 'the periodical burnings, which are the sole attention paid to the larger part of the soil of each farm' rendered tree growth impossible and destroyed those trees which 'originally adorned the kloofs'. Inspecting the unforested land, he noted that the cultivation of cereals was very small in the George, Knysna and Humansdorp region, with the remainder being used to graze cattle, 'the grass...renewed by setting fire to it periodically'. This land he regarded as 'waste land' on which tree growth was rendered impossible. Worse, 'from these useless pastures the fire spreads to the woody Kloofs surrounding the farms, each time burning a belt of forest of greater or less breadth, until the timber has been completely destroyed'.[28]

The Count noted the apparent anomaly that places that had been burnt were rapidly recolonised by 'keur bushes [*Virgilia* species which require fire to germinate] and brushwood'. This suggested there was a rich layer of soil underneath and excited 'the envy of those who have properties adjoining'. The Count was concerned that this apparent fertility of land after fires might encourage burning. He argued that the 'vegetative vigour' of burned lands was in fact due to 'the presence of vegetable matter accumulated during the period they were clothed with trees', and this fertility, accumulated over centuries, could be destroyed within a few years.[29]

Having asserted that 'the chief cause of the destruction [of the Crown Forests] is fire', the Count outlined the first comprehensive official policy for the protection of the Colony's forests from fire. Firstly, and based on Harison's practice at Knysna, forest borders would be cleared by burning grass in winter. Second, the debris of felled trees would be removed from the forest after felling. Third, paths would be cut at intervals of 200 yards to arrest fires, facilitate the removal of timber and enable proper surveillance and control of the forest. Whether such measures were legally applicable to Crown Forests situated in the midst of farms and locations remained to be determined, and he recommended that an act be introduced to ensure these measures were applied, to clarify responsibilities and enable 'repressive measures to be enforced'.[30]

Lister had noted in 1883 that 'fires on the mountains of the Cape Peninsula are yearly on the increase; Table Mountain and Devil's Peak are becoming ever more denuded'. He attributed these fires to woodcutters who burned in order to force proprietors to sell their trees for fuel. He argued that the result was that all the vegetation including seedlings was burnt, the humus was destroyed and the winds and rains carried away the ashes and then the soils. He feared the entire range would be reduced to bleak and baked clay and recommended legislation to 'compel private owners to lay out fire-belts along their boundary lines' for 'it is an established principle that as fast as forests are destroyed springs become dry'. The existing Herbage Act did provide a means of prosecuting igniters of bush fires, providing they could be caught.[31]

Practical measures were devised to protect the Peninsula's plantations, including the construction of miles of bridle paths and ditches and banks to keep out fires and trespassing livestock. New plantations were divided into compartments 'with numerous cross-roads and bridle-paths' and boundaried by 40-foot (12m) fire-belts. In 1896, belts of oak trees and bluegums (*E. globulus*, now considered a fire hazard) were planted as fire barriers on Peak Plantation (on today's Devil's Peak). Most of this work was achieved with convict labour. These protection measures paid off in 1887, when thousands of acres of private woodlands were destroyed by bush fires across the Cape Peninsula, but the Forestry Department's plantations remained untouched. Lister recommended that the Cape Town Council and suburban municipalities enact regulations compelling the proprietors of woodlands to keep their boundaries cleared, or plant them with deciduous trees like oaks and poplar, or succulent shrubs like aloes.[32]

The Forest Act No.28 of 1888 was the Cape Colony's first comprehensive forest legislation and provided the basis for all subsequent South

African forestry legislation. It was based on the Madras Forest Act of 1882 and the views of the Count, Hutchins and Fourcade on the importance of fire suppression were influential. While the colonial botanists had raised awareness of the threat of fire to forests, it was the foresters who had the legal backing to prosecute fire starters and the means and authority to undertake fire protection measures. The consequence was that there was a focus on forests and trees, which comprised a minute proportion of the region's vegetation. The attitude of the foresters to non-forested land (or 'waste land' in the Count's terms) is well put by Hutchins in 1893: 'Outside the forest the open country consists of barren moor land.... Unless heavily manured it will support only a useless vegetation of heather and rushes.' He attributed the poverty of the soil to the continuous rains, the absence of frost (which aids decomposition of rocks) and 'the veldt fires which periodically send seawards such plant food as the surface vegetation has formed'.[33]

An exception to this dismissive attitude towards the indigenous shrublands (fynbos and renosterveld) was the concern shown about coastal areas threatened by driftsands. Proscriptions on cutting, burning and trampling of the vegetation in these areas had long been in force, going back to the Dutch period. Ordinance 28 of 1846 specifically targeted 'the better preservation of the Cape Flats and Downs'. Exotic trees, notably Australian wattles, hakeas and casuarinas were imported from the 1840s as a means of controlling these driftsands, and drift sand reclamation received renewed impetus from the mid-1870s under Lister. In 1890 Christopher Harison noted with satisfaction that 'the wattle bark industry has become a means of livelihood to the German immigrants on the Cape Flats'. Many of these introduced species later proved invasive on the Peninsula, affecting the region's fire regimes.[34]

A challenge to foresters managing Crown Lands on the Cape Flats was that large unafforested areas were in effect commonages, used for harvesting a variety of useful plants and other raw materials. The solution was to issue permits, for example, for the grazing of stock, collection of shells, quarrying of limestone, harvesting of reeds and rushes, collection of wild fruits and clearing and removal of bush. This provided an additional source of income, but brought with it problems, notably concerning fire. In his 1884 report, Lister argued that issuing such permits should be reconsidered, 'not only to prevent bush fires, which may undoubtedly be traced to the carelessness of reed cutters and wood gatherers, but for the suspicious reason that many bush fires have taken place during one season in places on which licence to graze is requested in the next'. To give an indication of the scale of this problem, in 1884 forest rangers impounded 3242 cattle trespassing on Crown lands on the Cape Flats.[35]

Firefight over Table Mountain

The forestry philosophy encouraged by the Count was being fully implemented by the time his post was abolished in 1891 – and was soon to become a focus for public controversy. In 1892, Hutchins, now Conservator of Forests for the Western Conservancy, reported that the planting atop Table Mountain (confined to the lower plateau) was well established, comprising about 105,890 cluster pines *Pinus pinaster* and scotch pines *P. sylvestris* (the latter 'greedily devoured by buck'). The soils were poor, and he argued that planting 'opens the choice between a rich forest soil and a barren moorland'. Hutchins believed foresters were 'reforesting' the mountain, which would achieve the 'reconstitution of that vast vegetable sponge – the forest soil'. This would '[improve the] storage of water for the summer springs that supply Cape Town and the suburbs' and the brown water caused by peaty vegetation would become clearer. Trees would also reduce winds, precipitate 'drifting mist' and slightly increase the rainfall. In support of the idea of 'reforesting' he listed several indigenous 'well known forest trees' which occurred 'in a stunted form in the kloofs of Table Mountain', and argued that the occurrence of Silver trees (*Leucadendron argenteum*) on the open slopes 'show that good timber trees once covered the slopes of Table Mountain'.[36]

Unfortunately for the foresters, on 8 December 1892 a 'severe fire [spread over] the whole of the mountain slopes of Cape Town', and a further big fire raged above Rondebosch on 12 March 1893. These fires led to fierce criticisms of the Cape Town City Council and objections to plantations on Table Mountain. Hutchins complained that 'a sound public opinion has yet to be formed' on matters of fire and forestry and that in fact the fires had done more good than harm. The real problem was one of 'bad forestry' – private plantations which 'are dirty, irregular, and sparse ... are so common on the slopes of Table Mountain [and] simply invite fires'. The problem was containable by professional foresters: 'I speak with a 20 years' experience of the subject, 10 of which relate to India, where forest fires are an evil of terrible magnitude, but one that is successfully met and overcome.' Private plantation owners who allowed their plantations to 'degenerate to a jungle of scattered trees and rank ground herbage' were a 'nuisance and a danger to [their] neighbours' and should be penalised like the 'owner of a flock of scabby sheep'.[37]

Table Mountain was, in Hutchins' view, experiencing a process he had seen in many places in the Cape Colony, where irregular and excessive tree-fellings resulted in openings for herbage and bush, which allowed in fires, which further extended open patches until the forests were pushed back into dense (nearly) fireproof kloofs. Stopping this process was easy, and required: '(1) dense planting and maintenance; (2) fire-paths; (3) watching;' and the first of these was the most important. This three-point strategy would be implemented in the new planting scheme for the 'bare and stony slopes above Woodstock and Salt River' recently purchased from the Woodstock Municipality. Fire paths would be cleared and strips of oak planted as fire breaks if the poor soil permitted it. A forester's cottage would be built below the King's Block House, providing a vantage point from which any fires would quickly be spotted. It would be fitted with a telephone for summoning immediate assistance and a road was planned to link the plantation to Roeland Street in Cape Town. This work was soon completed.[38]

The public criticism following the big mountain fires of 1892/93 impacted significantly on the Forestry Department's workload. Hutchins complained that in 1893 'fire-paths and other work in connection with fire protection occupied a large proportion of the staff at Tokai for two months [which was] too large a proportion of the year's work'. In his annual report he included a cutting from the *Cape Times* newspaper (27 February):

> The picturesque slopes of the mountains of the Cape Peninsula are disfigured at frequent intervals along the entire sea frontage, from Cape Point, via Sea Point, to Hout's Bay [*sic*], by the brown patches left by this season's fires. It is years since there has been so large an area shorn of its verdant beauty at one time as there is now.

The 'severity of the season' had resulted in increased fire incidence and Hutchins fully expected there to be further fires and further public criticism.[39]

Drawing on his experience in India and South Africa, Hutchins argued that 'great misapprehension prevails on the subject [of fire], caused mainly by looking at it from the point of view of an inhabitant of Northern Europe'. Rather than seeing a 'veldt' fire as 'an incendiary disaster', people should understand it as 'a natural process, that usually is only dangerous when ignorantly interrupted or for some reason or other, too long deferred'. At Knysna, he pointed out, the Forest Department assisted farmers in regular velt burning to prevent fierce fires burning through overgrown vegetation. In contrast with hot countries like India where fires naturally burned through 'open forest' and required the forestry department to extinguish them, most forests in South Africa were dense and evergreen and not traversable by fire. Only the silver trees on Table Mountain and cedars on the Cederberg Mountains appeared to be naturally traversable by fire at intervals. The answer to the problem of veld fires on Table Mountain was to replace the velt with 'clean dense plantations'.[40]

The challenge foresters faced on the Cape Peninsula was that the public witnessed plantations burning on the slopes virtually every season. Hutchins lamented that the public did not understand that a 'close clean plantation' is 'as little liable to burn as a macadamized road'. He did caution that even the best plantations had to pass through a period of danger, from when the trees were young and the branches touched the ground until these were shed and all ground herbage was killed off. Ploughing to keep the plantation 'clean' was impossible on stony mountain slopes, but weeding, fire-paths and watching were adequate controls provided there were no neighbouring 'dirty' inflammable plantations. He acknowledged that 'considerable damage and much anxiety' had been caused by the Peninsula fires and he called for legislation to punish the owners of inflammable plantations.[41]

An additional immediate threat to the plantations on Table Mountain was the application of the Town Council for title to the ground and plantations on Table Mountain for the purposes of building a reservoir for the city. Hutchins used his annual report to make a case for the 'utility of forest', arguing that 'in many localities similar to Table Mountain, notably the French Alps, water storage is obtained at far more moderate cost and of purer quality from plantations than from dams'.[42]

Hutchins vigorously sustained his campaign to 'reafforest' Table Mountain over the next three years. He focused on the supposed benefits of afforestation, namely increasing water supplies and improving water quality. In 1894 he argued that clearing forests often resulted in the creation of 'pestilential marshes'. He reported that measurement of the flow of the springs on Mr Breda's property above the Molteno Reservoir in the Cape Town city bowl had showed that burning the bush around them had diminished their yield. He quoted George Perkins Marsh's pioneering environmentalist book *Man and Nature* (1864) on how the felling of forests in France, Germany and Australia had resulted in the failing of springs. Marsh had argued that classical Mediterranean civilizations (i.e. from a similar climatic region) had collapsed as a result of degrading their environments, notably through deforestation. Marsh however acknowledged that while fire may in some circumstances 'render the declivity of a mountain unproductive for a century', in others it favours and may even accelerate the growth of some trees, notably pines.[43]

Despite his best efforts, in his report for 1896 Hutchins had to report 'with very great regret... that at the close of the year tree-planting on Table Mountain stood suspended'. The City Council had convinced the Minister of Agriculture (forestry was a branch of Agriculture in the colony) that tree-planting was having an adverse affect on the city's water supply. Hutchins made his last stand on the basis that surface water captured in a reservoir could not compare with the clear spring water which forests would facilitate. He still maintained that only the afforestation of Table Mountain could provide enough water for the city and its future suburbs. If afforestation was stopped – 'in obedience to unreasoning panic' – then another masonry reservoir would soon have to be built 'at enormous cost... followed in after years by the colossal scheme required to bring water from the main land'. In terms of further costs and infrastructure projects Hutchins' prophesies were all fulfilled within a decade and the 'colossal scheme' by 1921. Nevertheless, the Forest Department and their afforestation programme lost out to the City Council's judgement that building reservoirs atop Table Mountain (and to an extent, removing plantations) was the best path forward for ensuring sufficient water supplies for the growing city.[44]

Conclusion

While many experts had accurately observed the regenerative effects of fire on the vegetation of the fynbos biome, their interpretive frameworks led them to argue that burning had negative effects over the longer term. Experts' arguments were often developed in the context of concerns over how 'lay' persons would interpret the effects of fire. Arguments from authority (rather than proof) were articulated and later enshrined in legislation with the intention of stopping or at least inhibiting veld burning.

This was not simply a process of Europeans imposing northern hemisphere ideas inappropriately. When formulating their ideas about the effects of fire and how it should be managed, Croumbie Brown, the Count, Lister and Hutchins drew on a wide range of sources and experience, including personal experience of fires and fire management in Mediterranean Europe and India. Hutchins regarded fire as a natural process in the region that got out of hand if poorly managed. It could be a good management tool if wisely deployed. He did retain a northern European bias towards closed forest, as did many others.

At the close of the nineteenth century, foresters were regarded as the colonial government's foremost experts on fire management on the Cape Peninsula. They maintained that the mountains of the Peninsula and surrounding region had been deforested, and the climate and soils were suffering as a result. The answer was to reforest with dense plantations of introduced trees, well protected from fire. From the 1890s, there was significant public opposition to this strategy in Cape Town, cohering around the fears that plantations of introduced trees sapped water and were a fire hazard. Within local government, these fears were shared by the Cape Town Municipality, and amplified by a water crisis in the early 1900s, resulted in the replacement of plantations atop Table Mountain with reservoirs. Planting continued on the slopes below, as the city urgently needed timber, and elaborate safety precautions suggested that plantations could be protected from fire. However, the fragmented nature of landownership on the Peninsula complicated fire management. The spectre of large destructive wildfires raging uncontrollably through plantations of inflammable introduced trees on the slopes above Cape Town continued to haunt the Peninsula's foresters.

Part II

Fynbos, Fire Research and Management, c.1900–99

3
Science, Management and Fire in Fynbos: 1900–45

> whole hill slopes being sometimes covered with Watsonias and other tall flower spikes [is a] sad spectacle [to the botanist] 'for he realises the tremendous slaughter that has taken place to produce this display'.
>
> Rudolf Marloth, 1924[1]

> a more realistic policy should now be adopted, which would favour controlled burning, except in particular cases where the need for total protection has been specially substantiated.
>
> C.L. Wicht, 1945[2]

South Africa has very little indigenous forest, with only 7 per cent of land cover suitable for forest and less than 0.1 per cent actually forested. In the southwestern Cape small patches of forest exist in fire-resistant refuges in shaded, moist montane areas, including the southern and eastern slopes of Table Mountain. Most of the indigenous vegetation (comprising the Cape Floristic Region) is virtually treeless. This was confusing for European botanical and forestry experts, particularly as some introduced trees grew very well. Many of them concluded that the region had been significantly deforested by humans.

The botanists and the burning bush

Botany was professionalised within South Africa following the country's political unification in 1910. Until that time, many of the acknowledged experts on Cape botany were amateur botanists and plant collectors. Prominent among these were Harry Bolus and Rudolf Marloth, both of whom were convinced that veld burning was resulting in terrible damage to the indigenous flora.

Harry Bolus (1834–1911) came to South Africa from Nottingham, England, to assist the plant collector William Kensit in Grahamstown. He took up plant collecting himself to help him assuage his grief at the loss of his eldest son in 1864. Bolus moved to Cape Town in 1875 where he joined a firm of stockbrokers. He undertook collecting expeditions across South Africa, and authored several botanical publications. Bolus became the major benefactor of botanical studies at the Cape in this period, founding a chair of botany in the South African College (now University of Cape Town) in 1902 (named after him from 1917, and still extant). In 1908, Bolus wrote to the director of the Royal Botanic Gardens at Kew, David Prain, that

> in dealing with Cape plants it should not be forgotten that the rarity of some of those sent by early collectors is doubtless due to the frequent bush fires. I believe hundreds of species have become extinct from this cause during the last century.[3]

Dr Rudolf Marloth (1855–1931) arrived in Cape Town from Germany in 1883. An analytical chemist by profession, he had some botanical training, and by the 1920s was South Africa's best-known botanist, having won the South African Association for the Advancement of Science's South Africa Medal and grant in 1914. Marloth travelled widely, and E.P. Phillips (Chief of the Union's Division of Botany and Plant Pathology, 1939–44) said of him in 1930: 'no one in South Africa knows the plants of the Cape Province as they grow in their native habitat as does Marloth'. His many publications include his six-volume *The Flora of South Africa* (1913–32) and *Das Kapland* (1908) on the ecology and plant geography of the Fynbos region of the Western Cape (Photo 3.1).[4]

Marloth was influential in his dire predictions about the threatened status of the vegetation of the Western Cape mountains. He blamed careless veld burning by pastoralists for this, arguing that 'hundreds of square miles of luxuriant *maquis* [fynbos] . . . have been devoured by the flames during a century or two of reckless burning'. For Marloth, the 'glorious sight' of 'whole hill slopes being sometimes covered with Watsonias and other tall flower spikes' is a 'sad spectacle' to the botanist, 'for he realises the tremendous slaughter that has taken place to produce this display, and the danger, no, the certainty, of the extinction of many species and perhaps even genera of plants'. The rich flora was being reduced to a 'monotony of Monocotyledons'.[5]

A wave of talented British botanists came to South Africa between the end of the South African War and 1930, including (Henry) Harold

Photo 3.1 Rudolf Marloth
Source: By kind permission, A.A. Balkema Publishers.

Pearson (arrived 1903) and Harold Compton (1919), both Cambridge-educated, and first and second directors of the Kirstenbosch National Botanic Garden in Cape Town. Illtyd Buller Pole-Evans (1879–1968), a Cambridge-educated mycologist, arrived in 1905, and in 1913 became Chief of the Union Agriculture Department's Division of Botany. John Bews (Photo 3.2) arrived in Natal in 1910, and in 1923 Robert Adamson arrived in Cape Town to become Harry Bolus Professor of Botany. Under their influence, botany became professionalised, and gained influence in land management sectors in the Union.

John William Bews (1884–1938) was a key figure. He grew up on a farm in the Orkney Islands (where he had seen heather burned as a boy), and was educated at Edinburgh University.[6] He came to South Africa in 1910 to take up the first professorship of botany at Natal University College in Pietermaritzburg (he was dismayed to discover on arrival that the college was still under construction).[7] Along with Professor Charles E. Moss, who founded the botany department at what became the University of the Witwatersrand ('Wits'), Bews was most responsible for introducing the ideas of the Nebraskan ecologist Frederic Clements

Photo 3.2 John William Bews
Source: By kind permission, University of KwaZulu-Natal Press.

(1874–1945) to botanists in South Africa.[8] Bews quickly met all the prominent academic and state botanists of the country at meetings of the Literature and Science Committees of the University of the Cape of Good Hope, which was then the examining body for all the country's college students. Bews later became Principal of Natal University College, was President of the South African Association for the Advancement of Science (the presidency lasts for a year), and led the South African delegation to the centenary meeting of the British Association for the Advancement of Science in London in 1931.[9]

In early papers on the vegetation of Natal, Bews argued that fire was an important factor in determining the distribution and character of woodland and grassland, and railed against 'primitive' and 'destructive' African farming methods including the use of fire to convert bush to maize fields. Bews argued this impoverished the soil and destroyed large areas of woodland. He maintained that grass burning favoured the less valuable veld grasses, impoverishing grazing resources. He also believed that burning was, together with deforestation, contributing to the desiccation of the country.[10]

Bews's 'An Account of the Chief Types of Vegetation in South Africa, with Notes on the Plant Succession' (1916), was the first national survey to apply an ecological approach informed by Clementsian ideas of succession. (Anker suggests that Bews adopted an ecological approach to assert his authority as botany professor in light of his ignorance of local plant species and lack of access to a laboratory, or space to establish a herbarium.) Clements's ecology, influential since the publication of his *Research Methods in Ecology* (1905) and summed up in his book *Plant Succession: An Analysis of the Development of Vegetation Change* (1916), might best be described as 'dynamic'. He argued that vegetation progresses linearly through a series of increasingly complex stages towards a stable climax plant community, which is in equilibrium with the prevailing environmental conditions. In the initial stages of succession, soil is more important, but ultimately it is climatic factors – rainfall, temperature and winds – which determine the nature of the climax community. Climate thus drives a linear unidirectional succession towards a single vegetation type (monoclimax), though a variety of disturbances (notably fire) may inhibit or temporarily reverse this progression. The concept of plant succession provided a powerful conceptual framework for landscape interpretation and manipulation which gave rise to rangeland succession theory as propounded by American rangeland theorists.[11]

In 1918, Bews published *The Grasses and Grasslands of South Africa*, which according to his student and successor as professor of botany at Natal, Adolf Bayer, stimulated the establishment of pasture research stations by the country's Division of Plant Industry.[12] According to George Gale, this book 'placed the study of South African grasslands upon a scientific basis'. It also alerted the country's statesman Jan Smuts (a key figure in the country's political life between 1897 and 1948, including stints as Prime Minister in 1919–24 and 1939–48) to the importance of the field. Smuts, a keen amateur grass scientist, later said of Bews: 'I have sat for years at his feet... I look upon him as one of the ornaments of science in South Africa'. Ultimately, Bews's research served as the foundation for the country's first chair in Pasture Management and Soil Conservation, established at Natal University College in 1948.[13]

Bews argued that in 'primitive' or semi-open grassland, burning stalls the plant succession, preventing the veld from becoming more nutritious. Because burning the prevalent pioneer grasses produces palatable shoots, farmers burn to get young growth early in spring. This prevents the grassveld from progressing towards a better type. In a more advanced, stable grassveld with the desirable rooigras (*Themeda triandra* species) dominant, Bews recommended grazing down the grasses. Burning them would retrogress the succession, replacing rooigras with more primitive types. Where the succession was advancing towards shrub and even seedling forest, with tall, coarse grasses dominating, Bews actually encouraged veld burning – should grazing be the desired land use. However, he argued that such land was best converted to timber production.

Ideas about vegetation and pasture science developed in the American Midwest proved influential in South Africa in the 1920s and 1930s, as the prairies were judged to be analogous to South Africa's grasslands. Researchers in both countries worried about the effects of poor farming techniques on soils, and following the Dust Bowl disaster on America's Midwest prairies in the 1930s, agricultural experts feared a 'dustbowl' scenario was imminent in South Africa. South African experts and officials engaged in modernising agriculture in the newly founded Union were much preoccupied with the notion that the country was drying up. There was a serious drought in 1919, amid a notably dry period from 1918 to 1929 in most regions of the Fynbos Biome. The influential Drought Investigation Commission's final report of 1923 concluded that this ongoing desiccation was not the result of declining rainfall, but rather of bad farming practices, including veld burning. The logical outcome, the report warned, was ' "The Great South African Desert" uninhabitable by Man'.

In the aftermath of the 1923 Drought Investigation Commission report, a special symposium focused exclusively on veld burning was convened in 1924, at which prominent Cape botanists lined up to criticise the practice. Bews had provided a theoretical framework for these sweeping statements about the effects of fire. Botanists then working on the fynbos, including Marloth, and Margaret Levyns and Professor Robert Adamson of the University of Cape Town, interpreted the ecological response of fynbos to fire as a linear decline resulting in an impoverishment of the flora, rather than a cyclical renewal through naturally recurring fires.[14]

Robert Stephen Adamson (1885–1965) moved to South Africa in 1923 to take up a professorship of botany at the University of Cape Town, where he remained until 1950. Born in Manchester, Adamson was educated in Edinburgh and Cambridge, taught botany at the University of Manchester, and collaborated on ecological studies with the originator of the ecosystem concept, Alfred Tansley, in England. Adamson has been credited with introducing plant ecology to Australia during a six-month stay at the University of Adelaide in 1922, immediately before moving to South Africa. His book *The Vegetation of South Africa* (1938) was the only one of the British Empire Vegetation Committee's planned series of monographs on the vegetation of the British Empire to be completed. He also co-edited *The Flora of the Cape Peninsula* (1950).[15]

In his 'preliminary account' of the vegetation of Table Mountain (1927), Adamson argued that the extremely diverse plant communities and their apparent lack of correlation with their habitats was a result of a high level of human disturbance. Burning, planting, grazing and the felling of timber had changed the nature of the vegetation almost completely, he argued, and of these factors, 'fire has been very much the most extensive and far reaching in its effects.' At this stage, Adamson regarded fire as an anthropogenic disturbance, rather than a natural agent of vegetation change integral to fynbos ecosystem functioning. In his discussion of the spread of alien plants on Table Mountain, he noted that fire was aiding the rapid spread of *Pinus pinaster* and species of *Hakea*.[16]

Adamson found evidence of widespread 'destructive change' to what he imagined to have been the original indigenous climax vegetation of Table Mountain. He was of the opinion that: 'by far the most effective agent of change has been the frequent burning of the vegetation. The effects of this need not be repeated; they result in the retrogression of the normal succession and an impoverishment of the flora'.[17]

The brilliant Margaret Michell (1890–1975) was similarly concerned over the effects of fire. A lecturer in botany at the University of Cape Town, Michell (hereafter Levyns: she published most of her work under her married name), had been convinced by Compton to abandon mathematics for botany. She won a scholarship to Cambridge, and later the John Innes Institute, before returning to research and teach botany at the University of Cape Town for the remainder of her career. In 1932 she was awarded the first doctorate in plant systematics in South Africa, and later served as President of the Royal Society of South Africa. Her contribution to Cape botany is honoured in the dedication of the book summing up the work of the influential Fynbos Biome Project (1992).[18]

Levyns conducted the first scientific research on fire in the Fynbos Biome, in the early 1920s. She focused on fire's negative effects in favouring the spread of the renosterbos ('rhinoceros bush', *Elytropappus rhinocerotis*). This invasive indigenous shrub, regarded as an agricultural pest because it is unpalatable to livestock, is characteristic of the renosterveld (giving its name to the vegetation type). Renosterveld predominates in areas of moderate rainfall and on more fertile and fine-grained soils, whereas Fynbos is found in areas with lower rainfall and nutrient-poor soils. Intriguingly, Levyns noted that while fires were said to have had eradicated proteas (Fynbos plants) from the slopes of Signal Hill, fynbos ('true Macchia') persisted on the slopes below the adjacent Lion's Head peak. She recorded that this area also burned frequently, but the soil was granitic. We now know that far from 'invading' Fynbos on Signal Hill, Renosterveld is the established vegetation type on the under-lying Malmesbury slates. Fynbos occurs on the granitic soils on Lion's Head. However Levyns, like Marloth (both were brilliant observers, but influenced by the conventions of their time), characterised renosterveld as a 'secondary' vegetation type degraded by human burning practices, and believed veld burning encouraged the invasion of fynbos by the renosterbos.[19]

This idea of 'secondary' vegetation and its outcomes well illus-trates how ecological theory has impacted on landscape management. Marloth had argued that Renosterveld was a 'Kunstformation', created by 'man's cultivation and interference', influencing Bews to characterise it as such in his 'Account of the Chief Types of Vegetation in South Africa'. This was chiefly the result of the identification of this vege-tation type with the eponymous renosterbos, which is actually much more widely distributed than renosterveld. Cape farmers had been burn-ing the veld to try to get rid of it since the eighteenth century. Bews similarly argued that 'man's interference and the continual burning of the grass' in Natal had resulted in a 'secondary succession' which created vegetation types he christened 'Changed Veld'.[20] Thus these botanists introduced a hierarchy within indigenous vegetation: of 'pri-mary' or 'natural' vegetation; and 'secondary' or 'artificial' vegetation degraded by the interference of humans. The latter was regarded as undesirable, with unfortunate long-term consequences for its survival. In a 1999 report by CAPE (Cape Action for People and the Environ-ment), lowland renosterveld areas were (belatedly) identified as 'top conservation priorities within the Cape Floristic Region'.[21]

The legacy of the idea of 'secondary vegetation' was continued by the agricultural botanist John Acocks (1911–79) in his extremely influential *Veld Types of South Africa* (1953). Acocks divided Fynbos (which he called Macchia) into lowland and montane Macchia, and the latter further into 'mountain' and 'false' Macchia. 'False Macchia' included all the Fynbos on the Cape mountains from the central Swartberg and Langeberg all the way to the eastern limits of these ranges near Grahamstown. It was 'false' because, he argued, it was derived from natural forest or grassland through human intervention – burning, overgrazing, and clearing. Fynbos ecologist Richard Cowling has since shown that this classification (and interpretation) is inaccurate and has proven harmful in its influence on the beliefs and land management practices of the region's agricultural community.[22]

Kirstenbosch and Compton

On the Cape Peninsula, botanists' efforts to protect and celebrate the local flora in the face of public indifference came to fruition with the founding of Kirstenbosch National Botanical Garden on the slopes of Table Mountain in 1913. In 1919, Cambridge-educated Robert Harold Compton (1886–1979) took over from Harold Pearson, becoming the second director of Kirstenbosch, and began editing the *Journal of the Botanical Society of South Africa*. Influenced by Marloth's ideas and those of his university colleagues Robert Adamson and Margaret Levyns, Compton developed a particular antipathy for veld burning, expressed in increasingly heated terms in the pages of his journal and in public lectures and papers. His views were the most widely disseminated of the Cape botanists of his time, reaching a rapidly expanding Botanical Society membership – which increased from 352 in 1914 to 912 by 1925 – as well as some of the growing numbers of visitors to Kirstenbosch. In 1925, the gardens recorded 56,312 visitors for the year's worth of Saturday afternoons, Sundays and Public Holidays.[23]

Compton's antagonism to 'promiscuous firemaking' in the wider region was fuelled by the danger that runaway fires on Table Mountain posed to the flora of his botanical garden (and that of the mountain in general). Discussing the wider region, he roundly chastised stock farmers in racist terms for using primitive 'Neo-Hottentot pastoral practice' (veld burning) learned from the pastoral Khoikhoi peoples. In pre-European times, Compton fulminated, the region's indigenous peoples burned the veld 'with that sublime indifference to ultimate consequences that characterises uncivilised races'. He lamented that

> it is remarkable how, when two races come into contact, each adopts the worst features of the other: so that while the black acquired the habits of wearing trousers and drinking spirits, the white learnt the gentle art of ruining the veld in the most certain and rapid way conceivable.

(It is not clear what he had against the wearing of trousers.)

In an attempt to shame white farmers out of burning, Compton argued that veld burning was 'Kafir pastoral practice...unworthy of Europeans'.[24]

Although Compton admitted that very little proper scientific study of the issue of veld burning had been done, he was nevertheless certain that 'serious degeneration as a result of burning actually exist[ed]. There [was] quite enough evidence and authoritative opinion that frequent veld burning [was] in the great majority of cases an evil of the first magnitude to justify immediate steps to put an end to it'. He offered a grand narrative of an evolutionary progression of vegetation, driven by a mutually reinforcing accumulation of nitrogen in the soil and vegetation, to a point of equilibrium at a climate- and soil-determined climax. Fire, Compton insisted, undid all this, and (anticipating recent attempts to quantify 'ecosystem services') he put a cash value on the loss: 'a quite moderate bush-fire of an area of a square mile...costs the country in combined nitrogen alone something like £2,000'. 'Neo-Hottentot pastoral practice' involved burning the veld as often as it would burn, and in the long run the land would return 'inevitably to a state in which it may be supposed to have begun, and the only hope for it is that man will disappear', he concluded, somewhat dramatically.[25] This aggressive moralistic rhetoric put him directly in the tradition of John Croumbie Brown, second Colonial Botanist at the Cape.

Compton was not an ecologist, but he was influenced by Bews's idea that burning or grazing reversed the succession of vegetation, and doing so resulted in the production of a more primitive vegetation type than had earlier existed. He drew clear parallels between 'primitive vegetation', 'uncivilised races' and 'backward and retrograde farmers' pursuing 'Neo-Hottentot pastoral practice' (i.e. burning the veld). Compton made the common assumption of the botanists of his era that the natural climax vegetation of the Southwest Cape was forest. The slow, but natural colonisation of scrub areas by evergreen forest species, he argued, was prevented by fires. Every bush fire burnt back the fringes of the forests and prevented their natural spread, and in these burnt fringes an earlier vegetation type, fynbos scrub, developed. Bush-fires negatively impacted on soil fertility and further set back the succession, with the result that:

after a series of such fires we reach a stage represented by hundreds of square miles in the Cape – white sandy or stony slopes, containing very little humus and occupied by a vegetation in which reeds [restios] largely predominate. The succession has been set back further still, and these areas are in the condition which they may be supposed to have left behind thousands of years ago before man came and reversed the natural succession.[26]

By seeing fynbos (despite its astonishing diversity) as a retrograde stage on the way to fire-induced desolation, Compton did not consider that fynbos regeneration and diversity might be *maintained* by fires.

In addition to proselytising against the use of fire in Fynbos, in the 1920s Compton together with the Council of the Botanical Society successfully championed 'the establishment of local nature reserves in the Cape to protect the natural flora and fauna from... flower-picking, veld-burning, grazing... and other forms of vandalism...'. These conservation efforts were important, but by the 1950s, policies of fire-exclusion on these flower reserves would achieve, in direct contradiction of their avowed purpose, a noticeable decline in flowers.[27]

The science and culture of agricultural research

After the South African War (1899–1902), British colonial attempts to create more efficient, modernised institutions guided by scientific research had been melded with progressive conservationist approaches from the United States. Following the American model, government-sponsored biological and environmental sciences were organised to research and improve land utilisation, and most (including forestry from 1934 to 1945) were organised as sub divisions of the Department of Agriculture. The Department of Agriculture became a key centre for fire research and management. John Bews had provided the intellectual impetus for pasture research as an economically vital research area for the country. The formalisation of pasture research on a national level was driven by a (belated) response to the recommendations of the 1923 drought commission, the economic woes of the great depression, and fears about waning returns from mining. The University of Pretoria and the African Explosives and Chemical Industry company (AE&CI, which developed fertilizers) put together a 'Grassland Research Committee' to tackle the Union's pasture problems. The committee's 1932 report urged further research, and under the leadership of Pole-Evans, the first government pasture research stations were opened in 1934. In this Pole-Evans was supported by his friend and plant-collecting companion Jan Smuts.[28]

The term 'pasture science' was a misnomer as most of South Africa is unsuitable for cultivating pastures and much of the research involved matters of veld (i.e. indigenous vegetation) management. Expert opinion was disseminated in the *Agricultural Journal of the Union of South*

Africa. The journal included correspondence from farmers, and the pros and cons of burning were hotly debated. Drawing on Bond et al.'s framing of the idea of Climate Dependent Ecosystems (CDEs) and Fire Dependent Ecosystems (FDEs), it is possible to summarise a complex situation by saying that fire tended to be favoured by farmers in FDEs, and less so in CDEs. Fire Dependent Ecosystems predominate in areas with mean annual rainfall above 650mm, and the plants and plant communities that dominate these ecosystems are well adapted to fire. The C_4 grasses dominate in the region's summer rainfall grasslands and savannas, including the much favoured grazing grasses in the *Themeda* genus in mesic summer rainfall areas. The Andropogoneae dominate sourveld (veld with seasonally nutritious grasses), and these grasses lose their nutritional value late in the growing season. They decompose slowly, and thus require defoliation through burning to allow their basal tillers to thrive. Fire exclusion results in a rapid decline of such grasses.[29]

The main reason given for burning by farmers writing to the agricultural journals was to get rid of old dry sourveld grass and force up new grass which was edible and nutritious. Farmers also argued that old dry grass was a fire hazard and should be burned off under safe conditions. The central argument of the anti-burning camp (mostly experts) was that veld burning led to the 'coarsening of veld grasses', resulting in reduced stock carrying capacity. Experts transposed arguments about the negative effects of fire on the species composition of (climate-driven) sweetveld (nutritious all year around), onto (fire-driven) sourveld areas. The 'natural succession' in the sourveld, the experts argued, was to scrub or forest, and thus sourveld should be developed for timber rather than burned for grazing. Most farmers resolutely ignored them for the remainder of the century. They could afford neither the labour nor the equipment required to implement the alternative methods to burning developed by agricultural experts.[30]

Agricultural researchers focused on frequency and seasonality of burning, with a view to manipulating vegetation succession to optimise land use for grazing and preventing soil erosion. Their linear Clementsian conception of vegetation succession, focus on desirable grazing grass species rather than biodiversity, and the rigidity of departmental policy on when (seasonally) it is acceptable to burn, would prove especially unhelpful in understanding the role of fire in fynbos.

The romance of the grassy vlaktes

Dubow has remarked (in a different context) on the curious amalgam of science and sentiment which characterised the articulation of Afrikaner nationalism in the 1930s. Louis Botha and Jan Smuts (who became South African, rather than Afrikaner, nationalists following Union in 1910), were the first and second Boer presidents of the Union. Both were farm boys and farmers who served as Ministers of Agriculture as well as Prime Minister. Both maintained a practical interest in, and a romantic attitude towards, the veld. They witnessed the closing of the South African frontier as increasing development, fencing and veterinary restrictions began to infringe on the practice of 'trekking' stock from place to place for seasonal grazing. Subdivision of bounded farms in accordance with Boer laws of inheritance threatened to make farms uneconomical in some regions. Veld management was required in a world in which it was no longer possible (or would soon no longer be) to 'trek on' to pastures new if the veld was over-utilised. In the 1930s, Afrikaner identity was being defined by recourse to a rural past in the period in which many Afrikaners were being urbanised, having been forced off the land by drought and economic hardship in the 1910s and 1920s.[31]

In the sphere of the governance of agriculture, there was an uneasy realisation that the much-vaunted *volk* wisdom acquired through experience by the Boers was, according to experts, leading to the wholesale degradation of land. The solution was to continue with the expert-led approach promoted by Alfred Milner's post-South-African-War reconstruction government, viewed as a patriotic undertaking which was both a practical necessity, and also a means of preserving the new nation's cultural heritage in the form of its developed landscapes and rural lifestyles. (Africans were not included in this vision, except as agricultural labourers.)

English-speaking South Africans including the author of the now classic South African novel *Jock of the Bushveld* (1907), Percy Fitzpatrick, were similarly motivated by this mixture of science and sentiment. Fitzpatrick evoked the romance of the veld in his best-selling novel, and dedicated his final years to experimental farming and irrigation schemes which ultimately gained the support of Smuts and the Minister of Lands, Deneys Reitz. In this period both Afrikaans and English-speaking South Africans looked to a technocratic and scientific vision of progress with enthusiasm and a sense of common purpose. An exemplar of this was the Afrikaner South African War hero and agricultural expert Heinrich du Toit. He argued that farming had become a scientific profession, not an occupation for peasants.[32]

Environmental parallels aside, for key Afrikaner role players like du Toit, embittered by their surrender to the British in 1902, the United

States represented an exemplary free nation which had thrown off the shackles of the British Empire. Du Toit, a *bittereinderer* who had not surrendered to the British in the field during the Anglo-Boer War, travelled to the United States in 1902 and again in 1907. In 1920 du Toit was appointed Chair of the Drought Investigation Commission, and the final report of the commission (1923) was to prove the most influential articulation of the ideology of impending environmental disaster in South Africa in this period. This report provided the ideological basis for the active interventions of the post-war period, enabled by the Soil Conservation Act of 1946. The South Africans noted that Roosevelt's New Deal government had taken direct responsibility for remedial measures following the disaster on the American plains in the 1930s, and Smuts was reported to have commented in 1936 that 'erosion is the biggest problem confronting the country, bigger than any politics'.[33]

Running through all of Smuts's engagements with environmental issues is both a strong scientific interest and a more romantic, at times spiritual response to the natural world. Ten days before his 80th birthday, Smuts would write: 'gradually I came to realize that the family of grasses was the most important of all ... give me the grasses, the rolling veld, the bushveld savannah ... this is the grass pattern of life, and there is no fascination like it'. The conservationist T.C. Robertson later recalled that, on their final walk in the veld together, Smuts picked the flower head from a grass and reminded T.C. that it had been named after him: *Digitaria smutsii*. Robertson mused: 'sometimes I think he was prouder of that than of all the many honours bestowed on him'.[34]

The pragmatic and romantic aspects of pasture (essentially grassland) research in South Africa in the first half of the twentieth century, together with the botanists' attempts to protect and publicise the new nation's indigenous flora, created a climate of opinion (for an influential clique) in which plantation foresters became the enemy. Foresters were, they would argue, converting the native veld and the grasses to aesthetically undesirable and environmentally destructive uniform plantations of trees introduced from other parts of the world. Foresters were also loyal members of the British Empire Forestry network.

State forestry

Under British colonial rule, the Cape of Good Hope Department of Agriculture oversaw forestry in the Cape Colony. Forestry was run by four conservators each overseeing a conservancy. Karen Brown has shown how at the turn of the century South African foresters participated in a crisis narrative of an impending timber shortage emanating from Gifford Pinchot's progressive forestry movement in the United States, and the colonial forester Professor Wilhelm Schlich at Cooper's Hill in Surrey, England. Amid an economic slump at the Cape following the end of the South African War in 1902, Leander Starr Jameson's Progressive Party came to power in 1904 in part through promoting the more efficient use and development of the colony's natural resources. Cape foresters Joseph Storr Lister and David Hutchins played a key role in promoting forestry as a means to achieving this. Once in power the Progressive Party reorganised the colony's civil service, making the forestry department an institution in its own right, independent of the Department of Agriculture. In 1905 Lister was appointed Acting Chief Conservator of Forests for the Colony, moving to Cape Town to take up his position. His brief was to institute and oversee a unified plan for the development of the forest resources of the colony. He based his approach on the programme for British imperial forestry outlined by Wilhelm Schlich, professor of forestry at the nerve-centre of British Empire forestry, the Royal Engineering College at Cooper's Hill. Working plans were devised, based on statistical calculations of growth and yield for plantations of single species chosen for their suitability to local environments and their marketability.[35]

This new management structure continued after Union in 1910, with Lister overseeing the conservancies created for the new Union Forest Department (he was succeeded by the scot Charles Legat in 1913). A Research Division was founded in 1912, later amalgamated with the Forest Products Institute in 1956, to eventually become the South African Forestry Research Institute. In 1931 the Department was Afrikanerised when F.E. Geldenhuys, a former undersecretary of agriculture, was appointed Director. Three years later his post was abolished and the department was demoted to a division of the new Department of Agriculture and Forestry, responsible to the Secretary in the Department of Agriculture and Forestry. Fortunately research and management were in the hands of well trained and dedicated Afrikaner foresters, Head of Forestry Johan Keet, and research officer Christiaan Wicht. This was an unhappy period for state forestry, with considerable friction between the agriculture and forestry sectors inside and outside of government (Photo 3.3).[36]

Two major branches of forestry developed in South Africa: plantation forestry and conservation forestry, and this book focuses on the latter. Plantation forestry focused on meeting the timber requirements of a developing country with large mines and a developing infrastructure.

Photo 3.3 Forestry Department Conference, 1931
Source: Seated (left–right): J.J. Boocock, A.J. O'Connor, P.C. Kotzé, J.D.M. Keet, F.E. Geldenhuys, J.E. Kaufman, B.R. Simmons, and N.B. Eckbo. William Watt is standing, second from left.[37]

Conservation foresters focused on conserving the country's limited water supplies through good catchment management. Their interest in the ecology of the indigenous vegetation of the region developed, at least at an official level, from their water conservation goals for catchments. While heavily exploited in the late nineteenth century, and policed by the department in later times, the indigenous forests were not regarded as viable for commercial exploitation long-term, mainly because the trees were difficult to propagate, and are slow-growing.[38]

State foresters working in the plantation forestry program imported introduced timber species which might grow well under South African conditions. Most significant were various species of pine tree, notably species from Mediterranean Europe and Mexico, followed by eucalyptus and acacia (wattle) from Australia. The plantation foresters perceived fire as a significant threat to their project, in part because their young trees could be devastated by surface fires entering the plantations from surrounding veld or farmland. New plantations were divided into compartments by cross-roads and bridle-paths, and were boundaried by wide fire-belts (30-foot belts were stipulated in the 1913 Forest Act), and in some cases by ditches or banks. In the Cape, belts of oak trees, eucalypts or acacias were planted as fire barriers. Within the plantations, dense planting and maintenance (clearing) were applied. In August 1913, Chief Conservator of Forests Charles Legat wrote to all conservators, recommending an article on fire protection by W.H. Miller in the journal *American Forestry*. He noted that while South African foresters were 'well ahead of American forestry as regards the use of living firebelts', in some cases (notably along railways) it might be useful to add trenches or cleaned strips, and create wider 'main firebelts'.[39]

Preventive measures included patrols, provision for firefighting and legislation. As noted by John Henkel in a memo to District Forest Officers in 1907, in advance notice of a new iteration of the regulations published in 1908, many undemarcated forests (including commonages, 'native locations', vacant Crown land) containing valuable timber, were open to the public, woodcutters and others. As draconian measures would be counterproductive, frequent patrols were necessary (draconian measures were applied to Africans in the Eastern Cape). Toolboxes containing firefighting equipment were placed along patrol routes, as early intervention was the key to controlling fires. Legislation differed markedly for undemarcated forests and demarcated (proclaimed) forests. Before 1913, punishments were based on the 1888 Act, amended in 1902, and for undemarcated forests included imprisonment up to 12 months, fines up to £20 or up to 25 lashes. For demarcated forests, imprisonment could be up to three years, fines up to £100 or up to 36 lashes. Damages could be awarded, and rewards were offered for identification of fire-starters. For undemarcated forests, rules applied only to lighting fires within forests (and only 'unsafe' fires). However, no one was allowed to set fire to the vegetation in the open air within a mile of a demarcated forests without giving prior warning to the Field Cornet of the ward, or a forest officer. Advance warning was required for the burning of firebreaks and both parties of adjoining properties had to agree on a date and be present, with labour.[40]

Before 1906 the country's foresters were trained in Europe or India. A school of forestry was opened at Tokai on the Cape Peninsula in 1906, where foresters could be trained with local environmental conditions in mind, but it was short-lived, closing in 1911. Until the 1930s, most of the country's senior foresters were trained overseas or drawn from the students who had attended the short-lived Tokai school. In 1932, Stellenbosch University opened a department of forestry, which became a faculty in 1956. Hutchins, who had come to the Cape from India, was unusual in regarding fire as a natural process in the region. He could have been influential, as he was the first professor at the Cape Town forestry school, but he left the school and the country in 1907.

This was because in 1905 Joseph Storr Lister had been appointed Chief Conservator of Forests for the Cape, rather than Hutchins.[41]

Whereas in the United States forestry research on fire began early in the twentieth century, driven and shaped by the need to prevent forest fires, in South Africa there was little research on forest fires. This was equally the case in many of the other commonwealth territories. What little research was done was undertaken by John Frederick Vicars Phillips (1899–1987). Phillips had an extraordinarily long and varied career, working in Southern, East and West Africa, and has often been described as the 'the father of ecology' in South Africa, though he gave priority to Bews. He began with a short period of intensely original and productive work as a forestry researcher at Deepwalls in the indigenous forests of Knysna in the 1920s. In this period he wrote his doctorate, through the University of Edinburgh, on 'Forest-Succession and Ecology in the Knysna Region'.[42] While doing so, he began to correspond with Frederic Clements, and like his friend Bews became an ardent advocate of the latter's succession theory.[43]

Phillips's research in this period included observations on the ecological consequences of the use of *Acacia melanoxylon* (Tasmanian blackwood) in the Knysna forests to kill weeds and accelerate the regeneration of indigenous trees destroyed by chopping or burning, and for firebreaks. He observed that while this Acacia did kill weeds, it also retarded the natural regeneration of indigenous species, and its numerous and highly fertile seeds were widely dispersed by animals and water. He feared it might prove invasive but did not think it could successfully invade fynbos as the soils were too poor. Phillips also worked to identify which associations of plants were indicators of past 'disturbances' by fire in indigenous forest at Knysna, and what the usual succession of plants was, following forest fires. He developed a narrative of degradation in which fires cleared (or were used to clear) the forests, and then together with overgrazing, kept down the tall fynbos species and prevented any improvement of the pasture through the natural processes of vegetation succession.[44]

After Phillips left Knysna in 1927, forestry research on the influence of fire on forests only got underway again with the work of Coert J. Geldenhuys in the late 1970s. Research on fire in fynbos would begin after 1935, but the first results became available in the late 1940s and were not to influence actual management until the late 1960s. Thus when state foresters were made responsible for conserving the nation's mountain catchment areas from c.1930, they initially relied on agricultural researchers who specialised in grasslands with grazing management in mind. The catchments were mainly grasslands, but included fynbos in the Cape. Foresters were aware of the antipathy to burning and grazing of Cape botanists. They were also influenced by the opinions and experiences of their colleagues working in other British colonial territories.[45]

The British Empire Forestry Conferences

Disruptions to shipping during World War I resulted in major international timber shortages, in response to which the British Forestry Commission was founded. The first of what proved to be a long series of British Empire Forestry Conferences was held in London in 1920. These conferences drew in foresters from almost every known forest ecosystem on Earth, on all the habitable continents, responsible for huge territories stretching across some 40 countries. One of the issues most of them seemed to agree on at first was the great danger posed by fires to forests and plantations, whether this was in the grassy savannas of West Africa or the closed forests of New Zealand and Canada.[46]

Volume Four of Dr Schlich's *Manual of Forestry*, 'Forest Protection', included a chapter on 'protection against forest fires', with a wide-ranging survey of causes, consequences and mitigation measures across Europe, the Americas and India. The chapter is included in the section 'protection against non-atmospheric natural phenomena', but the first sentence asserts that 'forest fires are nearly always caused by human agency'. Recommended protective measures included belts or admixtures of broadleaved species among inflammable pines, and fire-traces (cleared of vegetation by controlled burning) around and through plantations. State legislation and enforcement were judged essential.[47]

Firefighting and fire exclusion was a focus of the second conference in Canada (1923), and again at the third conference, held in Australia and New Zealand (1928). Australia's Inspector General of Forests, Charles Lane-Poole, argued that in Canada, the United States and Australia 'the Englishman' had brought 'no knowledge of the value of forestry with him, and faced with what seemed illimitable areas of timber...set to work and cleared and destroyed all before him'. In particular, 'the grazier...was now encroaching...on the forest areas, and his attack was always heralded by fire'. His sentiments were echoed by the Forestry Commissioner of New South Wales, N.W. Jolly, who noted that although early burning was being used to minimise the severity of incendiary or accidental fires, this was 'looked upon as the lesser of two evils and a first step towards complete protection in later years'. The Conservator of Forests for Western Australia, S.L. Kessell, was more cautious about the merits (and feasibility) of complete protection. He noted both the great dangers of the fire season in his region, and of leaving the indigenous jarrah forests unburned for long periods – but also the ecological importance of fire in the natural regeneration of forests in southern Australia. Kessell proposed that in his region, good management was more about fire control than fire prevention or suppression. In this he had one supporter at the 1928 conference in the person of E.O. Shebbeare, Conservator of Forests for Bengal in India.[48]

Shebbeare recounted that the head of the British forestry service in India, the German-trained forester Dietrich Brandis, had asked a Colonel Pearson to attempt fire protection in the Central Provinces in 1863. Pearson was almost universally expected to fail, but (in the face of considerable opposition from local peoples) by 1881 the protected area reached 11,000 square miles, and 32,000 by 1901. However, as early as 1883, a forester had complained in the pages of *The Indian Forester* that fire protection was reducing natural regeneration in teak forests (sal or *Shorea robusta*). In Burma an article published in 1896 by a forester named Slade started a movement against 'too much fire protection'. The matter was referred to Ribbentrop, Inspector-General of Forests, and he decided that fire protection should be continued. In their reports for 1902–03, two of the four Burma Conservators expressed doubts about fire protection, but the Lieutenant-Governor replied that 'the problem to be solved is not whether fire protection is advantageous, but how such protection can be made effective'. Interestingly, Brandis had advocated controlled early burning as a cheap way of preventing the worst outcomes of fire as early as 1877, and in 1895 Ribbentrop had advised controlled burning in some areas. The long and difficult fight to establish fire protection, however, made any proposal to either allow or use fire in certain circumstances very unpopular with the forestry authorities. In Burma (Myanmar), the argument in favour of stopping fire protection of some teak forests was finally won by Chief Conservator Beadon-Bryant in 1907, and by 1914 this was also accepted for Assam and Bengal. Thus in 1928, Shebbeare observed that in India the view was that 'fire is a good servant though a bad master, and, once it becomes possible to bring it under control, it is often more useful to use this power to direct how and where it shall operate than to suppress it entirely'. He concluded that 'in future...fire protection will tend more and more to become fire control'.[49]

These views were not shared by most foresters from other territories. At the conferences held in 1928, 1935 and 1947, representatives from Canada, Australia, New Zealand, several West African territories,

Nyasaland (Malawi) and Southern Rhodesia (now Zimbabwe) spoke trenchantly of the dangers of fire for forests, soils and water supplies. The chairman of the Forests Commission of Victoria A.V. Galbraith, asserted in 1947 that 'the primary internal consideration in Victorian forest policy is fire protection' (this followed the disastrous fires of 1939), and the aim for mountain forests was 'total exclusion of fires'. Until the 1950s, only foresters from Western Australia, the Indian Forestry Service and Southern Rhodesia (now Zimbabwe) had recommended using fire to aid the natural regeneration of useful trees. The general opinion was still that fire was a major (and usually anthropogenic) agent of environmental destruction, and total fire exclusion was the ultimate aim of good land management.[50]

Informed by both international and local opinion, then, South African foresters concluded that overgrazing and veld burning by farmers were the greatest threats to the nation's water supplies and soils. Farmers, however, believed that the afforestation of catchment areas with introduced trees threatened water supplies more than their own veld-burning. Introduced trees transpire a substantial amount of water, drawing it up from the groundwater table and losing it through their needles or leaves, thus diminishing local flows. Little research was available to clarify which practice depleted water supplies the most.[51]

Following a severe drought in Natal Province in 1933, at a time when the American dustbowl was much in the news, in 1934 Parliament debated a resolution to investigate the effects of veld burning on water supplies and erosion in the country's catchment areas (the effects of burning on run-off had not hitherto been explored scientifically). This was passed in the parliamentary capital in Cape Town, which experienced particularly high fire incidence in 1934 and 1935. As a result, the state began to buy up or expropriate private properties in mountain catchments, to be put under the control of the Division of Forestry. The Division took on extended protection measures in its catchments, and together with the Department of Lands surveyed crown lands in the Cape to decide where and to what extent water conservation measures (notably fire protection) should be applied. The resolution recommended an analytical survey of all of the country's mountain catchments, but this had to wait 20 years. In the interim, foresters impounded tens of thousands of the cattle farmers who were accustomed to grazing on mountain catchments, especially in times of drought, further inflaming the controversy.[52]

The perceived dangers of wildfires and desiccation were thus very much in the public eye in 1935 when the Fourth British Empire Forestry

Conference was convened in South Africa. At the opening session, the uncomfortably positioned Minister of Agriculture and Forestry Colonel Denys Reitz requested delegates to form a committee to advise him on the vexed question of veld burning in the Union. The foresters felt beleaguered, having been absorbed into the Department of Agriculture in 1934, and they knew that influential figures were briefing the Minister against their afforestation programme. In the run-up to the conference, the Chief of the Division of Forestry Johan Keet precirculated an edited selection of recent reports and speeches by important figures, including Pole-Evans and Smuts, in which they opposed afforestation with introduced trees.[53]

Keet had been angered when Pole-Evans took Reitz to visit a river catchment at Elgin and the Department of Forestry's Nieuwberg plantations (east of Cape Town) on 27 February 1935, ostensibly to see an area he wanted to develop as a nursery. Actually (as he minuted in a follow-up note to the Secretary of the Department of Agriculture) Pole-Evans intended to show Reitz the beauty of the indigenous vegetation and the healthy flow of local water courses – because he felt sure that afforestation with pines would destroy both. Keet was not notified of the visit, and a long and increasingly intemperate exchange of memoranda ensued between him and Pole Evans. Keet objected to being instructed by Reitz to create a wildflower sanctuary in the catchment of the Palmiet River at the Nieuwberg plantation, to protect a particular flower which he said was widely distributed in the region and was a fire ephemeral which would disappear in the natural course of succession of the fynbos flora. He pointed out that the Forestry Department's Annual Report for 1934 already listed a series of 'special areas set aside... for protection of wildflowers'. He was angered that the scientific competence of his division was in question, and equally that their dedication to nature conservation was in doubt: 'I, for one, stand back to no man in South Africa in my love for its flora; and the Forest Service has done more than any other... to preserve the flora. But the Nation will never countenance its Forest Service growing only flowers and no timber'.[54]

At the conference, Keet argued that the country's climate and soils were clearly drying out, and it was the responsibility of foresters to prevent this 'by preserving the vegetation and by increasing the area of forests'. He argued it was the yield rather than the quantity of water that mattered, and that forests afforded 'filtered storage' for the country's water supplies. He gave the example of 'French engineers who deforested their pine-covered catchments to increase runoff but had to reforest them when their reservoirs became silted up and their canals stood dry'. He argued against the 'retention and protection... at great cost, of uneconomic grass land or scrub vegetation [fynbos] on water sheds on sites where forests can be grown profitably'. Keet acknowledged he was up against 'many species of alarmists' who accused the department of '[ignoring] the ecological outlook' and planting trees in places where they were 'ecologically foreigners to our climate'. He did not see why acclimatised introduced trees should be 'any more desiccators of the soil and climate, than they are in their native habitat'.[55]

These 'alarmists' included the two major critics of afforestation who spoke at the conference: Jan Smuts and John Phillips, with the latter standing in for Pole-Evans. Pole-Evans' main argument against afforestation with introduced trees was that they 'transpire far more heavily than the indigenous vegetation and therefore make far heavier demands on the underground water'. At the conference, both Smuts and Phillips suggested that more care should be taken over where plantations were established, as in some areas particular eucalypts and wattles were proving detrimental to the soils and water supplies. Smuts argued that foresters had to accept that in some areas grasslands were the climax vegetation, not forest. Both men cautioned that more research was required before afforestation should continue.[56]

Phillips's critique of the South African forestry department's approach to afforestation was ecological. He blamed introduced trees 'for inhibiting the functioning of indigenous biotic communities' (his holistic conception of communities of flora and fauna). In arguing this, he was supporting what he viewed as the ecological turn in European forestry, as espoused by the Finnish forester Aimo Cajander. In addition to then-current arguments over desiccation and impoverishment of the soil, Phillips's own research suggested that introduced blackwood interfered with the regeneration of indigenous forests at Knysna. Smuts also raised considerations of national heritage: commercial plantations were 'sombre, regular, uninteresting in contrast with the riot of wild vegetation', and he noted public complaints about the plantations on the slopes of his beloved Table Mountain.[57]

Both Phillips and Smuts were criticised by the foresters present, for suggesting local foresters were not sufficiently informed of the effects of introduced trees, for mistaken ideas about eucalyptus biology (this from Australia's Lane-Poole), for wishing to prevent further afforestation in the country and for Smuts' remarks on aesthetic concerns. South Africa's P.C. Kotzé (a Rhodes Scholar trained in Oxford) spoke at length on erosion and desiccation and the urgent necessity of protecting the country's catchment areas: 'it is far better to have plantations than to have grazing plus veld burning, or agriculture on very steep mountain sides'. Erosion in plantations was simply a sign of a badly managed plantation. Turning to Smuts' comments, Kotzé responded that:

> it had been said that the time has come for a compromise between the foresters and a nature lover. I do not know what this means.... ours is not a tree country... at least two-thirds of our forest reserves will never be planted with trees... and there will be vast areas for those who love to see the South African countryside clothed with its own indigenous flora.... I think it was General Smuts who referred to compromise. I am sorry I have not got the picture with me, but I remember a little while ago in one of the illustrated papers, the General was shown on the slopes of Table Mountain on a hot day, sitting down on a bed of pine needles under the shade of *Pinus pinaster*. In the picture one could see what appeared to be a patch of wag-'n-bietjie ['wait-a-bit' thorn bush *Asparagus capensis*] and protea bush, but the General did not take advantage of the seat of the wag-'n-bietjie, nor of the shade of the fynbos, but sat under the pine trees. What better compromise could have been reached.[58]

Ultimately, the assembled imperial foresters concluded that their central challenge was to quantify 'the effects of forests on climate, water conservation and erosion', recommending a research programme 'of inestimable advantage both to [South Africa] and the world at large'. This research was deemed necessary in order to address the fears of 'the public' in South Africa about such negative effects resulting from afforestation with 'plantations of exotic pines, eucalypts and wattles'.[59] Veld burning would form a subset of these investigations.

Forestry fire research at Jonkershoek

Integral to the history of fire research and management in South Africa is the strength and persistence of a lineage of scientific research and management developed within state forestry that lasted some 54 years, from c.1935 to 1989 (and, through influential individuals, beyond). A key figure was Christiaan Lodewyk Wicht (1908–78), a local boy educated as a forester in Stellenbosch, before going on to Oxford and the Forest Academy in Tharandt, Germany. Through Wicht the international links and outlook of the department, moulded in the era when South Africa was part of the British Empire (later Commonwealth) forestry network, persisted despite the Afrikanerisation of state bureaucracies that began in South Africa in the 1930s, and the political isolation consequent on the government's Apartheid policies that began to take effect in the 1960s.

The Afrikanerisation of the Forestry Department had begun with the appointment of an agricultural economist, F.E. Geldenhuys, as Chief Conservator of Forests in 1931 (replacing Charles Legat). The talented Colin C. Robertson (1884–1946), first research officer for the department (from 1912), was a victim of this process. Robertson had been trained at Yale Forestry School and was important for his work on introducing tree species to South Africa from Mexico and Australia. He was Chief Research Officer and professional assistant to Legat from 1919, but following serious differences with Geldenhuys, he was retrenched in 1933. Wicht took up the position of research officer in 1934 and became Chief of Forestry Research in 1947, before resigning to become a professor of forestry at Stellenbosch University from 1950 to 1973, where he played a vital role as adviser, educator and mentor.[60] Wicht maintained wide scientific interests within and beyond forestry, introduced rigorous statistically informed experimental techniques to plant ecology and hydrological research in the country, and maintained an international network of contacts. On retirement he played an important role in the drafting of a new master plan for forestry research.

In December 1935, under Wicht's leadership, preparations began to establish hydrological experiments at the Jonkershoek Forest Influences Research Station. The station was established in the fynbos-covered upper catchment of the Eerste River just above Stellenbosch, a site already chosen in 1932. It was conveniently close to the subcontinent's only university-level forestry training school, founded within the Agriculture Faculty of the University of Stellenbosch in 1932. Purchased in 1933, the estate was combined with adjoining Crown land to comprise a 25,000 acre reserve. Wicht remarked that the Estate's 'beautiful old manor house and fine grounds' would be 'maintained in the traditional manner'. He felt it fitting that 'the creed of Simon van der Stel – the tree planter – should be appreciated and fostered on this old freehold, which he granted to Jan de Jonker in 1692' (Jan Andriessen was known as Jan the Jonkheer, or midshipman) (Photo 3.4).[61]

The overall aim of forestry research at Jonkershoek was to make a 'record of what happens to all the moisture deposited on mountain land, with due regard to the natural characteristics of the site, and to variations in treatments'. Veld-burning, with and without grazing, was included among the various experimental treatments designed to test the effects of various factors on the natural circulation of water. The fire component was in part a result of the conflict with farmers over the

Photo 3.4   Jonkershoek Forest Reserve, July 1970
Source: Fred Kruger, CSIR Database of Archival Landscape Photographs, University of Cape Town Libraries.

best use of mountain catchment areas, and foresters expected that the prescribed burning routine would show that vegetation burning by live-stock farmers was more detrimental to water absorption and stream flow than afforestation.[62]

Wicht decided to test an assertion by two Californian scientists that vegetation removal through burning could actually increase water sup-plies. Viehmeyer and Johnston had argued that soil moisture loss is mainly due to transpiration by plants. This idea that burning could *increase* water supplies through reducing transpiration would later be one of the arguments used in favour of prescribed burning as a catchment management practice. However, at the outset of the pro-gramme, Wicht cautioned that 'the long point of view is a *sine qua non* for success in... forestry... we must be prepared to pave the way so that our grandchildren may complete the good work, and benefit by it'.[63]

At this stage Departmental policy in unafforested catchment areas was to prohibit grazing as far as possible and annually burn protective fire-belts 50–100 yards wide around areas 2000–4000 acres in extent with the aim of limiting the spread of veld fires. Belts were burned around the perimeters of indigenous forests, annually in grassland areas and every two to three years in fynbos. It was difficult to control the kindling of fires in forests under the department's control as the extant Forest Act, No.16 of 1913, did not make kindling a fire an offence in itself, even in dangerous localities or weather conditions. The Act only allowed for prosecutions where fires caused damage to forest produce or could be proven to have been left unattended in an irresponsible fashion. In his paper 'Forest Protection from Fire in South Africa' delivered at the 1935 British Empire Forestry Conference, William E. Watt (another Rhodes Scholar) had revealed that the department was spending £28,000–£30,000 per annum on fire protection (including extinguishing fires) for its 3,000,000 acres of Forest Reserves (approximately 5 per cent of average total expenditure for the division in the period 1931–35). In 1935, a little over 90 per cent of this was being spent on protecting the 287,000 acres of plantations.[64]

Wicht's watershed report

Conservation concerns over the deterioration of the country's river catchment areas and soil erosion, coming after a long cycle of low rainfall (1921–49), became prominent in the 1940s. In 1945 Wicht completed a report commissioned by the Royal Society of South Africa on protecting the montane vegetation of the south-western Cape. Wicht worked on the report with the retired forester John Henkel, Cape botanists Robert Adamson and Harold Compton, and the entomologist and conservationist Sydney Skaife. The result was the most important and prescient account of the effects of fire in fynbos, and the management implications of this, published in the first half of the twentieth century.

The Royal Society commission was stimulated by a public furore about fire and a memorandum in response to this by the forester and botanist John Spurgeon Henkel (1871–1962). Born in the Eastern Cape, Henkel was educated at the Royal Indian Engineering College at Cooper's Hill. He worked as a forester in the Eastern Cape, at Cape Town (briefly replacing Hutchins as director of the Tokai forestry school), in Natal and Rhodesia, before retiring to Pietermaritzburg where he remained active as a botanist and author. Henkel wrote his memorandum in response to the public outcry about mountain fires reported in the Cape press in the summer of 1942/1943. Henkel feared that the called-for attempts at fire suppression in the fynbos-clad mountain catchments of the southwestern Cape were doomed to failure, dangerous, and ecologically harmful. His intimations on the role of fire in fynbos, his perspectives on the human dimension of burning, and the clarity and boldness of his management recommendations (notably prescribed burning) are outstanding. Henkel's memorandum also came at the right time: scientific forestry research on fires was well underway, fire incidence was high, and public opinion was threatening to dictate environmental policy. In a major departure from accepted thinking and practice at the time, Wicht et al. concurred with Henkel in many respects.[65]

The important second chapter of Wicht's report, on 'characteristics of vegetation', was written by Robert Adamson of the University of Cape Town (Wicht chaired the commission and wrote all the other chapters). It is striking how Adamson's thinking on fire had developed since the 1920s. He now explicitly linked the diversity and structure of the vegetation to fire, noting that these were only partly correlated with local habitat conditions, and that 'the frequency with which fires have occurred often tends to obscure the more direct relations of the vegetation to habitat'. In most areas 'the vegetation exists... in the form of a mosaic of patches of varying stages of re-growth after fire destruction, and the most obvious boundaries between communities are those dependent on the date and extent of fires'. His own early work in the Mediterranean and Australia enabled him to make comparisons between Mediterranean-climate-type ecosystems, and he noted that 'the vegetation in all these regions is liable to fire damage in the dry season' and 'many of the phases in re-growth after destruction are very much indeed like those here'. While the plants were different, 'the parallelism in behaviour and in the life-form of the plants is very close'.[66]

Adamson examined six major processes alleged to cause the vegetation to deteriorate: 'veld-burning, pasturing, erosion, spread of undesirable species, conversion of veld to other uses, and gathering of flowers'. With regard to burning, he noted that after fires, 'black ruin remains, and casual observers are impressed by the completeness with which the vegetation has apparently been destroyed' and 'they are usually led to condemn veld-burning as an unmitigated evil'. This, some 20 years previously, would have been a reasonable description of Adamson's own view. However, 20 years' on, he continued that

> those who observe the redevelopment of the vegetation for a year or two after the fire, on the other hand, are often equally impressed by the rapid regrowth and the wealth of flowers produced on some sites, and they frequently arrive at the diametrically opposite conclusion, that burning is good and essential to preserve the vegetation.

Adamson did caution that 'it is doubtful whether the evidence is available at present to judge the full effects of burning'.[67]

Regarding management, Adamson repeated Henkel's suggestions on veld burning. Henkel argued that veld burning promoted the growth of grasses and certain characteristic fynbos species, did not result in erosion on mountain slopes, and nor did it affect water conservation adversely. Excluding fire eliminated species which would otherwise be present, and resulted in devastating fires. Fire should in fact be considered as an integral part of fynbos ecology which had been so for centuries. It was wrong to consider fire as solely caused by humans, because under the right conditions lightning and falling rocks could ignite extensive natural fires (noted in forestry department annual reports from 1917). Regarding human influence, deliberate burning to create grazing had been practised long before Europeans arrived at the Cape.[68]

Adamson identified four main strategies evolved by fynbos plants to survive fires: regeneration from soil storage organs (geophytes); resprouting from rootstocks after the aerial parts have been killed by fire; growing from dormant buds protected by insulating bark on stems and branches; and regeneration from seed by woody shrubs. He concluded plants that exhibited the first three of these strategies were unlikely to be threatened by fires. However, woody shrubs that regenerate from seeds (e.g. many proteas and ericas), required four to five years between fires in order to set seeds again.[69]

Marloth had argued that several Penaeaceae species not seen for a century had disappeared in consequence of a series of fires at short intervals. However, one of these (of the genus *Glischrocolla*) was collected again in 1943, and Adamson suggested that some plants which had their vegetative parts destroyed by fire might survive several subsequent fires without flowering, before reappearing. This has remained a vexed question in fynbos ecology, with a number of apparently 'extinct' species reappearing after very long intervals and several fires. A classic example is *Mimetes stokoei*, which had apparently died out in 1945. One seedling was found in 1966, but it died without setting seed. Despite deliberate controlled burns in its last known locality in the Kogelberg, the plant failed to reappear and was regarded as extinct – until 24 seedlings popped up after an intense fire in December 1999.[70]

Discussing the former forester Henri Fourcade's interesting historical comparison of species collected in four forestry divisions in 1843 and 1943, Adamson was sceptical of Fourcade's findings on declining diversity, attributed to burning and pasturing. Adamson had discovered that many alleged extinctions were in fact the product of mistakes in identification, or locality records, or were invented by commercial collectors aiming to sell new or rare plants. Most plants believed lost on the Cape Peninsula were originally found on the Cape Flats, and here construction, the spread of introduced plants and cultivation were 'much more important factors in the extermination than burning or pasturing'.[71]

Adamson pointed out that 'the frequency with which fires re-occur in the same area is often over-estimated. The expression "annual burning of the veld," is loosely used.' In fact, he argued, very few, if any, areas of fynbos can burn annually. Usually it requires three to four years to attain an inflammable state, and longer if it is grazed. Burning that occurs after the minimum interval is often patchy and burnt areas can be reseeded from adjacent, unburned areas. He also questioned Henkel and others' contention that very intense fires generated by older, denser vegetation were more destructive. Research at Jonkershoek was suggesting otherwise.[72]

In sum, Adamson concluded that while there was much to learn about fire in fynbos, 'the destruction may not be as serious as has been generally accepted in the past. Noteworthy and lasting harm is apparently done only on unfavourable sites, and when the burning is repeated at short intervals over a long period of years'. Lasting damage to vegetation by burning was most likely on 'shallow, lean, Table Mountain sandstone soils' on mountain slopes in areas of low rainfall. However, while sanguine about the effects of fire on plants, Adamson maintained the accepted view that fire impacted negatively on soils and water supplies. He argued that 'burning entails the continual interruption and renewal of succession', the most serious consequence of which is the failure of the mature vegetation to build up deep soils rich in humus, capable of storing water.[73]

The greatest threat to soils and vegetation involving fire, argued Adamson, was deliberate burning followed by grazing. No research had been done on the effects of browsing on fynbos, but Arthur Sampson's experiments in Californian chaparral were quoted to suggest that grazing of steep, newly burned slopes measurably increased soil erosion. Photographs of South African landscapes were included in the report as local evidence of this process. The mountain areas were only really suitable for pasturing goats, and these had a disastrous effect on the vegetation. Pasturing wool-sheep on fynbos required burning for grazing, and resulted in the dominance of the unpalatable Renosterbush.[74]

In discussing 'Measures for Preservation', Wicht stated unequivocally that 'the signs of rapid deterioration are unmistakable' and concluded that scientifically-informed management intervention was necessary. Conservation could not be achieved by 'merely excluding all human influence', but should aim at the 'perpetuation of particular conditions or the restoration of conditions known to have existed in the past' (or what is now known as restoration ecology). He acknowledged that current management measures might be proven unnecessary or mistaken by subsequent scientific research.[75]

Wicht tackled fire first, as a major influence requiring management. He pointed out that Europeans had been trying to prevent veld burning since Van Riebeeck's arrival in 1652, without success. Recent attempts by the Department of Agriculture and Forestry to control fire through the use of firebreaks and comprehensive legislation had failed. Indeed, Wicht argued that 'the opinion that total protection is impossible, or nearly so, and that controlled burning is preferable has consequently gained ground and is now widely held'. Neither discouraging deliberate uncontrolled burning, nor using propaganda to discourage carelessness, was effective in preventing the burning of fynbos. Rather, preventing fires required the construction of access paths, wide firebelts and lookouts (manned day and night in fire season), and labour available for firefighting around the clock. This would be too expensive to implement over the large areas of mountain veld involved. Further, successful total protection was not necessarily desirable: the attainment of so-called climax vegetation would result in the disappearance of many species characteristic of interim phases of fynbos succession, notably the striking geophytes. Thus for both financial and ecological reasons, total protection was 'not the universal remedy it has often been supposed to be'. It should only be tried in valuable catchments (providing important water supplies), experimental areas, and where indigenous forests were to be encouraged (he doubted forests of any utility for timber would result).[76]

Wicht's important recommendation was that deliberate controlled burning should be applied to all other areas. On most sites, this would be to 'prevent the over-luxuriant development of the scrub vegetation [Fynbos] in order to decrease its inflammability'. Drawing on Adamson, Henkel and others, he concluded that fire damage to fynbos was not as bad as had been generally supposed, so long as burning was not too frequent and the area to be burned was chosen with care. He suggested dividing areas selected for burning into patches of 40–50 acres in accessible areas, and 200–300 acres on rugged mountains. These should be burned in rotation, and subsequent burns should not lie adjacent to one another, so a patchwork of young and old veld would be achieved. Periods between successive burns should be varied, so as not to disadvantage particular species which may suffer from recurrent burns at fixed intervals. These he identified as species regenerating from seed, which require four or five years between burns to set seed. Burning should also not be delayed for too long, as this would increase the chances of extensive accidental fires. Controlled burning should not be difficult, providing that intelligent use was made of weather conditions, the natural topography and areas of young less inflammable veld, to control the spread of fires. He totally opposed the pasturing of domestic animals on fynbos, especially on mountain slopes.[77]

Wicht and his fellow commissioners concluded that, considering the extent to which the indigenous vegetation of the region was being modified, replaced and destroyed, the only long-term prospect for its protection was the establishment of nature reserves. Drawing on the reports 'Nature Conservation in Britain' by the Society for the Promotion of Nature Reserves (1943), and the British Ecological Society's 'Nature Conservation and Nature Reserves' (1944), the commissioners stipulated that nature reserves (five were recommended) should be scientifically managed by a single authority in accordance with plans which should be revised periodically. The major management interventions to be implemented on these nature reserves would include the application of controlled burning and the removal of introduced (he used the term 'exotic') species.[78] These recommendations anticipate the management plans for the Table Mountain National Park declared a half century later.

In a confidential departmental memorandum to the Conservator of Forests, Cape Town, in 1944, Wicht had already communicated the Royal Society committee's findings that veld burning did not harm fynbos, and he urged him to formulate a statement of policy on prescribed burning. This was despite the fact that the Director of Forestry had replied to the Royal Society commission's call for comment on fire in fynbos in favour of 'the traditional policy of complete protection'. The context for this was a period of heightened interventionism by the state in environmental affairs: a draft forestry bill had been submitted in 1940, and an 'important innovation' was heralded, relating 'to the expropriation of land, or the suspension of the owners rights therein, where reclamation measures or the protection of vegetation have to be undertaken at public expense in the national interest'. A further aim of the new bill was to consolidate the country's laws on forest and veld protection from fire (most from the pre-Union era): there was widespread pressure on government to bring these 'into line with the needs of the State as reflected in national sentiment and demand'. The Department acknowledged that:

It is recognized that in most part of South Africa farming cannot be carried on without veld-burning. Expert opinion, based on empirical observation as well as on scientific investigation, moreover supports the view that burning is not necessarily deleterious to the pasturage if done at the correct season, in proper rotation and if followed by the judicious regulation of grazing.

The aim of the new legislation, enacted the following year as 'The Forest and Veld Conservation Act – No.13 of 1941', was thus not to forbid burning, but rather to 'ensure that veld burning is undertaken with precautions'. However, it was asserted that: 'somewhat different is the argument versus veld burning in important catchment areas in mountain regions, especially in the "bossie"-veld [fynbos] of the southern and southwestern Cape Province, which must be protected, in contrast to the grassveld of the summer rainfall region'.[79]

Wicht 'respectfully' urged that 'a more realistic policy should now be adopted, which would favour controlled burning, except in particular cases where the need for total protection has been specially substantiated'. He felt emboldened to make this recommendation as 'from discussions with you I have concluded that you have also come to conclusions similar to these'. Wicht noted that even Compton, 'a strong opponent of deliberate burning, admits: "In actual practice...it is impossible to exclude fire for ever from the mountain vegetation. There are in fact many indications that fire must be regarded as one of the essential ecological factors in this type of vegetation".' For the Jonkershoek Forestal District, Wicht proposed only attempting total protection in the catchment of the Eerste River, as a trial, and undertaking prescribed burning in the Banhoek, Helderberg and Hottentotsholland regions.[80]

Wicht was well aware that such a change in policy would 'involve the somewhat delicate task of re-enlightening a public, which has been taught to condemn all burning'. He argued that the present moment, when past attempts to totally exclude fire had been obviously proven to be failures, was an ideal moment to publicise 'a clear exposition of the problem and [draw] the obvious conclusions viz. controlled burning or protection at very high costs'.[81] Likewise, the aftermath of the 2000 fires should have been an ideal moment to educate the public about the positive ecological role of fire within natural limits of variation and the merits of prescribed burning. Alas, as in 1945, so in 2000 – no such outcome was to materialise.

4

Science, Management and Fire in Fynbos: 1945–99

> South Africa is drying up on a large scale...Most of this has resulted from long-continued reckless burning-off of mountains and hills that formerly nurtured indigenous water-holding growth.
>
> Hugh Hammond Bennett, 1945[1]

> The basis of sound forestry is management to produce sustained yields. Good farming is permanent farming, deriving sustained yields from the land, without deterioration of soils and water supplies. Conservation of soils and water is therefore common ground for farmers and foresters. It has been one of the chief objects of forestry in South Africa for seventy years.
>
> C.L. Wicht, 1948[2]

In 1948, F.H. Wroughton speculated in the pages of *The Journal of the South African Forestry Association* that fire exclusion might lead to the extinction of fire-adapted species where fire was successfully suppressed for long periods, and harmfully intense fires when fires did finally break out. He advocated experiments in using fire sensibly as a management tool, seeing that 'fire, wind and irresponsibles will always be with us' and bush fires therefore impossible to eliminate. He acknowledged that his views might 'appear unspeakably heretical', but argued that 'when we set out to assist nature we should not discard as utterly harmful a factor which has influenced nature in developing the covering of our soil...merely because fire is absolutely destructive in exotic plantations'. He encouraged South Africans to learn from colonial foresters in India who had had to overcome their European anti-fire prejudices and discover that fires were essential for the regeneration of some indigenous Indian forest trees. In the same issue of the journal, Wicht recommended 'controlled broadcast burning' to 'prevent the luxuriant development of the sclerophyll scrub vegetation in order to decrease its inflammability'.[3]

Based on Wicht's 1944 memo and the 1945 Royal Society report *Preservation of the Vegetation of the South Western Cape* (see Chapter 3), and informed by research conducted at Jonkershoek, in 1948 the Forestry Department decided that controlled burning was acceptable in fynbos catchments. Circular No.15 of 1948 formulated a policy of burning: controlled burns would permanently isolate sensitive areas such as valleys encompassing the headwaters of significant streams, from the wildfires that inevitably burned through the vast inaccessible mountain regions from time to time. This policy was not implemented, however.[4] In fact, forestry management of fire remained inherently defensive, and prescribed burning remained minimal in fynbos until the late 1960s: why?

The campaign to save the soil

In August 1944, Dr Hugh Hammond Bennett, Chief of the Soil Conservation Service of the US Department of Agriculture, toured the Union at the invitation of the Department of Agriculture and Forestry. His report, based on a visit of just over two months, was intended to 'advise the Government on matters of policy and organisation' concerning 'the all-important problem of erosion control'. Bennett condemned 'indiscriminate' veld burning, and was of the opinion that overgrazing and burning were contributing to sheet erosion even in 'good Bushveld' (savannah areas). Crucially, he concluded that the country was drying up on a large scale, mostly as a result of 'long-continued reckless burning-off of mountains and hills that formerly nurtured indigenous water-holding growths'. The state should acquire the most critical of these catchments and set about restoring them to their original condition, notably through adequate fire control. Bennett's parting shot was 'a prophecy of threatening calamity': the 1923 Drought Commission Report 'had some good suggestions', he allowed, 'but these have not been carried out except to a ridiculously small degree'.[5]

After the War, the National Veld Trust vigorously pursued its mission, which was couched in the military phraseology of the time. Edward Roux's booklet *The Veld and the Future* (1946) included a foreword by the chairman of the Veld Trust, ex-High Commissioner in London Charles te Water, in which he wrote:

In the Union of South Africa a national catastrophe, due to soil erosion, is more imminent than in any other Country of the

World... like locusts, ten millions of people swarm upon our good earth and are eating it bare, so that its bones are uncovered and its soil is washing to the sea... We of to-day cannot plead the innocence of our Ancestors. Their careless ignorance has been replaced by our conscious understanding of the soil.

His remedy was characteristic of post-war modernist planning: 'The Conservation of our Soil calls for total planning; the prevention of the erosion of our lands for total war'. Roux's book was a call to arms aimed at 'the Men and Women of South Africa of all races, young and old alike, of the Towns and of the Country'.[6]

The most remarkable figure in the Veld Trust was Thomas Chalmers (T.C.) Robertson (1907–89), who as director from 1950 to his retirement in 1968 almost single-handedly directed a crusade to conserve the country's natural resources. A farm boy who grew up on the grassy plains near Middelburg in the Transvaal, as a teenager T.C. won an award for a collection of 30 grass species. His mother Eliza was a keen botanist who related that her Dutch ancestor Jan Andries Truter had arrived in the Cape in 1722 and became master gardener for the VOC. He was remembered for his efforts to plant heath (fynbos) and grasses to stabilise the sand dunes on the Cape Flats. By his mid-20s, T.C. had forged a successful career as a tough-talking, risk-taking, hard-drinking investigative journalist and editor. In 1939, he accepted Smuts' invitation to become his chief of (pro-Allied) propaganda during World War II. He was chosen for his extensive political contacts and acuity, and writing and campaigning skills. In the post-war period, a political career, or a return to the newspaper world beckoned, but he chose instead to dedicate the remainder of his career to what he believed was the most important issue facing the country and the continent: veld conservation.[7]

As editor and major author of the Trust's magazine *Veldtrust*, Robertson was the ideal person to implement the Trust's mission to produce 'research, propaganda and agitation' that would bring about a new state of mind, a national will for environmental conservation. A barrage of articles were published in *The Farmer's Weekly*, *Libertas*, *The Outspan* and other magazines and newspapers, and the Trust backed the filmmaker C.J.J. van Rensburg whose propagandist anti-erosion film *South Africa in Danger*, edited by the prominent conservationist and Trust member Eve Palmer, proved a great success. The Trust acquired a four-ton mobile cinema which toured the country, H.C. le Page showing films on the dangers of erosion and measures to prevent it. Pamphlets were distributed to schools and school competitions and projects were a mainstay of the campaign. Money was raised through a series of national Green Cross stamp campaigns, the first of which was opened with an address to the nation by the leaders of all three of the country's political parties, D.F. Malan (National Party), J.G.N. Strauss (United Party) and H. Davidoff (Labour). On a policy level, the 1946 Soil conservation Act was modelled on a draft supplied by the Trust. However, officials at the Department of Agriculture resented this perceived meddling with their remit, and in 1953 it was decided the Trust's activities should be restricted to education and information dissemination.[8]

At the heart of Robertson's narrative for the nation was the importance of nurturing the right grasses. Drawing on the martial language of the War years, he described the ecology of the country's vegetation in terms of a 'war of the floras'. The 'southern fynbos' and forests of the winter rainfall areas were pitted against the tropical forest, savannah and grassveld of the summer rainfall areas. Over the course of shifts in climate over millennia, the grasses had driven the fynbos south – 'but there was always trouble lurking behind the lines where there was a botanical fifth column: the woody shrubs and thorn trees'. Drawing on the writings of John Acocks, Robertson argued that it was 'Man [who had] unintentionally helped the thorns in their battle against the Grasses' (much ink was spilled by agricultural experts on the subject of 'bush encroachment'). Over-grazing and incorrect burning were resulting in the replacement of good perennial grasses by less nutritious annual species, which would further deteriorate to leave only bare ground under thorn thickets. Robertson described fire as 'a most powerful weapon' in the 'natural battle of plant succession', which could be used for good or ill.[9]

1948–68: A failure of nerve?

The three major departments with land use oversight in this period had distinct positions on controlled burning. Forestry researchers had concluded it was acceptable in some areas in fynbos and grassland catchments if properly applied. Agricultural researchers condoned properly applied controlled burning in grasslands but not fynbos, and disapproved of its use in catchment areas. The Soil Conservation Board opposed all burning in catchments.

In this portentous period in which the Veld Trust was disseminating a narrative of ongoing environmental collapse across the Union, supported by the Department of Agriculture – with whom foresters were in competition over the most responsible use of catchment areas – foresters were reluctant to make a radical departure in their fire management policy in catchments. They were also unsure about how best to burn fynbos.

Foresters also faced significant practical challenges in this period. The Department was focused on an ambitious afforestation programme, which required the strenuous protection of these plantations of young fire-vulnerable trees from fires. The vital importance of the forestry sector in South Africa had been reaffirmed by a timber famine during World War II, and in October 1945 the Department of Forestry was reconstituted, with William Watt as Director. By 1948, the Department had afforested 163,000 ha of grassland and scrub with exotic timber trees. They aimed to afforest 300,000 ha, amounting to about 21 per cent of the land under their control. Of the remainder, 60 per cent would be kept as natural vegetation managed (primarily) to conserve soil and water. The total was only just under 1.2 per cent of the area of the Union, but included large regions of the vital coast-facing mountain ranges which extend from the Cape Peninsula through KwaZulu-Natal to Limpopo Province, and are the country's major catchment areas.[10]

Increased fire incidence in the forest reserves, and the challenges of implementing fire management on large, remote catchment areas, contributed to the Forestry Department's reluctance to implement controlled prescribed burning in the fynbos as recommended by Wicht. Fire incidence in state-managed plantations and forest reserves increased across the country as a whole from the late 1940s. Annual fire incidence had averaged 45 fires from 1915/16 to 1939/40 (no reports were published from 1940/41 to 1944/45 for reasons of state security), rising to an annual average of 153 fires for 1945/46 to 1949/50, and this increase continued, peaking in the mid-1960s. This was in large part the result of the expansion of state forestry in area afforested, into the inflammable grasslands of the summer rainfall region of the country. By 1956, the Western Cape conservancy had the smallest afforested area of the forestry conservancies, but was second only to the Cape Midlands in total area under departmental control because it included large catchment areas.[11]

Expansions of settlement, cultivation and transport infrastructure contributed to increased fire incidence in forest reserves. The Department was up against human psychology, technology and natural causes (see Appendix 2 on causes). The country's mileage of bituminous roads trebled between 1948 and 1961, opening up large areas of previously inaccessible land to hikers and picnickers, and opportunities for roadside ignitions were greatly increased. Steam locomotives emerged as a

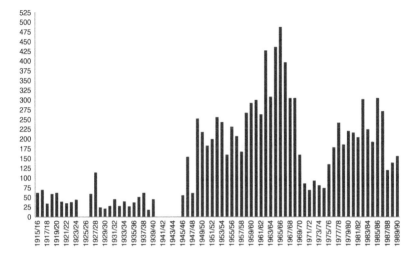

Figure 4.1   Total number of fires on state forestry lands per annum, 1915/16–1989/90

Notes: Afforestation was expanded into eastern South Africa from the 1950s, with afforested area more than doubling from 1930 to 1950, reaching a plateau in the mid-1960s. Increase in afforested area thus partly explains increasing fire incidence from the 1930s to 1960s, but not the dramatic fall during the early 1970s. The 1960s were a peak period for fires caused by steam locomotives and arson.

significant cause of fires from 1958/59 to 1965/66, after which they were gradually phased out, being replaced by electrification or diesel. Property owners, farmers and state and private foresters whose lands were expanding ever deeper into the remaining areas of untouched veld, felt threatened by these fires (Figure 4.1).[12]

Labour shortages made implementing fire protection plans difficult. Forest rangers employed 'coloured or native' labourers to carry out fire protection, detection and firefighting duties. The work was dangerous, difficult and poorly paid, and often undertaken in remote regions with poor transport links. Although labour was always an issue, shortages were noted in the annual reports in the mid-1950s. By 1958, the Department's total staff complement was 2,335 officers and employees (including desk-bound and field workers) and 19,544 labourers, overseeing an estimated 804,271 acres (325,477ha) of forestry land (afforested and unafforested).[13]

With catchments being added piecemeal to the Forestry Department's responsibilities from the 1930s, additional prescribed burning must have seemed daunting. In 1955 the Department decided to formulate a comprehensive fire protection plan for its catchment areas in the Western Conservancy in the Cape Province. Planner P.J. Le Roux formulated a defensive system, advocating rotational burning of very wide breaks on the mountains' lower fringes. An exception was the Cederberg, where he recommended broadcast burning at four-year intervals, but this was only implemented once in this period, in 1959.[14]

The use of fire was restricted (at least in theory) for purposes of soil conservation by the Soil Conservation Act, No.45 of 1946. The Department of Agricultural and Technical Services' advisor on soil conservation, J.C. Ross, wrote that 'veld burning is in many cases inseparable from good veld management practice, but it should be restricted to a minimum and carried out at the right time'. He was particularly concerned about the impacts of bad veld burning practices on catchment areas, which he blamed for 'the drying up of local water sources' and 'disastrous floods' downstream.[15]

The Soil Conservation Act of 1946 provided for fire prevention on private mountain catchment land through the creation of Fire Protection Districts, to be managed by Fire Protection Committees. According to forestry planners Armin Seydack and Stefanus Bekker, this contributed to a significant drop in veld burning for grazing purposes in catchments.[16] The management emphasis was once again on fire prevention and suppression. Foresters believed that the only permissible fires were those burned as firebreaks to protect plantations, an attitude shared by most state foresters in the United States, Canada and Australia in the period. So, what led South African foresters to change their minds in the 1960s about prescribed burning in fynbos, particularly considering the very high fire incidence recorded in forestry plantations and reserves during the decade?[17]

Hydrological controversies and conflagrations

Developments in hydrological research, safety concerns, and doubts over the feasibility of fire exclusion added to the momentum for a reconsideration of the role and use of fire in fynbos catchments. The first results of the Jonkershoek fire experiment had been published in 1947. Assistant Forest Research Officer Brian (H.B.) Rycroft (later Director of Kirstenbosch) concluded that late autumn burning of fynbos could cause flooding and erosion, because the vegetation could not recover in time to check runoff during the winter rains. Spring or early summer burning would allow the vegetation to re-establish itself before the onset of the rains.[18] This seemed to confirm criticisms of veld burning by farmers in the region, who burned in February and March (late summer) to force some fresh grazing before the return of the rains in winter and remove long, hard grasses.

The question of burning in catchments was finally addressed in depth by an interdepartmental committee appointed by the Soil Conservation Board in December 1952. They undertook the survey of the country's catchments first recommended in 1934. This was the country's (and, surmised the commissioners, possibly the world's) first comprehensive national survey of mountain catchments. The resulting report of the Ross Commission of 1961 (J.C. Ross chaired it) identified and ranked the country's 109 major catchments in order of importance as water sources, investigated the technical and administrative challenges in conserving them, assessed the adequacy of existing legislation, and formulated a comprehensive national policy on mountain catchment conservation. They concluded that the burning of mountain slopes, destruction of indigenous forests, overgrazing and trampling of vegetation, and vertical ploughing of upper catchment areas was resulting in 'destruction and devastation'. Of these, 'unsound grazing management on private land associated with injudicious burning' was 'the greatest single factor responsible for catchment degradation in the Union'. Urgent action was required to achieve the 'restoration of the earth's protective covering to hide the shame of its nakedness'. The Commission rehearsed detailed arguments from the 'two schools' on veld burning: the 'total elimination' school and the 'controlled burning' school, concluding that in drier regions with vulnerable soils there should be no burning, but in some humid catchments veld burning was acceptable, and even advisable for the prevention of large accidental fires. They took an ecological perspective: decisions on burning would 'have to be determined for every individual catchment ... with due reference to prevailing conditions ... and local experience'.[19]

In a 1958 technical report on the management of water catchments, Wicht criticised the use of costly engineering remedies at the expense of what he called 'bionomic catchment management'. This report gives us an insight into the theoretical basis for his scientific approach at this time. Synthesising elements of the theories of Clements, Tansley and (for good measure) Smuts, Wicht advocated a holistic ecological approach to understanding and managing mountain catchment areas. He used the term 'biocoenose' as a synonym for Clements' and Shelfords' biome concept, to refer to the 'living community of plants and animals which develops under specific ecological conditions' on a specific site. (The term and concept 'biocoenose' was coined by the German ecologist Karl Möbius in the 1870s.) Wicht argued that catchment management required an understanding of the functions of living organisms, their functions when associated in biocoenoses, and the functions of these biocoenoses understood in interrelation with their physical environments. This would inform a Smutsian holistic approach to catchment management, which synthesised 'organic or vital' and 'inorganic or mechanical' approaches and recognised that the whole was more than the sum of its parts. Wicht deployed Tansley's ecosystems concept to sum up this 'combination of relationships char- acterised by the reciprocal dependence of all the organic and inorganic components upon the whole'.[20]

Wicht maintained that biocoenoses directly impede or delay the progress of the water cycle, and indirectly do so through their inter- relations with climate and soils. The key impacts on water cycles are: affecting evaporation through precipitation–interception and shading; retarding water movement, which promotes infiltration and reduces flooding; and creating deep, stable, absorbent soils. He focused on the importance of transpiration, meaning the transfer of water to the atmosphere by plants. This, he related, had only become clear to researchers in the past 10–15 years, notably from studies in the arid west of the United States. For all but the most denuded South Africa catchments, he estimated, evapo-transpiration accounted for the disper- sal of 'at least half, and often much more' of the local rainfall into the atmosphere.[21]

The key lesson for catchment management was that the loss of water to evapo-transpiration and precipitation–interception was directly related to vegetation density. Wicht identified three main approaches to management, with different outcomes in mind. First, complete protec- tion of the biocoenoses from fire and other disturbances would result in a progression to climax communities: good for flood-control and retaining water in the catchments, but bad for catchment water yield and prolonged flow. Second, in the Southwestern Cape's winter-rainfall catchments, modifying the biocoenoses to reduce their density would reduce interception and transpiration losses, so improving groundwa- ter discharge in streams in quantity and duration. This could be done through slashing, cutting or controlled burning. While the former two methods were safer and retained surface matter (humus), they were often impractical and too expensive to implement in montane areas. Thirdly, maximising water yield could be achieved by totally removing the biocoenoses, but this would result in denudation and erosion, fol- lowed by flooding and irregular stream flows. All three approaches could be supplemented with structural engineering works.[22]

In 1960 Wicht gave his expert opinion on the effects of controlled burning and afforestation on the hydrology of mountain catchments in the winter rainfall region. He modified his three management options to total protection, controlled burning with no grazing and afforestation. He argued that total protection of the fynbos from fire would create unnatural conditions because the vegetation had burned, from time to time, for centuries. Many of the animals which had had important impacts on the vegetation had disappeared, so it was impossible for managers to turn the clock back to a 'natural state' through protection from fire. Total protection would, he believed, result in a climax vegetation of trees and shrubs. This would maximise water loss and trap more water in the catchments – but reduce flood risk. Controlled burning without grazing would prevent vegetation reaching its climax state, so less water would be lost and streamflow prolonged. There would be more of a flood risk (in 1964, C.H. Banks showed that burning fynbos resulted in only short-term increases in runoff following storms). Wicht thus believed veld burning offered the most effective means of controlling water resources in catchments. While afforestation with introduced trees could quickly stabilise denuded catchments, and bring economic benefits, plantations used more water than control-burned fynbos.[23]

There was high fire incidence in the Cape's mountain catchments in 1961, linked to the drought-related granting of extensive grazing permissions in the Cape mountains. Then in July 1962 a fire burned through over 20,000 acres (8,094ha) of veld and 12,274 acres (4,967ha) of plantations in the Outeniqua Mountains in the Southern Cape, and the annual fire damage estimate exceeded ZAR1 million for the first time. The Department concluded that prescribed burning: 'now requires intensive study and a clear formulation of policy. The present policy [of fire protection] is only successful until such time as the vegetation matures, when these fire disasters seem to be uncontrollable despite any protective or preventive measures which may have been taken'. This observation seemed to be borne out by the extremely high overall fire incidence of the 1960s (see Figure 4.1). The Department decided to abandon its system of intensive firebreaks within plantations. Experience had shown that open firebelts, or even the traditional planted belts, did not check fires under really unfavourable conditions. Fires could be dealt with even within planted stands, under less extreme conditions. Further, fire control capacity had been improved through better communications by radio, new road networks, and well-equipped vehicles.[24]

As a result of these developments, in 1965 experimental research focused on the hydrological effects of fire was begun in four fynbos river catchments in the southwestern Cape. A promising young Stellenbosch forestry student named Frederick Kruger was hired to work on fire in fynbos. Kruger had four generations of family involvement in forestry, and shared with his grandfather Johan Keet (Chief of the Division of Forestry in South Africa from 1934 to 1942) an interest in nature conservation. Kruger undertook forestry training to get into nature conservation and his aim was to work in South West Africa (now Namibia). However, Hilmar Lückhoff, then Deputy Director of Research, had other plans for him. Lückhoff had a passion for the Cape – his father James and brother Carl were well known Cape plant collectors and friends of the prominent Cape botanists Rudolf Marloth and Harriet Bolus (Harry's daughter) – and Hilmar wanted Kruger to work on the problem of fire in fynbos.[25]

Hydrological controversies revived

A further spur to a change in burning policy was a hydrological report compiled in response to a resurgence in the controversy over the effects of afforestation of catchment areas. The Forestry Department afforested extensive areas of pastoral and arable land in the 1950s and 1960s, during a period of prolonged drought culminating in 1966. From 1960, the Department received 'numerous complaints' to the effect that 'afforestation was ... directly responsible for the drying up of many streams'. These complaints came in the form of a series of resolutions adopted at congresses of organisations including agricultural unions, the South African Water Catchments Association, local chambers of commerce and the National Wool Growers' Association. In response, an interdepartmental inquiry was assembled to investigate 'afforestation and water supplies in South Africa'. The committee included Forestry, Agricultural Technical Services, and Water Affairs officials, and was chaired by H.L. de W. Malherbe, Secretary for Forestry.[26]

The committee reported in 1968, observing that most complaints came from areas where afforestation had been rapidly expanding in recent times, chiefly the Natal midlands and eastern Transvaal. They concluded that conflict was an inevitable consequence of competition for water resources between established agriculture and commercial forestry interests, and the state should consider adjudicating. The commission argued that protecting natural vegetation from fire reduced stream flow because (as Wicht had suggested) fires were believed to lower average veld age and reduce evapotranspiration. However, this was not regarded as a licence for uncontrolled burning. In order to limit the spread of accidental fires in catchment areas, Section 21 (4) of the Forest Act, 1968, was amended to make landowners or occupiers responsible for clearing firebelts to prevent fire spreading from their properties onto adjacent land.[27]

In 1970, administration of the Fire Protection Committees, which controlled extensive areas of privately owned land in mountain catchments, was transferred from the Department of Agriculture to the Department of Forestry. This was important because many Department of Agriculture officials and Soil Conservation Committees remained vehemently opposed to all veld burning, despite 20 years' of veld burning research at the University of Natal which suggested that correctly implemented prescribed burning was an acceptable agricultural practice. The Forestry Department was further empowered to carry out this mandate through the passing of the Mountain Catchment Areas Act 63 of 1970.

In December 1970 the Forestry Department set out a new policy for catchment area management in the Western and Southern Cape Forest Regions, specifying 'controlled rotation burning in humid catchments in a checkerboard pattern', on a minimum rotation of 12 years. Finally, prescribed block burning of mountain veld was fully accepted as the major management tool for these areas and implemented from the mid-1970s.[28]

Charismatic species and conservation

Wicht had led the Royal Society Commission into the conservation of indigenous vegetation in the southwestern Cape (1945), and state forestry researchers had become increasingly intrigued by the ecology of the unique indigenous fynbos. They came from an established if mostly informal tradition of nature conservation within the Forestry Department, encouraged in the forestry training offered at Stellenbosch University. A series of discoveries about fynbos species reproduction, alongside a noticeable decline in certain species following fire exclusion, encouraged a rethink of fire management from a wildlife conservation perspective.

In 1949 Pieter Jordaan, then senior lecturer in botany at Stellenbosch University, used a study of the life cycle of *Protea mellifera* Thunb (the common sugarbush, now *Protea repens*) to make deductions about when it was safest for these plants to experience fires. He concluded that to ensure sexual reproduction, January–March (late summer)

Photo 4.1   Foresters at Jonkershoek, 1971
Source: Seated, left–right: Diek van der Zel, Christiaan Wicht, Philip Boustead. Standing, left–right: Abel Coetzee, Ross Haynes, Danie van Wyk, Fred Kruger, A.P. Botha, J.D. Kotzé, J.C. Myburgh; Image courtesy of the CSIR, CSIR/SAFRI Jonkershoek collection.

was a safe time for these proteas to experience fire, April–June (late autumn/winter) was unfavourable and July–December (winter, spring) was dangerous. Basing his calculations on the time it took for *Protea repens* to reach reproductive age, Jordaan estimated that the minimum interval between fires should be eight years. He argued that the same calculations could be made for other plant species. Knowing the various safe periods (season), and ideal fire intervals between safe burns, for a range of species, would enable managers to calculate when an area could be burned without drastically reducing the numbers of various species in that area.[29]

In 1953, the Forestry Department acknowledged that fire exclusion from flower reserves was resulting in a dearth of flowers, and in 1957 they resolved to conduct scientific research on the problem. It is ironic that it was particularly the Marloth Flower Reserve on the Swellendam Mountains that occasioned this rethink – Marloth had been one of the most vocal proponents of the idea that fire harms fynbos. In 1959, University of Cape Town botanist Anthony Hall published a paper on the distribution and ecology of orchids in the Muizenberg Mountains of the Cape Peninsula. His five-year study focused on the effects of fire and fire exclusion on these geophytes. Hall realised that fires triggered the flowering of all kinds of rare previously unrecorded species, speculating that this was either because fire stimulated their dormant tubers, or created ideal conditions for germination of seed distributed from elsewhere after fires. As a professor of botany at Stellenbosch, Wicht was sure to have been aware of Hall's findings.[30]

Brian van Wilgen, selected by Kruger to work at Jonkershoek in 1974, argues that the realisation that fire was ecologically necessary in fynbos 'was brought home by the spectacular failure of fire protection policies to prevent the decline to apparent virtual extinction of two rare and charismatic plants – the Marsh Rose *Orothamnus zeyheri* and the Blushing Bride *Serruria florida*'. Discovered and described by Carl Thunberg in the Franschhoek Mountains in 1773, the Blushing Bride was not recorded again until the botanist Peter MacOwan rediscovered a population in the same region in 1891. In the 1920s, the plant collector Thomas Stoekoe and others feared fires would result in the extinction of both of these species. Fire protection of the declining population of Blushing Bride near Franschhoek from 1930 resulted in its total disappearance by 1962. A few plants were then found nearby at Assegaaiboskloof in that year, and the site was cleared of litter as this was believed to hinder germination, with little result. Then, following an accidental fire, underground seedbanks were stimulated and the plants reappeared. This clearly demonstrated the role of fire in the reproduction of some fynbos species.[31]

Fire protection of one of the best known populations of the Marsh Rose, in the Kogelberg Mountains, had similarly resulted in a decline. The coastal region below the mountains had been a grazing farm, burned every year by the Walsh family until in the mid-1930s they sold it to a consortium of Johannesburg businessmen led by the architect Harold Porter. These businessmen developed a series of little holiday resorts named Betty's Bay, Sunny Seas and Pringle Bay, and sought to restrict burning so as to revegetate the extensive sand dunes and protect the buildings they were erecting. The Forestry Department became responsible for the Crown Lands on the mountains above in 1937, and similarly attempted to exclude fires. John Rourke, former Curator of the Compton Herbarium at Kirstenbosch, speculates that it was these restrictions on burning that changed the fire regime of the area and eventually led to the decline of the Marsh Rose. By 1967, there were only six plants left, and Rourke recalls walking the area with the head of forestry research at the time, Hilmar Lückhoff. Lückhoff thought the plants needed fire, but he was worried that his career was on the line. It didn't help that his brother Carl, a well known Cape plant collector and conservationist, had vehemently opposed veld burning and expressed this forthrightly in his well-known book *Table Mountain* (see Chapter 6). Fortunately, Hilmar had the courage of his convictions and authorised a burn. The result was the reappearance of numerous seedlings of the Marsh Rose.[32]

It was helpful to Lückhoff that the rise in environmental awareness internationally and nationally had contributed to a climate of opinion in which the protection of the country's indigenous flora had become a legitimate priority for the Forestry Department. Although the Department had been involved in protecting indigenous flora in catchment areas since the 1930s, before 1972 the word 'conservation' appears in the annual reports always in connection with soils or water supplies. From 1972, for the first time the word was explicitly linked with the conservation of the indigenous fauna and flora. The Forest Amendment Bill (adopted in 1972) provided for 'the separation of wilderness areas in State forest and for more effective protection of such areas so that they can be preserved as they are', and the first three wilderness areas were proclaimed in 1973, two in the Natal Drakensberg mountains and one in the Cederberg. Amendments to Section 21 (4) of the Forest Act of 1968 were now justified as necessary because 'accidental fires take an annual toll by destroying the fauna and flora of the Republic'. In 1973, work began on a 'master plan' for forestry research in South Africa, and this included conservation forestry research focused on: the effects of current management measures on water supplies in catchment areas; ecological research on catchment ecosystems to better conserve biological diversity in natural environments; reclaiming drift sand areas; and planning ecologically sensitive outdoor recreation facilities for the public. The Forestry Research Institute established a separate department of Conservation Forestry Research in 1974.[33]

The new emphasis on nature conservation in the Forestry Department was part of a wider movement in forestry in South Africa. This motivated the 'Our Green Heritage' campaign organised by the South African forestry industry and launched in the Western Cape by State President

Jacobus Fouché (a farmer and former Minister of Agricultural Technical Services and Water Affairs) on 11 February 1973. The aims and values informing this campaign were set out at a symposium organised by the South African Forestry Association (established 1938) in Stellenbosch in May 1973. The President of the Association was Johan Keet, and the Chairman of the Executive Committee was Hilmar Lückhoff. Key papers were published in a special issue of the Association's journal entitled 'Our Green Heritage'.[34]

In their 'Green Heritage' article, Wicht and Kruger described the challenges facing the Department of Forestry following the recent consolidation of responsibility for catchment management to the Department. Catchment management had formerly been divided between the Forestry Department for Crown land, Agricultural Technical Services for private land and Bantu Administration and Development for the so-called African Homelands concocted by the Apartheid state. The authors noted that the cooperative efforts between these departments had been limited to protective measures to improve water supplies, and had not included management for nature conservation and outdoor recreation. The primary goals of catchment management were now water and nature conservation. Nature was to be protected for scientific, aesthetic and social reasons. Facilitating outdoor recreation, the appreciation of mountain beauty and relaxation were identified as important secondary goals, of great social and economic (notably overseas tourism) value. Wicht and Kruger encouraged young scientists to take up the challenge to research and manage the '*terra incognita*' of the mountain veld: this was an exciting opportunity to combine a stimulating and satisfying career with valuable *landsdiens* ('service to the nation/the land': a Christian agricultural youth movement of this name was started in South Africa in 1914, and still exists).[35]

A renaissance in fire research

Once prescribed burning was again recommended as a water and nature conservation management tool in the 1970s, researchers needed to work out an ecologically sound method of applying it to fynbos. The still dominant Clementsian equilibrium-based approach to understanding the role of fire in vegetation (and the allied idea that fire was an unnatural disturbance) was unhelpful. More than this, there was still widespread popular and expert disapproval of burning to overcome.

The wider environmental context, certainly for managers but also the public, was that the 1960s had been a period of exceptionally high fire incidence in forestry areas nationwide. Section 12 of the Forest Act, 1968, had conferred powers on the Minister of Forestry to forbid the making of fires within five kilometres of the boundaries of state forests, and the fire risk was judged so severe in the Western Cape Forest Region in the summer of 1970/1971 that on 26 February the Minister issued a government notice enforcing this for the first time. There were numerous fires in the region in 1972, including large mountain fires. Fire protection committees were active, and reconstituted Committees were appointed for the Cape Peninsula and Hottentots Holland mountains. Very dry conditions persisted into 1973, with 500 fires reported in the region overseen by the Cape Peninsula Fire Protection Committee, 43 of these serious. Controlled burning was hazardous in these conditions, and a burn carried out by the Fire Protection Committee in the Central Langeberg raged out of control.[36]

Wicht had long experience of all of this, and in 1973 he gave a series of talks on fire in fynbos intended to promote the necessity of controlled prescribed burning. In his first lecture, Wicht acknowledged the urgency of doing something about the 'destructive, uncontrolled veld fires' which lit up the mountains with 'red chains of fire' at night. However, he also pointed out that humans had used fire in the region for centuries, and the fynbos had burned for thousands of years. The plants which had survived over the long term all had strategies for coping with fire, and he identified five main strategies. All fynbos plants fitted into one of these, and he argued that fire (without grazing) was not as big a threat to these plants as was generally believed.[37]

In his second lecture, 'Fire: Evil or Asset', he maintained that total fire protection over a long period would result in the development of climax vegetation. However, this would result in the loss of interesting and attractive interim stages of the fynbos plant succession, with their characteristic plant and animal species. Total protection in areas with more than 30 inches of annual rainfall could result in the development of indigenous forest, but for drier areas, total protection had more mixed results. The climax vegetation would be dense fynbos scrubland which was the most inflammable vegetation type in the southwestern Cape, and used up a lot of water. Research showed that reducing the density of vegetation would improve water supplies, and in the drier fynbos regions, controlled burning should become standard practice. He acknowledged that he was 'daring' to go against the 'ingrained belief' that fire was always bad.[38]

In his final lecture Wicht noted that efforts to prevent fires through propaganda and punishment, from the days of Simon van der Stel in the late 1600s, to the present, had failed. Total protection was just about possible around plantations, but much too expensive to apply extensively. The solution was a system of controlled veld burning, and he offered four pointers for 'healthy' burning. First, the fire return interval must not be too short or too long: 10–12 years was recommended. Second, size and shape of burn must be determined by local circumstances: in accessible well-populated areas burn units should be small (43ha would do), whereas in remote mountain areas burns could be ten times that size. Smaller burns offered the chance of reseeding from plants bordering on burned areas, and allowed fauna to more easily escape burning plots. Very extensive units risked causing erosion as large areas would be denuded of vegetation. Burn units should be shaped by the local topography, following ridges and using natural barriers as borders. Third, the best season to burn was early summer. Late summer burning left the ground bare before the winter rains, risking erosion and flooding, and spring burns could damage flowering species. Fourth, fires should be burned downhill as such fires burn more slowly and are easier to control. Still, cool days were best. Wicht advised on how to plan the burn cycle for an area, to ensure all units were burned within the chosen fire return period. In conclusion, he admitted that knowledge of the ultimate outcomes of controlled burning was incomplete, but argued that the problem was too urgent to delay action. Existing knowledge was sufficient for practical recommendations.[39]

The more detailed scientific research on the effects of fire Wicht said was still necessary was hampered by a lack of adequate theoretical tools to tackle the complexity of fynbos ecology, and the lack of empirical data on the life cycles of individual plant species. Developments in theoretical ecology and interactions with the international group of ecologists cohering around a comparative study of Mediterranean-type ecosystems (later named MEDECOS) gave South African fynbos ecologists the tools necessary to rethink fynbos ecology.

International scientific influences

In 1953, the American ecologist Eugene Odum published his textbook *Fundamentals of Ecology*. He addressed the criticism that ecology was not a proper science by proposing the ecosystem concept (Alfred Tansley's idea) as the organising principle of a scientific approach to ecological questions. Odum's book quickly became the major textbook in ecology, remaining a core text into the early 1970s. The ecosystem concept, as developed in the United States by Raymond Lindeman, focused on nutrient cycling and energy flows between biotic and abiotic (living and non-living) dimensions of the environment. This offered a flexible and general approach to explaining both very small and very large aspects of the biosphere and the full range of environmental problems. A novel feature of Odum's book, which showed the influence of his father, the sociologist Howard Washington Odum, was an emphasis on applied ecology and the social role of ecologists. Odum argued that ecosystems ecology offered the means to comprehend and mend the planet's damaged systems (much American environmentalism in the first two decades after World War II focused on pollution). Following his father's New Deal values, Eugene Odum emphasised cooperation and interdependence in nature and human society. Ecologists should manage nature for the benefit of humankind, an all-embracing task for the discipline which he sometimes described as 'human ecology'. They should provide government with expert technical guidance to ameliorate or avoid the harmful side effects of new technologies like nuclear power.[40]

The problem was that just as public (and expert) calls for an engaged, applied ecology were emerging in the period of rising environmental awareness beginning in the 1960s, serious splits were emerging among theoretical ecologists. Where theorists like Eugene Odum put a positive spin on the ecosystem concept, with cooperation and harmony achieved through group selection, others argued that evolutionary adaptation can only happen on the level of the individual organism, and not at group level. The concept of linear succession with its implications of direction and destination was challenged by Drury and Nisbet, Buzz Holling, Robert May and others, who instead emphasised chaos and complexity and what Holling called the 'resilience of non-linear ecological systems'.[41]

Amid this ferment of environmental concern, technological optimism and theoretical fragmentation, in 1961 the International Union of Biological Sciences (IUBS) resolved to undertake an International Biological Programme (IBP). The theme was to be 'the biological basis of productivity and human welfare'. The IUBS were confident that scientists would soon be able to manipulate 'the global equilibrium and balance sheet of organic materials and resources, so as to improve the balance sheet by increasing production and reducing losses'. This flagship for big biology ran from 1967 to 1974, focusing on large-scale studies of the functioning of ecosystems.

Nature was presented as a system which should be controlled and managed in such a way as to remain in a state of equilibrium. This would remove the need for the natural disasters which would otherwise occur in order to return its (particularly anthropogenically) perturbed ecosystems to their natural steady state. The cybernetic model of nature, and of society, represented them as closed systems that could be controlled and manipulated by experts from the outside. This positive view of the potential of the ecosystems approach to reveal the structure of the ecosystem as machine and specify the conditions under which it would remain in good running order (stable) had obvious appeal to scientifically inclined bureaucrats. It also chimed with the conceptions of nature presented in key texts of the environmental movement such as Buckminster Fuller's *Operating Manual for Spaceship Earth* (1969).[42]

MEDECOS

The IBP had been 'the first attempt to look at the Earth's ecosystems in a functional manner', which was necessary in order to develop ecosystem models. Also in the late 1960s, a small group of ecologists began making comparative studies of Mediterranean-type ecosystems. They realised that the matching climates and apparently similar flora of these far-flung regions in Australia, Chile, California, the Mediterranean basin and South Africa provide a remarkable opportunity for studying the evolution of ecosystem function. The sclerophyll ecosystems (characterised by plants with tough, evergreen leaves) of Chile and California appeared to be a result of climate-driven convergent evolution. Here, two very different and distant ecosystems had evolved similar physical and functional attributes. The IBP California–Chile Convergent Evolution Project was set up to investigate this further, and the meeting which initiated this collaborative research program in March 1971 was the first of what has become an on-going series of conferences called MEDECOS from 1984 (the term Mediterranean Type Climate regions (MTCs) is now preferred to Mediterranean Type Ecosystems).[43]

According to the American ecologist Harold Mooney, a founder of MEDECOS, basic (as opposed to applied) ecologists traditionally sought to understand the interrelationships of organisms with their environment through evolutionary time, seeing human activities as 'disruptions of these relationships which resulted in disequilibrium conditions'. During the IBP, ecologists had discovered that 'virtually all ecosystems have been influenced by man' and these impacts were increasing. An important realisation linking basic and applied ecology more closely in this period, was that factors which disturb so-called equilibrium conditions, such as fires, storms and droughts, are often important *ongoing* influences on the structure and functioning of ecosystems.[44]

The second MEDECOS meeting, held at Stanford University in 1977, included scientists from Mediterranean-type ecosystems in Mediterranean Europe, South Africa and Australia. It was a 'symposium on the environmental consequences of fire and fuel management in Mediterranean-climate ecosystems', bringing together basic and applied ecology at the level of ecosystems. At this meeting Fred Kruger, by now the dynamic coordinator of South Africa's Conservation Forestry research program, presented a pioneering paper on the 'ecology of Cape Fynbos in relation to fire' – a summary of what was known (or suspected) at the time.[45]

The third MEDECOS was held in 1980 in Stellenbosch, despite South Africa's pariah status at the time as a result of its Apartheid policies. It seems that fynbos, as one the world's few, isolated Mediterranean-type ecosystems, was of sufficient scientific interest to override such considerations. A guiding question from the previous meeting was 'the degree to which ecological principles, derived from research on particular populations, communities or ecosystems, are generally valid, and hence also to what extent resource management principles are general'. The Australian botanist Raymond Specht had challenged the northern-hemisphere notion of Mediterranean-climate biome types first advanced by the nineteenth century European biogeographers August Grisebach and Andreas Schimper. He showed that the ranges of Australian and South African sclerophyllous shrublands extend well beyond the Mediterranean-climate regions, and correlate more closely with (nutrient-poor) soil types. Thus this third MEDECOS meeting focused on 'the role of nutrients in species and ecosystem convergence in Mediterranean-climate ecosystems'. Another organising question was, why are these ecosystems so rich in plant species, and how is this diversity maintained where nutrients are scarce?[46]

A number of answers were proposed to these latter two questions in particular, and in relation to fynbos, fire was attributed an important role. Forestry researcher William Bond advocated the 'humpbacked model of diversity' to explain fynbos diversity patterns. Here, species diversity is low in 'stressed, arid habitats' and also 'high biomass sites with strongly expressed dominance' (e.g. dense stands of tall shrubs). Diversity is high in 'intermediate conditions'. In this context, fire acts like a non-selective herbivore consuming all plant types, thus reducing dominance by any one of them. Diversity will therefore be lower in communities that experience very frequent fires, and also in those that experience very infrequent fires.[47] This neatly explains the deleterious impacts for fynbos diversity of human-induced changes to fire regimes, through either increased sources of ignitions (e.g. by farmers, hikers or picnickers) or fire suppression (by foresters).

In a wide-ranging and prescient paper, Kruger synthesised the available information on 'the dynamics of Mediterranean-climate vegetation in relation to fire regimes'. Following the dominant trend in ecological studies, he focused on the relatively (anthropogenically) undisturbed MTCs of South Africa, Australia and California. Despite the concerns expressed at the 1977 meeting on the consequences of different national fire management strategies for presumably very similar ecosystems, ecologists still preferred to avoid 'confounding' human activities.[48] In South Africa, the focus on Forestry-run mountain catchment areas resulted in the neglect of the lowland fynbos and renosterveld areas, with serious long-term consequences for their conservation.

Kruger articulated the emerging view that fire and its effects was one of the 'major features common to these different shrublands' and that the general succession model advanced by Clements and others did not apply. Rather than developing from simple to complex species-rich climax communities following fires, in fynbos, species diversity *decreased* with age after fires. Van Wilgen had demonstrated that fynbos shows increased species richness in the first years after a fire, with the rapid disappearance of short-lived 'fire ephemerals' being followed by a gradual loss of species during succession resulting from the suppression of understorey (lower-growing) species by overstorey species.[49]

Kruger's discussion of the degree of convergence between the Mediterranean-type regions in responses to fire was framed by the Australians Malcolm Gill and Richard Groves' work on fire regimes. It has since been recognised that this was a major breakthrough in how ecologists understand fires and fire effects: for the first time researchers and managers had a sound framework for understanding the complex and dynamic role of fire in vegetation. The components of fire regimes as understood at the time were 'fire type, intensity, seasonality, and frequency'.[50]

When fires occur, they can result in markedly different impacts on plant species. In 1965, Jordaan had shown marked seasonal effects of fire on the regeneration of seeding species in fynbos, with summer and autumn fires favouring their regeneration, and winter and spring fires resulting in reduced populations. Forestry researchers found that to manage fires with seasonal effects in mind required them to pay attention to regional variations in fire seasons. In the Western Cape these were found to vary considerably within a six-month range, for example 52 per cent of fires in the Cederberg occurred in summer and 32 per cent in autumn, 51 per cent of fires in the Groot-Swartberg occurred in summer and 24 per cent in spring, and the southern coastal mountains showed a higher incidence of big winter fires. This had broad ecological implications for regional prescribed burning programmes.[51]

Because South African forestry researchers had shown little interest in forest fires, fire intensity was not understood. Fire intensity is an amalgam of temperature (release of heat energy) and how fast the fire moves. Variation in plant responses to fires is in large part determined by variations in species' sensitivities to heating, and thus researching fire intensity was vital for understanding the effects of different fires on biodiversity. In 1974 Kruger spent a sabbatical in Australia where he studied Alan McArthur's work on measuring fire behaviour. Kruger later favoured the more technical approach to fire modelling developed by Richard Rothermel at the Intermountain Fire Sciences Laboratory in Missoula, Montana. Also in 1974, Kruger encouraged Brian van Wilgen to work at Jonkershoek, where his postgraduate studies focused on fire in fynbos. Brian was later sent to the United States and Australia to study fire behaviour (Photo 4.2).[52]

Kruger summed up the current thinking on fire and diversity in Mediterranean-type ecosystems by observing that these systems experience fires of intermediate intensity and frequency. He speculated that this might account for the diversity of modes of regeneration displayed

Photo 4.2   Brian van Wilgen monitoring vegetation on a firebreak, Jonkershoek, c.1980. The fire lookout station is just visible above the measuring board
Source: Image courtesy of the CSIR, CSIR/SAFRI Jonkershoek collection.

(a range of seeding and sprouting strategies). These fire regimes are pre-dictable, yet display variance in aspects of fire behaviour, frequency and seasonality, which prevented the competitive exclusion of any species following a particular survival strategy. Spatial variation in fire inten-sity created a diverse patchwork of microhabitats. Kruger argued that: 'The research interests of ecologists and resource managers are largely coincident now. Alternative fire-management options must be selected on the basis of the predicted effects of different regimes on community structure and diversity.'[53]

For Kruger and his researchers, it was vital to work out an ecolog-ically sound method of applying prescribed burning to fynbos. Fire regimes were believed to differentially affect populations of fynbos shrubs that reproduce by seeding, and those that reproduce by resprout-ing. Sprouters were regarded as usually subordinate to seeders in fynbos, and only markedly favoured over seeders by very low (>40 years) or very high (<10 years) frequencies of fire. Management-oriented research was thus focused at this time on the population dynamics of sprouting species. In the interim, a rule of thumb was applied where the slowest maturing species in any fire management compartment was allowed to mature to the point where 50 per cent of individuals had flowered three times or more, before burning was allowed.[54]

Kruger presumed that the principle of convergence in plant life-history strategies would make it possible to generalise on the basis of detailed knowledge of a few species. To facilitate this, he urged further research on the demography of 'representative' plant species, which he hoped would 'reduce apparent vegetation complexities to intelligible patterns'.[55] For this, new intellectual tools were required.

New ecological approaches

A South African who studied ecology at Exeter University in the 1960s, William Bond came to forestry from a background of plant and soil surveys. Bored with the pattern-finding nature of South African ecol-ogy in the 1970s, he had been energised by the dynamics of lake ecology while working on a project at Cahora Bassa in Mozambique. In 1976 the Forestry Department offered Bond a job because of his soil and vegetation mapping experience (he was interviewed by Wicht and Lückhoff). The Department had been given control over all State catchment areas, and wanted to inventory what was there. Bond was asked to find a method of classifying fynbos that would enable managers to work knowledgeably without having to master the full complexity of fynbos taxonomy. While working on this, at Saasveld near George, he became interested in the dynamics of fynbos vegetation, specifically the way fires could result in large shifts in vegetation structure and composition.[56]

In 1976 the Forestry Department's burning policy for fynbos was still focused on burning in spring to maximise vegetation cover in winter, with the aim of reducing runoff and erosion in winter. However, Bond's experimental work in the southern Cape showed that burning successively in spring (or winter) could drastically reduce some *Protea* populations. This work contributed to a switch to autumn burning – but then it was realised that fire season varies regionally in the fynbos.[57]

In this period, Bond was impressed by a meeting with Martin Cody, an American ecologist of the mathematically-oriented MacArthur school and a founder of the MEDECOS project. According to Donald Worster:

> MacArthur was convinced that the structure of any ecological community, the fauna and flora living together in a place, was determined by the interaction of organisms, not by fortuitous external factors like climate. If he was right, these interactions could be plotted and predicted, and the community structure described likewise.[58]

Cody visited the Cape doing research on how clutch size varied in birds between tropical and temperate climes. The way ecological theory equipped him to see structure and relationships in fynbos even though he was seeing it for the first time convinced Bond of the value of acquiring such an intellectual toolkit. The Forestry Department funded his doctorate at UCLA with Cody as his supervisor, where Bond developed an interest in evolutionary questions including the influence of fire as a disturbance (he worked in the chaparral). A year after returning to South Africa, Bond joined the botany department of the University of Cape Town, where he dedicated a significant proportion of his research to understanding the role of fire in various ecosystems including fynbos, co-authoring a monograph *Fire and Plants* with Brian van Wilgen (1996).[59]

Another influential alternative theoretical perspective to Clementsian succession theory was supplied in this period by John Harper of the Department of Agricultural Botany, University College of North Wales, Bangor. In his 1967 paper 'A Darwinian Approach to Plant Ecology', Harper pointed out that in striking contrast to zoological research, plant population biology had hitherto been woefully neglected. The challenges were considerable, but nevertheless it was necessary, he argued, to 'trace the fate of individuals within populations'. It was necessary to examine the responses of numbers of plants, and of individual plant size, to changes in the density of populations. Populations living in environments with recurrent natural hazards, he wrote, 'may spend most of their time recovering from the hazards'. Population size thus might be a 'function of the magnitude of the last catastrophe and the time available for recovery'.[60]

Focusing on plant-to-plant variation revealed dominance and suppression occurring within a single species stand (a population). The self-limitation of populations of a species might prevent a fight to the death between different species inhabiting the same space at the same time (a plant community), permitting diversity. For plants, Harper puzzled, it was 'not easy to see how a group of species which all require the same basic food requirements ... may possess sufficiently diverse biologies to prevent a best species from excluding all others'. Research in fynbos would later suggest that the different responses of plants to particular fire events could provide one answer to this question. Harper presented a fully-worked out version of his ideas in his book *Population Biology of Plants* (1977). The book bursts with ideas concerning the kind of demographic questions then being tackled by ecologists working on fynbos.[61]

The fynbos biome project (1977–89)

Despite South Africa's political isolation, South African scientists remained well connected to international scientific networks into the early 1980s. This was particularly true for forestry researchers, where strong imperial links forged across the Anglophone world were maintained. Communications between the British Commonwealth forestry networks and South Africans (South Africa was out of the Commonwealth from 1961 to 1994) were continued via the Oxford Forestry Institute. Further, the International Union of Forest Research Institutions (IUFRO) opposed any ban on race, creed or politics, and the category 'special guest of IUFRO' was developed by Professor Jeffrey Burley (Director of the Oxford Forestry Institute until its closure in 2002) to enable South Africans to attend their five-yearly meetings. The International Council of Scientific Unions (ICSU) also opposed the exclusion of scientists on political grounds, and South Africans participated in the Scientific Committee on Problems of the Environment (SCOPE) program on ecological effects of wildfire initiated in 1979. This exposed

South African foresters to the pioneering fire behaviour work of a local agricultural researcher, Winston Trollope.[62]

South Africans were excluded from some major international initiatives, notably the IBP. They responded by setting up their own program of biome projects coordinated by the Cooperative Scientific Programs secretariat of the country's Centre for Scientific and Industrial Research (CSIR), led by Brian Huntley. The Fynbos Biome Project was initiated in 1977, and over the next decade a new generation of ecologically minded researchers and land managers revolutionised the understanding of this biome. The extraordinary outpouring of cooperative research completed during the project's lifespan (1977–89) would be synthesized in *The Ecology of Fynbos: Nutrients, Fire and Diversity* (1992), edited by one of the project's major contributors, Richard Cowling, then a botanist at the University of Cape Town. Fred Kruger was instrumental in setting up the Fynbos Biome Project, and forestry researchers including William Bond, Brian van Wilgen, David Le Maitre and David Richardson played a key role in unlocking the complex effects of fire in fynbos. Regional Directors of forestry in the Western Cape (Gerhard Wagner and later notably John Fenn) instituted regular meetings to sustain continuous exchange between researchers and managers, in order to integrate current knowledge and management in fire management, water resource protection and control of invasive plants. Planners like Armin Seydack parsed research for work with management implications, and managers had input on the practicalities of implementing ideas. This institutional innovation was later named MAREP (Management, Research, Planning), and it facilitated the close coordination of research and management of the fynbos catchment areas.[63]

In contrast to the ecosystems approach of the other biome projects, the fynbos researchers adopted a population–community ecology approach. (They did attempt an ecosystems approach to their catchment hydrological programme, but according to Kruger this proved unexpectedly challenging in complex, highly diverse fynbos ecosystems.) The population–community ecology approach was based on the newfound understanding of the ecological role of disturbances – notably fire. Also important was the MacArthur school's theory that ecological communities are determined by the interactions of their fauna and flora and not by external factors like climate, and John Harper's methods for studying the interactions of plant populations and communities.[64] Numerous studies were made of the effects of different intensities and frequencies of fire on the dynamics of particular fynbos plant populations and communities. They confirmed that 'variable fire regimes maintain coexistence by preventing the long-term dominance of any single species'. Individual fires impact differently on the various reproduction strategies of individual fynbos species.[65]

Management implications

Linking plant diversity with fire regimes offered a useful approach to managing the complexity of fynbos ecosystems in that fire regimes could be used as a proxy for biodiversity. The effects of management interventions on biodiversity are notoriously difficult to monitor. In effect, pyrodiversity (within safe ecological thresholds) could maintain biodiversity. The challenge, as Richard Cowling succinctly put it, is that there is no one optimum fire regime for conserving diversity.[66] Having significantly extended their theoretical understanding of fynbos ecology, then, researchers set out to establish ecological guidelines to inform management practice.

Deciding the frequency of burning in such a complex biome was perhaps the major management challenge tackled by the fynbos researchers. The Australian scientists Ian Noble and Ralph Slatyer's 'vital attributes scheme' (1980) was used as a means of developing generalisations about the responses of plant communities to variations in fire frequency. A number of 'indicator species' were classified according to how they respond to fires. For instance many serotinous *Protea* shrubs showed local extinctions under frequent fires, and regenerated poorly after fire intervals of more than 40 years. Intervals of 8–30 years were recommended for them. *Mimetes splendidus* matured and shed seeds (after fire) at five to six years, giving a minimum period for this class of species. The suppression of understorey species as a result of long inter-fire intervals, and patchy regeneration and possible erosion following fires in old fynbos, were used to suggest an upper limit of 25–30 years. Because sprouting species are longer lived and can survive fires even before maturity, non-sprouters were used to dictate the recommended frequency of prescribed burns. This approach resulted in a general recommendation of fire frequencies of 10–25 years, and by the mid-1980s this was the approach adopted by most managers.[67]

A survey of fynbos species led to the recommendation that 'the late summer-early autumn period is the best time to burn' (March/April in the region), generating intense fires. This revised management practices based on agricultural and early Jonkershoek research, and perhaps, vindicated the practice of farmers in the region. These intense fires favour serotinous proteas, the dominant shrub in most fynbos areas, the presence of which also correlates well with overall fynbos plant diversity. Serotinous proteas hold their seeds in fire-proof fruits in their canopies, releasing them when hot fires burn up the parent plants. However, variation is required as some fynbos species do better under less intense fires. For managers, delivering intense fires proved challenging because this is the most dangerous time of year to burn with a high risk of runaway fires. Dense stands of invasive plants further complicated the picture, because they caused abnormally intense fires which have adverse ecological effects. This hampered prescribed burning in invaded areas because these plants had to be removed before burning.[68]

The difficulties faced by managers who could only safely burn in March/April, with limited suitable fire weather days within that period, stimulated research on fire danger rating systems and fire behaviour prediction models for fynbos. With the cooperation of researchers at the Northern Forest Fire Laboratory in Montana, United States, Jonkershoek forestry researcher Brian van Wilgen adapted the US National Fire Danger Rating System for fynbos. The sophisticated US approach was preferred over the field-oriented Australian approach, but it was highly technical, and by 1990 was judged not to have been a success with managers in the field.[69]

Threats

Aside from fire, the major challenge identified for management of fynbos with conservation of species diversity as a goal was invasive plant control. Invasive introduced (or more pejoratively, 'alien') plant control in nature reserves was first noted as a management goal in the Forestry Department's annual reports of the mid-1950s. In 1960 'plants/weeds' appears for the first time under 'Harmful influences'. By the late 1970s invasive plant control was official policy and included in management plans for Fire Protection Committees. Strategies included mechanical, biological and chemical control. Trees and shrubs of the genera *Pinus*, *Acacia* and *Hakea* were believed to threaten many fynbos species with extinction. Fynbos Biome Project researchers credited plant invasions with causing the extinction of 26 fynbos species, and 750 species were regarded as 'at risk' should existing burning and invasive plant control programs break down.[70] (I have already noted the difficulty of predicting the extinction of fynbos species by fire.)

The rapidity of spread of invasive species in, and transformation of, fynbos ecosystems unmodified by humans was regarded as unique. A puzzle in fynbos is that in its natural state it is virtually treeless, and yet bioclimatically well suited to tree growth. David Richardson

and Fred Kruger argued that many fynbos shrubs are focused on sur-
viving fire and exploiting the brief post-fire period, lacking capacity to
develop dense fire-resistant stands later on. It is the capacity of invasive introduced trees and shrubs to most effectively exploit the brief
invasion windows caused by fires (which kill non-sprouting tall fynbos
shrubs), that enables them to invade fynbos communities. While this
kind of local extinction and invasion is a normal part of fynbos succession, the difference is that these introduced invasive species are
more fire resilient. They mature quickly, well within the usual fynbos
fire return period (around eight years), and establish persisting dense
stands which exclude competing indigenous species. These stands burn
more intensely than fynbos, but are less inflammable. Thus the non-
equilibrium system usually maintained by fire in fynbos is arrested and
the only way to reintroduce it is to fell the adult invasive plants and
burn the area following seed release before seedlings can mature.[71] This
became the management practice of the time.

Ironically then, contrary to the views of early Cape botanists and ecologists, it is not fire disrupting the equilibrium of fynbos that destroys it,
but rather the disequilibrium maintained by fire that maintains fynbos.

Fynbos species diversity was also being threatened by attempts to
keep large mammals in fynbos nature reserves, for example the Cape of
Good Hope and De Hoop nature reserves, and Bontebok National Park.
In 1971, the small fynbos- and renosterveld-clad Helderberg Nature
Reserve boasted 3 eland, 1 rooihartebeest, 4 springbok, 7 bontebok,
2 duikers, 25 steenbok, 21 grysbok and 3 wildebeest. Large grazing
mammals require grasses, which meant that managers turned to unnaturally frequent burning to promote grasses over shrubs. Agricultural
researcher Winston Trollope had shown that in the Eastern Cape you
could (in his interpretation) reverse the succession of fynbos to grassland
through sustained frequent burning. A problem for fynbos is that
grazers and browsers concentrate on post-burn areas, consuming a
disproportionately high quantity of fire-stimulated reproductive parts.[72]

Amidst this ferment of activity, the CSIR's Cooperative Scientific
Programs began to be phased out. Researchers in the pure sciences complained that funding was being spread too thinly, and indiscriminately.
From 1985, Jacobus ('Jack') de Wet (1913–95), an Oxford-educated
mathematical physicist, engineered a funding shift from the collective
model towards a system which rated and rewarded individual excellence. This brought to a close a period of sustained excellence and
productivity in ecological research in South Africa which has yet to be
replicated. The Fynbos Biome Project was wound up in 1989.[73]

There had been a revolution in fynbos ecology, but what of management? Overall, the implementation of the new scientifically informed management policies had proved challenging. Although fire incidence on forestry lands plummeted in the early 1970s, the inclusion of block burning from the mid-1970s did not prevent large fires. In 1979, some 38,257 acres (15,482ha) of catchment were burned in the Cederberg Mountains, and 13,640 acres (5519ha) in the Groot Winterhoek Mountains. Two forestry employees were killed, and two seriously injured, fighting mountain fires. Block burning in rugged, remote areas was dangerous. In 1982 a block burn got out of control in the Du Toitskloof, and workers and equipment had to be airlifted out. A fire in the Hex River Mountains burned for 18 days.[74]

In 1983 and 1984, dry and difficult conditions hampered block burning, notably in the Western Cape Forest Region. In 1984 there was an increase in uncontrolled fires in South Africa, one of which killed nine Forestry Department employees. Difficult weather conditions and labour shortages again restricted fire protection and block burns in the Western Cape in 1985. Van Wilgen and Richardson calculated that, for the Western Cape, there were only 12 days per annum of weather suitable for burning within the season deemed ecologically acceptable by researchers (March/April). Thus fires often had to be burned under ecologically undesirable conditions, and in dangerous conditions with a risk of runaway fires. Veld fires burned 76,158 acres (30,820ha) of mountain catchments in the Western Cape Forest Region in 1986 (665ha of plantation were burned). A shortage of funds restricted fire prevention measures, and firebreaks were only burned where absolutely necessary. Combined with unfavourable weather, this meant many prescribed burns were again postponed, with the result that the mosaic (or 'checkerboard') effect intended to buffer against accidental fires was not realised. It is unclear whether this contributed to the unprecedented number of large veld fires recorded in the southwestern Cape in 1988, burning an estimated 249,121 acres (100,816ha).[75]

The demise of state forestry

In the face of a falling gold price, sinking exchange rate and international sanctions, budget cuts bit deep in South Africa in the 1980s, proving fatal for state forestry as a research and management organisation. State President's Minute 1109 of 18 November 1986 dictated the devolution of the management of mountain catchments to provincial authorities as of 1 April 1987.[76] The catchments were handed over to inadequately resourced and (when it came to large-scale use of fire as a management tool) inexperienced provincial nature conservation departments from 1992. Many experienced forestry staff did not transfer to provincial departments. The long-term mountain catchment experiments were terminated in 1990, and the Forestry Research Institute was moved to the CSIR to become the Division of Forest Science and Technology. Its remaining government-funded projects expired in 1995. Key management figures like Fred Kruger left (to private consultancy, in his case). Those who remained in the CSIR struggled to find funding for research on fire in fynbos.[77]

In 1993 the Assistant Director of Conservation Forestry Armin Seydack, and head of mountain catchment management Stefanus Bekker, produced a review of how the research-led policies and management guidelines of the mid-1980s had played out. It makes for depressing reading. While some managers had burned to rejuvenate the veld, 'for most it had effectively degenerated into a fire hazard reduction operation'. Fires were burned in the wrong season under the wrong conditions, sometimes under pressure from neighbouring farmers experiencing drought who mistakenly believed burning would increase water supplies (possibly a misinterpretation of forestry hydrological research). Rigid application of controlled block burning removed the variability required to maintain co-existing species and thus species diversity.[78]

By the late 1980s the Department's budget for prescribed burning and invasive plant control was being diverted to meet the booming demand for recreational facilities in forest reserves. The rise of popular environmental awareness in the 1970s had been registered, and encouraged, by the Forestry Department's 'Our Green Heritage Campaign'. Ironically, in the period in which prescribed block burning was introduced with conservation aims in mind, an increasingly vocal, conservation-minded public pressurised the department to exclude fire from fynbos. The Department was also concerned about the safety of the rapidly increasing numbers of people using their outdoor recreation areas.[79]

The budget for prescribed burning and invasive plant control was also being sapped by fire protection measures made necessary by the expansion of farms into mountain land. In 1989 fynbos flower harvesting generated an estimated ZAR29 million and there had been 'a flurry of claims for compensation, running into millions of rand, for lost income after runaway fires'. In 1991, the State paid approximately ZAR2.8 million in compensation to landowners, after farms used for flower harvesting were burned after a lightning fire crossed onto their lands from the Kogelberg State Forest. This inhibited managers responsible for prescribed burning. Social upliftment became a necessary priority in the region, further boosting labour costs (for state-funded organisations) to well over the rate of inflation (15 per cent), a rate not met by budget increases. Researchers feared that the consequent reductions in prescribed burning and invasive plant control could result in fynbos species extinctions (Figure 4.2).[80]

Finally, managers contested the effectiveness of prescribed burning to prevent large fires. They noted that from 1981 to 1989 far larger areas of the Western Cape catchments were burned by wildfires (943,698 acres/381,901ha) than by prescribed fires (330,817 acres/133,877ha). The introduction of rotational prescribed burning in the Cederberg National Forest had reduced fire frequency, but on average, wildfires had doubled in extent.[81] This may have been the result of failing to implement block burning to create mosaics of veld ages, intended to limit the spread of large wildfires. Nevertheless, as a result of all these problems, the Department's strained resources were refocused on invasive plant control. By the early 1990s, the controlled block burning system had collapsed and in effect, 'no systematic fire management [was] being carried out over a large proportion of mountain catchment land'.[82]

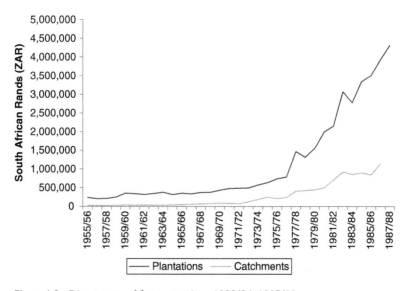

Figure 4.2 Direct costs of fire protection, 1955/56–1987/88

Source: From 1986 to 1989 total costs (damage, extinguishing costs and protection costs for plantations and catchments) was ≥R6m, and in 1990, the total cost was R28,758,197 (9581ha of plantation burned).

In addition to prescribed burning of catchments, much defensive burning had been undertaken by state foresters responsible for plantations of introduced trees (they opposed the use of fire within plantations throughout). In 1990 however the Government decided to devolve its commercial timber activities, creating a parastatal, the South African Forestry Company Limited (SAFCOL), to take over most of its plantations. Discontinuities in staffing and training impacted negatively on managers' and labourers' willingness and capacity to burn.[83]

This collapse in the nation-wide management of fire in key catchment areas of South Africa was exacerbated by the fragmentation of land management that occurred in the mid-1990s. Under the new African National Congress (ANC) led government from 1994, development was devolved to the level of regional government. There were no longer any centralised national organisations to maintain fire control measures and undertake firefighting. Environmental management became a provincial responsibility under the new, smaller provinces. The new 'joined-up' municipalities were now responsible for all the lands around their built-up areas to the borders of neighbouring municipalities. Many did not have the competency or staff to undertake fire prevention and firefighting measures. In contrast to the situation in France or Australia, for example, in South Africa fire brigades were only trained to deal with structural fires and were out of their depth when confronting veld or forest fires. The provincial conservation organisations similarly lacked the necessary training. In terms of fire control, this was a potentially chaotic situation. Extremely high fire incidence in fynbos catchments in 1998 was attributed to 'a steady decrease in prescribed burning'. Further experienced staff were lost in the Western Cape when many were offered, and accepted, severance packages in the late 1990s.[84]

In 1998, the fractured legislation on wildfire regulation and oversight of fire management was finally unified under the National Veld and Forest Fire Act No.101 of 1998. Reflecting the management vacuum, this Act made landowners responsible for forming Fire Protection Associations, formulating fire protection plans, undertaking fire prevention measures, monitoring wild fire occurrence and developing capacity for firefighting on their lands. Some interpreted this as an abnegation of responsibility on behalf of the state. Tax-paying citizens were required to organise their own fire prevention and firefighting, both skilled and dangerous activities. They were not provided (as was the case in Australia, for example) with state-funded training and equipment. The Fire Protection Associations system was not a great success. According to Zane Erasmus, Programme Manager for Fire for CapeNature, in many areas people still expected foresters to come and burn as they always had done. Some commercial foresters still did so, if they had timber in the area, but not if there were no assets to be protected. When South Africa became a democracy, state forestry was folded into the Department of Water Affairs and Forestry, with enabling and regulatory rather than a management or research roles. At the time of writing, it is a branch of the Department of Agriculture, Forestry and Fisheries.[85]

In terms of the 1998 Act, it was the responsibility of the Minister to implement a national fire danger rating system (to be tailored to local conditions). It was the responsibility of local Fire Protection Associations to keep their members constantly apprised of fire danger ratings. Unfortunately, no such fire danger rating system existed, and local conditions vary significantly across the country. An ecologically informed national fire danger rating system would only be ready in 2005.[86]

The period of institutional collapse and transition that was the fallout of South Africa's political revolution of the late 1980s and 1990s left environmental research and management in a state of collapse, completely unable to support the legislation carefully designed in an attempt to cope with the mess. In 1999, a staggering 182,006ha of veld burned, including two fires of 30,000 and 70,000ha apiece that each burned for a period of 17 days, in February and March, respectively.[87] While this may not have been an ecological disaster, and historically big fires like this may not have been unusual, by the late twentieth century exurban settlement and the development of vineyards and orchards on mountain slopes had significantly increased the magnitude of the fire risk. The scene was set for a major fire disaster.

Part III

Fire on the Cape Peninsula, 1900–2000

5
Fire Geography and Urbanisation on the Cape Peninsula

Along the Twelve Apostles open veld festooned with various attractive indigenous species including watsonias has been destroyed by the building of hideous flats and houses. Milnerton, Blouberg [and] Camps Bay [have been] ruined by unsightly buildings. It would seem that this 'progress' and beauty...cannot live together.

> G.P. Blake, 'Peninsula's Ugly Face',
> *Cape Times*, 27 October 1971

The environmental lobby has been increasingly active and has kept us all very aware of the sensitivity with which our surroundings and heritage are to be treated. It is essential however that a balance be maintained between preservation and development.... I am sure my colleagues... will apply themselves... to 'direct the great sources of power in nature for the use and convenience of man; being that practical application of the most important principles of natural philosophy which has, in a considerable degree, realized the anticipations of Bacon, and changed the aspect and state of affairs in the whole world'.

> D.G.D. Riley, City Engineer, City of Cape Town, 1989[1]
> (citing the English engineer Thomas Tredgold)

The rich flora of the Cape Peninsula has evolved in the context of a complex mountainous topography formed over millennia by a range of geological processes, of a winter-rainfall summer-drought Mediterranean-type climate, and fire. The Peninsula's fire regimes are shaped primarily by vegetation types and age, weather patterns and events, and the incidence of ignitions. Human interventions have

117

influenced all of these. For centuries, vegetation cover has been destroyed, fragmented and transformed through the actions of hunter-gatherers, the development of livestock farming, agriculture, forestry and urban areas and infrastructure. Anthropogenic climate change has affected weather patterns, increasing the incidence of conditions conducive to fires. Humans have introduced new sources of ignitions progressively over the past 2000 years, with an unparalleled escalation over the course of the twentieth century. Part I traced some of the major environmental impacts of human settlement and land use from 1652 to 1900, and this chapter provides a framework for thinking about the impacts of urban development in the twentieth century.

The geology of the Peninsula

The sphinx-like profile comprising the 669m-high peak known as Lion's Head and the 2.5km ridge running north to the 'rump' of Signal Hill conceals a clean break between the two major geological formations underlying the Peninsula. The Malmesbury Group shales to the north are remnants of the late Precambrian Adamastor seafloor, and the Cape Granite to the south intruded into these shales as molten rock 630–540 million years ago, crystallising 8 km below the surface. Lion's Head is an isolated remnant of Peninsula Formation Sandstone sitting atop a narrow band known as the Graafwater Formation. These two topmost geological strata on the Peninsula were formed from sands eroded by rivers from adjacent and long-disappeared high mountains, which were washed down onto flat alluvial plains and gradually compacted into sandstone rocks between 500 and 450 million years ago.[2]

The upper northern slopes and cliffs of Table Mountain that loom sheer above Table Bay – joined by the narrow ridge of the 'saddle' to Devil's Peak to the east – are exposed Peninsula Formation Sandstone. This thick sandstone crust is separated from the underlying granite, and in the case of Devil's Peak shale, by a narrow band of Graafwater Formation rock comprising sandstone, pink siltstone and maroon-coloured shale. These pinks and purples are the result of oxidised iron, and are easily distinguished from the ochres and greys of the Peninsula Formation sandstone above. This is the structure of the Table Mountain chain extending southwards along the fissured sandstone towers of the Twelve Apostles towards Llandudno and the isolated outcrop of the Karbonkelberg above Hout Bay on the West Coast, and from Constantia Mountain in a dogleg through to Muizenberg on the False Bay coast. The sandstone peaks of the southern Peninsula, running down to terminate in tumultuous seas churning beneath sheer cliffs at Cape Point, are separated from those of the northern Peninsula by a sandy gap running west to east between Noordhoek and Fish Hoek.

The Peninsula's sandstone-derived soils are acidic and very poor in nutrients. The soils derived from shale or granite are less acidic and somewhat richer in nutrients. This combination of a variety of relatively nutrient-poor soils, and complex montane topography interspersed with sandy flats, has contributed to the extremely high level of plant diversity and endemism on the Peninsula. The renosterveld favours the more fertile Malmesbury shales and granitic soils, while the fynbos thrives on the nutrient-poor sandstone soils.[3]

Climate and biotic influences on biodiversity

The Peninsula has an extremely high level of floral diversity with 2285–2572 indigenous species (tallies vary) and over 400 naturalised introduced species. Endemism is high, with 194 near-endemics (≥80 per cent of populations or numbers of plants exist only on the Peninsula), and 90 endemics (occur nowhere else). In addition to the complex topography and soils, this is also associated with the significant variations in rainfall and temperature experienced across the region. The Table Mountain plateau receives on average 1000–2000mm of rainfall per annum; Cape Town city bowl gets 600–800mm, the Southern Suburbs 1000–1500mm, the Cape Flats 500–600mm and the southern Peninsula 400–700mm per annum. Aspect and winds account for some of this variation: the summer southeaster provides significant precipitation at higher elevations on the southern- and east-facing slopes. In general, the wetter higher and eastern slopes are moister and experience fewer fires, though in extreme conditions almost everything burns. The Peninsula's few patches of Southern Afrotemperate Forest inhabit ravines on the eastern and southern slopes of the mountains, notably above Newlands and Kirstenbosch and at Orange Kloof, which are moist and sheltered from the winds, so keeping fires at bay. In winter, northwesterly winds frequently reach gale force, as do summer southeasterlies. Both can quickly dry out the fine-leaved fynbos, rapidly fanning a small fire into an unmanageable blaze.[4]

The role of insects and birds in pollinating specific plants has also been important in evolving a complex flora on the Peninsula. Cape sugarbirds *Promerops cafer*, endemic to the Cape Floral Kingdom, are particularly adept pollinators of *Protea* and *Mimetes* species, and overly frequent fires has reduced their numbers as the larger, slow-to-seed shrubs decline. Ants play an important ecological role by burying the seeds of hundreds of species, so protecting them from predators and fires, until heat or smoke triggers their germination.[5]

Vegetation and fire

Cowling et al. completed the first comprehensive vegetation map of the Peninsula in 1996, recognising three major plant formations as follows: Cape Fynbos Shrublands (92 per cent), Renosterveld (5 per cent) and forest and thicket (3 per cent). Their map recognised 14 vegetation types including ten varieties of fynbos. The fire-prone parts of the Peninsula fall mostly within their categories of 'renosterveld and grassland', and 'mesic proteoid fynbos'.[6]

Subsequent vegetation mapping – in most cases reconstructions of what was believed to exist in now-developed areas, generated from proxies like soil types – has generated slightly different categories and varieties of vegetation. It is well for the layperson confronted with authoritative maps and statistics to remember that these are all (well-informed) approximations based on assumptions about historical vegetation distributions and varying approaches to defining vegetation types (Map 5.1).

Mucina and Rutherford's (eds.) vegetation map of 2005 shows ten vegetation units on the Peninsula, including five units of fynbos. Examined alongside fire incidence data, it appears that the most fire-prone vegetation types are Peninsula Shale Renosterveld and Peninsula Sandstone Fynbos. Peninsula Sandstone Fynbos is described as 'a medium dense, tall proteoid shrubland over a dense moderately tall, ericoid-leaved shrubland', characterised by proteas, ericas (heaths) and restios (rush-like plants).[7]

The Peninsula's fire regimes

Rebelo et al. venture fire return intervals of 10–20 years for fynbos in the region. Kruger's rule of thumb (see Chapter 4) proposed that the minimum interval between fires should be based on how long it takes the slowest maturing species in any fire management compartment to mature to the point where 50 per cent of individuals had flowered three times or more. Based on this ecological rule for fire-return intervals, Van Wilgen and Forsyth (2008) recommend a minimum period of 10–12 years between fires on any one area of fynbos on the Peninsula. Fire-free intervals of longer than 35 years may result in senescence (reduced vigour and reproductive capacity) of some fynbos species, notably the endemic silver trees. The currently recommended season for burning is summer or autumn.[8]

Peninsula Shale Renosterveld is described as 'tall, open shrubland and grassland, typically with Renosterbos not appearing very prominent'. Rebelo et al. ascribe the grassiness to 'frequent fires and lack of grazing', though they note that on parts of Devil's Peak the renosterveld is mowed for grazing. They commented in 2005 that 'this vegetation burns every 3–5 years to the consternation of Cape Town citizens', though large areas on Signal Hill have been protected from fire for as long as 25 years. Estimates for 'an ideal fire frequency' in renosterveld vary from two to seven to three to ten years. Apparently, increases in fire frequency since 1970 have not had negative ecological effects on Renosterveld.[9]

If as Seydack et al. argue, fuel (inflammable vegetation) age only significantly affects fire size in drier fynbos areas, then fire size on the Cape Peninsula depends primarily on fire weather conditions and ignition rates, rather than fuel loads.[10]

No reliable scientific data exists on which to base accurate predictions of the study area's pre-1970 fire regimes. More data is available for the Cape of Good Hope (southernmost) section of the Table Mountain National Park, a nature reserve since 1965. I have largely excluded this area from my study as my focus is on the urban–wildland context of fire on the Peninsula. CSIR researchers Greg Forsyth and Brian van Wilgen conducted a study of fire regimes of the parts of the Peninsula now incorporated into the Table Mountain National Park over the period 1970–2007. They concluded that, based on a recorded 373 fires covering 45,757ha in that period, the mean fire return interval for the whole area was 22 years. However, in fynbos, mean fire return intervals declined from 37.9 years in the 1970s to 12.6 years between 2000 and 2007. The area of fynbos experiencing short-interval fires (≤6 years) during 1970–89 grew nearly threefold in 1990–2007. For renosterveld, the increase was nearly fivefold. Overall, more than 90 per cent of fires burned in summer or autumn.[11]

If we rely solely on data about variations in the inflammable nature of vegetation types at different fuel levels, and relatively short-term fire incidence data, we will fail to understand the fire regimes of the Cape Peninsula. Using this kind of evidence it has been assumed that the Peninsula's fire return intervals fall between 12–25 years for Fynbos and five to ten years for renosterveld. In reality, fire weather (optimum atmospheric conditions for fire ignition and spread) is dynamic, and fynbos and renosterveld respond quickly to this. Human ignitions are also unpredictable.

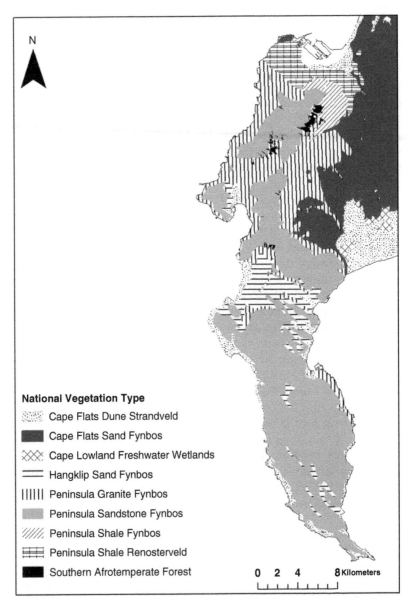

Map 5.1 Indigenous vegetation of the Cape Peninsula, past (above) and present (opposite)

Source: These maps are derived from A.G. Rebelo, P.M. Holmes, C. Dorse, and J. Wood, 2011, 'Impacts of urbanization in a biodiversity hotspot: Conservation challenges in Metropolitan Cape Town', *South African Journal of Botany*, 77, 20–35, since updated and groundtruthed, and supplied by courtesy of the City of Cape Town's Environmental Resource Management Department.

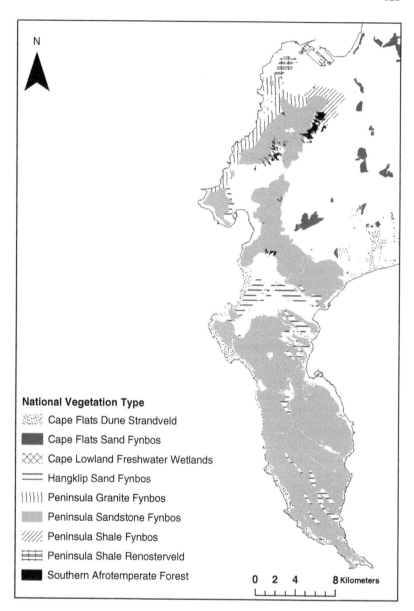

National Vegetation Type
- Cape Flats Dune Strandveld
- Cape Flats Sand Fynbos
- Cape Lowland Freshwater Wetlands
- Hangklip Sand Fynbos
- Peninsula Granite Fynbos
- Peninsula Sandstone Fynbos
- Peninsula Shale Fynbos
- Peninsula Shale Renosterveld
- Southern Afrotemperate Forest

0 2 4   8 Kilometers

Map 5.1 (Continued)

Diane Southey has argued that fires in the Western Cape region are not primarily driven by fuel accumulation, but rather ignition incidence and weather patterns. In winter, the South Atlantic high pressure system is at its northernmost and polar frontal (low pressure) systems move from west to east over the Western Cape, bringing cool temperatures and rain. In summer the South Atlantic high pressure system is at its southernmost, blocking the advance of polar fronts. The system moves from east to west, causing dry, hot sunny conditions and onshore southeasterly winds. Southey used self-organising maps (SOMs) and sea-level pressure charts to generate 12 discrete synoptic states typical of circulation patterns over the Western Cape. She analysed these alongside fire incidence records (1970–2007), revealing a significant relationship between fire incidence and synoptic states. Most fires occurred in summer in periods, when the alignment of the South Atlantic high pressure system (at its southernmost) and a low pressure system over the western mainland, brought high temperatures and desiccating southeasterly winds to the Western Cape. The frequency of these synoptic states increased over the period, and region's climate is predicted to become warmer and drier in future.[12]

The other dynamic variable is anthropogenic ignitions, but ecologists and managers have very limited data on this. They do assert that in the later decades of the twentieth century, areas were burned more frequently than their minimum recommended fire return intervals, with 11.4 per cent of the fynbos area of the Table Mountain National Park burning at intervals of ≤6 years since 1990. To begin to understand the long-term dynamics of anthropogenic influences on the Peninsula's fire regimes, we must look to historical sources.

In the course of the twentieth century, humans began to dominate the fire regimes of the Peninsula. Large-scale transformations of the vegetation cover as a result of increased afforestation with introduced trees (and plant invasions) played a role, but most significant were unprecedented rates of urban and infrastructural development, and of the diversity, distribution and frequency of ignitions.

Rural Peninsula, c.1900–40

Until the late 1940s, the Cape Peninsula retained significant undeveloped areas used by livestock owners for pasturing livestock. Livestock were kept for transport, meat and dairy products. Until the early

1930s the municipal authorities recorded large numbers of sheep and cattle trespassing on their plantations and open areas. The Colony's foresters impounded 1210 cattle on the Cape Flats in 1903, and 245 head of livestock were impounded on municipal lands. Throughout the 1920s figures remained high on municipal lands, dwindling to double figures on the northern Peninsula after 1933. This decline is linked to (a) the electrification of Cape Town and development of refrigeration, (b) improved rail transport linking Cape Town to its hinterlands and (c) the replacement of animal-drawn transport with motorised vehicles, trams and trains, notably from the 1930s.[13]

The pasturing of livestock on the commons and the mountain slopes above the City raised fire incidence because graziers burned to create fresh grazing. There was a dairy farm in Upper Gardens in the City Bowl until the early 1900s, Rondebosch Common and the grassy renosterveld on Signal Hill were used for grazing, as were the slopes above Muizenberg. A map surveyed in 1932 shows a dairy farm and municipal stables at Muizenberg, and still showed Salt River Outspan (an area kept for pasturing draft animals) north of Maitland, now Ysterplaat Aerodrome. As late as 1940, eight bush fires on City Lands were attributed to burning for grazing, and 357 cattle were impounded at Silvermine in 1942. In an account of a 'great bush fire' on Devil's Peak which threatened Vredehoek Estate just above the City Bowl in 1944, we read that 'a dairyman at the corner of St James and Derry Streets removed his cattle out of danger'. In 1951, Lückhoff related that frequent fires kept areas near Kasteels Poort, Hout Bay Nek and the seaward slopes of Signal Hill in an 'unnaturally' grassy state, and 'many of the larger heaths [*Ericaceae*] are today represented by dwarf shrubs less than a foot high'. In the second half of the century, small herds were still kept on the southern Peninsula, notably around Noordhoek and Kommetjie.[14]

Rhodes' estates and amenity areas

Urban development on Table Mountain's eastern slopes was significantly checked by the efforts of Cecil Rhodes (1853–1902), who made a fortune on the Kimberley diamond mines in the 1870s and held a seat in the Cape House of Assembly for the last 20 years of his life. In the 1890s, Rhodes purchased extensive properties on the eastern slopes of the mountain in order (reported his architect Herbert Baker) to save it 'for ever from the hands of the builder' and to educate and ennoble the people 'by the contemplation of what he thought one of the finest views in the world'. Rhodes bought the old Cape Dutch Groote Schuur residence in 1893 and had it remodelled by Baker, filling the gardens with a riot of exotic flowers. Large areas of natural vegetation on the estate were cleared to form park-like grazing lands for imported deer, and he created grazing lawns for a range of large African antelope. When he died he left the estate to the nation, stipulating that it could not be developed to provide suburban residences. Land on the Estate was later allocated for the establishment of a university (now the University of Cape Town), and for the Kirstenbosch National Botanical Garden.[15]

In addition to Rhodes' extensive estates, several categories of areas were protected from urban development during the course of the century, managed by a host of different management authorities. The major categories of those important for terrestrial biodiversity were Council land on the slopes and plateaus of Table Mountain set aside for water conservation, suburban parks and commons set aside for outdoor recreation, privately owned agricultural lands (notably vineyards and orchards), defence forces land, state, municipal and private lands earmarked for afforestation with introduced timber trees and municipal and provincial nature reserves.

Urbanisation and infrastructure

Following the discoveries of diamonds in the 1860s, and gold in the 1880s, Cape Town boomed, the population of the city and its suburbs growing from an estimated 45,000 in 1875 to 171,000 by 1904. There was a major influx of British troop as well as refugees from the Transvaal following the outbreak of the South African (or Anglo-Boer) War in 1899, but many left following the peace in 1902. There would be similar spikes in population during World Wars I and II as troops passed through the city.[16]

By 1900, previously outlying villages had become substantial municipalities, and new suburbs were developed. The forester David Hutchins commented that inappropriate materials were being used in the building boom following the end of the South African War. In 1903 he noted that 'Cape Town during the year has been subject to disastrous fires, so disastrous that the whole community is suffering from enormously increased fire insurance rates; yet, the inflammable foreign Pitch-pine continues to be used in preference to the non-inflammable British Jarrah'. Hutchins combined safety concerns with colonial solidarity, advocating importing 'British' colonial timber (Australian eucalypts) over 'foreign' pines (from southern United States).[17]

By 1904 Cape Town was running out of fresh water every summer. Despite the creation of Molteno Reservoir in the City Bowl, a reservoir at Newlands, and of five reservoirs atop Table Mountain between 1896 and 1907, supplies were still insufficient for the growing city. Waterborne sewerage was not possible for many southern suburbs until after World War I. The realisation that economies of scale were required to solve these service provision challenges drove the unification of the city under one municipality in 1913. As Hutchins had predicted, in 1918 the unified city sponsored a £1.5 million plan to create a dam in the Steenbras Valley and pipe water some 40 miles (65km) to the City Bowl, a project completed in 1921.[18]

The city was gradually electrified, notably during the 1920s and 1930s, and by 1945 most of the study area had electricity. Prior to this, open flames were universally used for cooking, lighting and warmth, with the attendant risks of fire. People setting the veld alight to create dry firewood (*Protea nitida* or 'wagon tree' was favoured) provided an additional fire hazard on the Peninsula. Electrification reduced the risk of fire in the city and its more affluent suburbs immediately surrounding the Peninsula's mountains. In the townships and informal settlements further out on the Cape Flats (and in District Six in the City Bowl until it was demolished in the apartheid removals of the 1960s), open flames were used throughout the century.[19]

The first railways were laid on the Peninsula in the 1860s, and by 1890 a railway line linked the city with the southern suburbs and Simon's Town. There was a Cape Flats line, and another to Sea Point. Sparks from steam locomotives occasionally caused fires, peaking between 1965 and 1970, dwindling fast thereafter until steam was phased out in the late 1970s. Wynberg Rifle Range Plantation, defunct by 1949, was most at risk as the Cape Flats Railway passed through the middle of it. By 1902 an electric tram linked the City Bowl to Camps Bay via Kloof Nek, and in 1905 foresters completed a contour bridlepath linking Kloof Nek to the King's Blockhouse on Devil's Peak. This contour path linked up with another path all the way to Constantia Nek, part of Hutchins' original plan to facilitate firefighting on the eastern slopes of Table Mountain. From the Kloof Nek tram station, it was a short walk up to join this path, and this link opened up the front of Table Mountain to outdoor recreation. Ultimately a road was built to the cable car station (opened in 1929), and extended along the bridlepath, and popular picnic spots including barbecue areas were established along it (Tafelberg Road). The tramway was closed in the 1930s, after private motor vehicle ownership became more widespread.[20]

In combination with new roads – built as relief labour schemes in the late 1930s, and in aid of postwar reconstruction in the late 1940s and 1950s – car ownership opened up undeveloped parts of the Peninsula to hikers, campers and picnickers.[21] From the late 1930s, cigarettes thrown from cars were a frequently cited (though unproven) cause of bush fires. In February 1938, local resident Reg de Smidt wrote to the *Cape Times*:

> As I write the night sky over the Muizenberg Mountains is lurid flames that have been raging since last night. There has been a veritable inferno of fire over the Peninsula this weekend. Veld fires have been said to be caused by coloured people who stand to gain from the fuel they can collect after such a fire. Actually there can be little doubt that they are caused by cigarette-smoking Europeans [who] ... bowl along the Peninsula in their cars over a weekend (significantly most of these fires occur in the weekends) and pitch their cigarette-stubs through the windows onto the dry grass, and the South-Easter does the rest.[22]

The increase in traffic also provided complications for firefighters. Chaos ensued during a fire above Clifton on 20 January 1934 after 'hundreds of motor cars arrived from all directions'. Besides choking up the main road, one car caught fire, and another crashed into a street light after the driver was blinded by smoke while trying to negotiate the tram-line. This was a problem for firefighters from this period on, exacerbated by huge influxes of holidaymakers in the Easter and Christmas periods, with onlookers and traffic congestion hampering access and safe operations.[23]

From the 1940s light industry spread eastward through Woodstock and Salt River and onto the Cape Flats. There was an influx of labour during the wartime boom in industrial activity, but housing was inadequate, with Langa being developed as a township for Africans on the Flats in 1943. There had been sporadic attempts to remove non-Europeans from the city bowl from early in the century, but this changed following the (narrow) victory of the Afrikaner-dominated Nationalist Party in 1948. They began to implement their policy of Apartheid in the 1950s (enabled by the Group Areas Act of 1953), and in the 1960s so-called 'non-whites' were systematically exiled from the city and its suburbs, to the sandy Cape Flats. The conditions were harsh, with no shelter from the sun, high winds and driftsands in summer, and flooding in winter. The rare flora was cleared to make way for townships, woody plants were gathered and burned as fuel and wildflowers were sold in the city. Invasive species introduced to stabilise driftsands proved welcome sources of building materials and fuel.

In the 1960s and 1970s Cape Town's city engineer Solomon (Solly) Morris oversaw a period of high modernist town planning. Morris, born and educated to university level in Cape Town, spent most of his working life in the city. He was promoted to City Engineer in 1950 (the youngest, and first South African, appointee). In 1958, Morris enthused that 'South Africa stands today on the threshold of great development', proposing that 'in the full exploitation of our natural resources and in the creation of a sound national economy civil engineers must play a vital role. [It is not possible to] ignore the ineluctable march of progress'.[24]

The opening of De Waal Drive and the Table Bay Boulevard in 1959 provided two multi-lane freeways linking the City Bowl with the Southern Suburbs and Cape Flats. In January 1960, the Cape Divisional Council approved a £548,000 scheme to create scenic drives around the Peninsula, deemed 'essential for a healthy tourist trade'. These included roads to link Constantia Nek with Lakeside, to open up the Silvermine plateau, a coastal drive along the shores of False Bay to Somerset West, a circular drive to open up the Cape of Good Hope Nature Reserve and its beaches, and a West Coast road north of Cape Town. In 1968 the Eastern Boulevard was smashed through District Six and together with the Western Bypass (10 years later) came to dominate the City Bowl landscape, cutting the City off from the sea. A huge tanker basin was built in 1963, expanding the harbour. New 'scientifically' planned townships were built on the Cape Flats, next to new industrial estates like Retreat West Industrial Area and Epping Industria. A freeway link between Somerset West and Cape Town was opened in 1971, and there were plans to build a freeway through Bishopscourt, cutting off a corner of Kirstenbosch National Botanical Garden (see Chapter 6), to link Union Avenue in Newlands with the new Simon van der Stel freeway to Muizenberg. The Ou Kaapse Weg was built over the Muizenberg mountains, providing a direct link between the southern suburbs and Noordhoek on the southern Peninsula. Altogether, the area of constructed roads recorded in the report of the City Engineer in 1936 had doubled by 1966, and trebled by 1976.[25]

Capetonians were alarmed at the pace and scale of this transformation of their city. In October 1971 a Mr G.P. Blake wrote to the *Cape Times* bemoaning the scarring of the Peninsula's landscapes with new roads, and the ruination of its scenic drives by the resulting urbanisation along the new routes. In just 20 years, he noted, several prominent wildflower beauty spots had been buried beneath concrete:

> The *Flora of the Cape Peninsula*, edited by Adamson and Salter in 1950 mentions Paarden Eiland as the habitat of various genera of wild flowers. How many of them can you find to-day? The beautiful splashes of pink and white Drosanthemums...a pleasing sight as one drove in spring towards Milnerton along the Otto du Plessis Drive are replaced by huge and none-too-attractive warehouses and other buildings. Blouberg, once an unforgettable site in late winter and spring with species of *Arctotis* and other daisylike flowers and *Lachenalia pendula* has little floral beauty now...Along the Twelve Apostles [open veld festooned with various attractive indigenous species including watsonias has been] destroyed by the building of hideous flats and houses. Milnerton, Blouberg, Camps Bay [have been] ruined by unsightly buildings. It would seem that this 'progress' and beauty...cannot live together.[26]

The effects of settlement and urbanisation on fire regimes

The City Bowl and Camp's Bay

In the early 1900s the City Council developed its lands on Signal Hill for outdoor recreation. The bridlepath linking Kloof Nek to Devil's Peak opened up the front of Table Mountain to outdoor recreation – and more ignitions. In 1909 the Cape Town city bowl was still confined to the flatter land from the base of Signal Hill east along Table Bay to Woodstock, and not as far upslope as Molteno Reservoir or the prison (now the Provincial Archives). Green Point Common was a large open area where animals were grazed and grass fires were occasionally a problem. The Atlantic coast below Signal Hill and Lion's Head – and extending all the way south to Cape Point – was sparsely settled.

Thirty years later, suburbs extended from Bantry Bay on the west coast below Lion's Head to enfold the 'lion's rump' as far upslope as Tamboerskloof. East of that was the new suburb of Oranjezicht. Another 30 years later, the suburbs had expanded to fill the City Bowl's lower slopes to virtually their current extent, extending east from Oranjezicht to Deer Park, and north along the lower slopes of Devil's Peak to Vredehoek. Lion's Head and Signal Hill were virtually encircled by suburbs, with nearly continuous settlement along the Atlantic coastline south to Bakoven below the Twelve Apostles.

Fires on the slopes of Lion's Head and Signal Hill now directly threatened housing on the slopes below. New suburbs forming the upper perimeter of the city bowl were similarly exposed. High winds spread big fires east or west through the wide belt of fynbos separating the suburbs from the north face of Table Mountain, even spreading on occasion over Kloof Nek into Camps Bay, or up the saddle between Table Mountain and Devil's Peak. The mountain fires watched with such equanimity and fascination by early inhabitants of Cape Town now burned at the city limits, threatening lives and homes. This may explain the disappearance of aesthetically appreciative descriptions of mountain fires from newspaper reports in the postwar period.

From Table Bay to the Southern Suburbs

Wetlands around the lower reaches of the Salt and Black rivers where they approach Table Bay had long formed a natural barrier to the spread of fires from the Cape Flats onto the northern Peninsula. South of this, urban development along the railway to Simon's Town progressively cut off open veld on the Flats from the Table Mountain chain. The large forestry plantation developed at Uitvlugt until 1918 was both a screen from and a location for fires, but by the early 1920s was being replaced by residential areas (including Pinelands). By the 1939 this separation of the city and its suburbs from the Cape Flats was more or less complete, with small industry replacing the remnants of Uitvlugt and Epping Forest Reserve in the north, and the suburbs of Observatory, Mowbray, Rosebank, Rondebosch, Newlands, Claremont, Kenilworth, Wynberg, Plumstead, Diep River and Heathfield linking up along the railway to the south. A series of wetlands and the Sand, Spaanschemat and Keysers rivers flowing into the Sand Vlei separated the mountains at Muizenberg from the Flats, and east of the Sand River much of the land was driftsand and marsh. Thus the Cape Peninsula became a largely closed system from the perspective of fires in vegetation.

In the second half of the century, new freeway links were built to the southern suburbs of Cape Town alongside the Black and Liesbeeck rivers, which were canalised and had their banks cleared of vegetation. New roads built in the late 1960s and early 1970s further aided the development of the lower mountain slopes from Bishopscourt west to Constantia Nek, and southwards across the Constantia Valley and Tokai to the railway line and beyond. Settlement closed in around the remaining vineyards and plantations at Constantia and Tokai. In Hout Bay, the fishing industry was developed and industrialised from the late

1930s, largely based on exports of frozen lobster tails to the United States. A hard road was built, and the area later became a favoured peri-urban settlement for wealthy Capetonians and expatriates. Settlements expanded up the mountainsides. Introduced plants used to stabilise the dunes and driftsands invaded the slopes above.[27]

Throughout the century, the major locations for large bushfires were the Peninsula's undeveloped mountain slopes. Turbulent winds and rugged topography meant fires could spread in several directions at once. The wetter and colder plateaux atop Table Mountain were little troubled by fire until late in the century, and the wetter, shadier, upper eastern and southern slopes of the Table Mountain chain with their deep forested kloofs (ravines) were likewise relatively untroubled. The development of (Devil's) Peak Plantation from 1894, first primarily to prevent erosion, later for timber, meant that fires originating on Cecil Rhodes's park-like Groote Schuur Estate (frequent in the early 1900s), were combated by foresters and prevented from spreading further ups-lope. However, as this plantation was developed, particularly after World War I, when fires did take hold they became uncontrollable. Invasive pines spread up the cliffs above the plantations, extending the risk of severe fires to the upper slopes.[28]

The development of Newlands Forest, Kirstenbosch Botanical Gardens (from 1913) and Cecilia Plantation on the southeastern slopes of Table Mountain, all protected by firebreaks and firefighters, helped interrupt the spread of large fires along these slopes. On the other hand, some introduced plants invaded the surrounding Fynbos, creating higher fuel-loads. Antipathy to burning fynbos above Kirstenbosch remains a fire-hazard to this day.

The dogleg of fynbos-covered mountains from the Vlakkenberg south of Constantia Nek, via the Constantiaberg and Noordhoek Peak to Muizenberg Peak, provided space for large fires to develop throughout the century.[29] Strong, desiccating southerly winds primed these exposed slopes. The plantations on Tokai Forest Reserve, already well developed by 1909, formed a kind of screen between the Constantia and Steenberg mountains and the vineyards of Constantia and the southern suburbs. These plantations were often threatened by fires sweeping down from the mountains. As late as the 1960s, large undeveloped areas remained around the plantations, with the danger of fires spreading from them. The plantations were also a source (and sometimes target) of fires, which subsequently spread up into the neighbouring veld, occasionally crossing the mountains to threaten Hout Bay to the northwest or Noordhoek to the southwest.[30]

Some of the century's most extensive wildfires have originated in the farms and veld around Noordhoek and on the mountain slopes and plateaux to the north and east.[31] Improved access to the mountains, particularly following the opening of the Ou Kaapse Weg over the Muizenberg Mountains, and of Silvermine Nature Reserve in 1964, exacerbated the fire risk.

The Southern Peninsula

The southern Peninsula is separated topographically from the northern by a flat sand movement corridor that once linked the beach and wetlands of Noordhoek on the west coast with the wetland and beach at Fish Hoek on False Bay to the east. This formed a natural barrier to fire, but it became vegetated after the movement of sand was interrupted by the development of Fish Hoek from the 1920s. With the exception of the port of Simon's Town, the area south of the corridor remained a rural and remote region of the Peninsula until road improvements in the 1960s and 1970s allowed Capetonians to commute more easily between the southern Peninsula and the City, and residential areas were developed in what had hitherto remained farmland or open veld. The settlements at Silvermine, along the Fish Hoek-Kommetjie road, at Da Gama Park and at Glencairn (developed in the 1980s) were very exposed to wildfires. The risk of severe fires was exacerbated by invasions of Australian *Acacia* species introduced to stabilise driftsands. These urban developments did further separate the undeveloped lands on the southern Peninsula from the Silvermine and Kalk Bay mountains.

The southern Peninsula has remained relatively undeveloped, and covered with inflammable fynbos. There have never been significant settlements south of Scarborough and Simon's Town, and from 1939 this area began to be developed as a nature reserve. The Cape of Good Hope nature reserve was established in 1964. Cape Coloured residents were removed from coastal villages to Ocean View (formerly Slangkop) township by Apartheid planners in 1968, and some fires have been attributed to wood collectors around it.[32] One of the big January 2000 fires was started near the informal settlement at Red Hill above the navy base and settlement at Simon's Town, and was quickly spread by high winds through mature fynbos and invasive plants. The other big fire was ignited near the picnic area at Silvermine.[33]

In 2011 Rebelo et al. estimated that urbanisation had transformed 57 per cent of the original extent of Peninsula Granite Fynbos, 2.5 per cent of Peninsula Sandstone Fynbos and 87 per cent of Peninsula Shale

Renosterveld. Both fynbos types are regarded as threatened, with the renosterveld classified as critically endangered. To fully illuminate the human dimensions of the Peninsula's fire regimes in the twentieth century, it is now necessary to investigate cultural responses to the local environment.[34]

6
Conserving Table Mountain

[Cecil Rhodes'] intense and genuine love of the big and beautiful in natural scenery prompted him to buy as much as he could of the forest slopes of Table Mountain, so that it might be saved for ever from the hands of the builder, and the people, attracted to it by gardens, wild animals, and stately architecture, might be educated and ennobled by the contemplation of what he thought one of the finest views in the world.

Sir Herbert Baker, remembering his patron Cecil Rhodes[1]

When we reach the mountain summits we leave behind us all the things that weigh heavily down below on our body and our spirit.... The mountains of our lovely land will make a constant appeal to us to live the higher life of joy and freedom. Table Mountain, in particular, will preach this great gospel to the myriads of toilers in the valley below. And those who... make a habit of ascending her beautiful slopes in their free moments, will reap a rich reward... it will make us all purer and nobler in spirit and better citizens of the country.

Jan Smuts
Oration on Table Mountain,
May 1923.[2]

After 1910 Table Mountain was suggested as a unifying symbol for the newly created Union of South Africa. In 1923, one of the chief architects of that Union, General Jan Smuts, argued that ascending the mountain, which he frequently did, could literally enable the diverse citizens of this new and divided nation to '[become] purer and nobler in spirit and better citizens of the country'.[3] However, approaches to the environmental management of the mountain showed no such symbolic coherence. State foresters planted it with exotic species and aspired to clothe its 'bleak and forbidding' slopes with lush woods.[4] Municipal authorities created reservoirs, planted exotic trees for shade and shelter from the wind and gardened with indigenous species to 'improve our native flora'.[5] Botanists were determined to preserve the mountain in its 'natural state'. While some celebrated the Fynbos for its floristic diversity, and because it was indigenous, for others it was featureless, dangerously inflammable, 'sun-scorched, snake-ridden bush'.[6] The one point of agreement for these diverse approaches to the mountain environment was a vehement antipathy to fire.

Contrasting visions of Table Mountain

The disagreements over the management of Table Mountain stem from quite different visions of and for the mountain. These emerge strongly in the decades of transition from British colonial to Afrikaner nationalist government, and Van Sittert's psychogeography of Table Mountain (1891–1952) offers a stimulating account of this. He provides a nuanced portrait of the contested and protean nature of white South African identity in the Cape in this period. Simple notions of imperial or national identity are inadequate categories for analysing a period of South African history which Beinart has characterised with the phrase 'a state without a nation'.[7]

Although Van Sittert refers to a seemingly unified 'bourgeois eye', he shows that European settlers' views of the mountain were refracted through the lenses of class, ethnicity and gender. European-led fashions for alpinism and orchid collecting were expressions of romanticism and a taste for empirical enquiry which combined at the Cape to make Table Mountain 'a site of scientific and romantic pilgrimage for Cape Town's itinerant intellectual and administrative elite'.[8] The precipitous north face with its views over the city from on high were the focus for this attitude, inducing 'godlike feelings of "superiority and command" ', which stimulated climbers to sing hymns and 'God save the King' on the summit – rejoicing in the echo from the 'loyal mountains'.[9]

The Mountain Club was formed in 1891, growing its membership from 60 to 490 by 1908. Van Sittert argues that its middle class members pressured the 'utilitarian' city engineers and foresters into taking their more sentimental approach to the montane landscape into account, allowing them access to restricted areas on the mountaintop and relocating buildings which spoiled skylines. As we shall see, however, we should be cautious about linking foresters with utilitarian municipal officials. Key Forestry Department officials had a strong aesthetic vision for their work on the mountain.[10]

This appropriation of the 'domesticated' mountain by Cape Town's bourgeoisie represented a shift from a predominantly utilitarian perspective to a kind of 'Christianised romanticism', van Sittert argues. This shift was political in that the bourgeoisie sought to assert their vision of the mountain in an exclusivist fashion. Litter louts, fire bugs and flower pickers (of the lower classes) were to be excluded, and fences were erected across unofficial paths and set up to channel climbers onto policed routes. The City Council's early attempts (from 1907) to make the summit more accessible to mass tourism were decisively blocked.[11]

This bourgeois reorientation towards the mountain as an elevated space above the 'humdrum' world of the city resulted in a reorientation away from the north face of Table Mountain, with its views over the city. Rather, they turned to the forested eastern face, with its views over 'European colonialism's primal scene in Africa' (early Dutch farms and estates), and the suburban villages where the Anglo middle class were relocating to in order to escape the 'vulgarity of the city'.[12]

Van Sittert links two key political figures of this period with this Anglo middle class attitude to the mountain: Cecil Rhodes (prime minister of the Cape Colony from 1890 to 1896), and the Afrikaner general and statesman Jan Smuts, a key architect of the Union. Rhodes affected 'the local veneration of the "spirit of the mountain" ', buying an estate below the eastern slopes which he developed scenically and architecturally to combine mountain romanticism with the cultural legacy of Dutch settlement. This was a gesture to facilitate a political union with the Afrikaner Bond.

Rhodes' creation of a 3700-hectare (9143-acre) estate on the eastern slopes of Table Mountain, which he landscaped and opened (in part) to the public, certainly provided a focus for local pride and identity. In his will, Rhodes bequeathed it to a future unified country. He commented in 1900 that contemplating the mountain put petty human politics into perspective, and inspired noble thoughts about the betterment of humanity. His love of place and strong identification with particular mountain landscapes cannot solely be attributed to political expediency at the Cape, however. In accordance with his will, following his death in March 1902 Rhodes' body was transported by rail from Cape Town to Rhodesia (now Zimbabwe), where he was buried on the summit of Malindidzimu Hill in the Matobo Hills.[13]

Science, nature and South Africanism

The ideology of South Africanism emerged in the aftermath of the South African War of 1899–1902, with science and nature providing potentially neutral grounds for building a shared national identity. Science was used to express universality and progress (within the imperial 'chain of civilization'), and was regarded as a means of uniting moderate Afrikaners and British colonials in a positive and progressive patriotism.[14]

The union of South Africa in 1910 left the formerly dominant Anglo elite scrambling to reorient themselves toward the new nation. Fortunately, a key figure of the new government provided an inspirational figure. Jan Smuts was both a central figure in the South Africanisation of science, and fond of extolling the spiritual virtues of nature and mountaineering in particular. He hymned Table Mountain for its botanical and cultural significance. While not exactly 'an indigenous reincarnation of Rhodes' for the Anglo middle class, he certainly provided leadership for those who saw Table Mountain as an exalted space to be protected from the vulgar and the mundane.[15]

Two important developments following Union linked science, the indigenous flora, South Africanism and Table Mountain. These were the formation of a national botanical society and the creation of a national botanical garden on the mountain slopes. The political union of South Africa was seized upon by Harold Pearson, Harry Bolus Professor of Botany at the South African College, as an opportune moment to motivate for the establishment of the country's first national botanical garden. In his presidential address to a meeting of the South African Association for the Advancement of Science on 4 November 1910, Pearson laid out the case for doing so in Cape Town. He pointed out that while Cape Town had the oldest municipal garden in South Africa, established in 1848, it had not been organised scientifically and by the 1890s had declined into 'but a town pleasaunce of flowers and shady walks'. The distinguished Indian botanist and forester J. G. Gamble had been sorely disappointed when he visited Cape Town in 1890, complaining that the municipal gardens contained none of the 'beautiful and interesting "bush" or "veldt" vegetation … not even a single silver tree'. Embarrassingly, the Director of Kew had noted in 1895 that the Cape was 'the only important British possession which does not possess a fully-equipped botanical institution'. Seizing the historical moment, Pearson argued that establishing a national botanical garden:

cannot merely be an economic undertaking; it must also be an expression of the intellectual and artistic aspirations of the New Nation whose duty it is to foster the study of the country which it occupies, to encourage a proper appreciation of the rare and beautiful with which Nature has so lavishly endowed it.[16]

The Kirstenbosch National Botanical Garden was established on the eastern slopes of Table Mountain in May 1913, to protect, explore and display the indigenous flora of the Union. A National Botanical Society of South Africa was founded in June, albeit with a decidedly Cape focus. The symbolic importance attached to the Cape flora in this new nation is apparent in that in 1921 'a fine wreath of Cape Flora embodied in a decorative tribute' was sent to Europe by ship 'to be deposited on the grave of the American unknown soldier'.[17]

In 1923 Smuts had delivered a famous address on the summit of Table Mountain in which he spoke of 'the religion of the Mountain'. In his last year, he wrote that Table Mountain was 'the greatest national monument of South Africa' and that 'to interfere with it is to desecrate what should be our national temple, our Holy of Holies'. Van Sittert suggests that, for the Anglo bourgeoisie at least, the mystique of the mountain waned after Smuts's death in 1950. The increase in private car ownership, road development and the opening of the cable car in 1929 had already delivered the Peninsula and Table Mountain into the realm of mass tourism. The location of the cable car station, and the opening of Van Riebeeck Park in 1952 to mark the tercentenary festival of the first European settlement at the Cape, resulted in a reorientation back to the iconic north face of the mountain.[18]

'A tremendous slaughter': Botanists versus fire

The most enlightened of the groups contesting the management of Table Mountain, when it came to the effects of fire, should have been the botanists. As related in Chapter 3, however, they came from a tradition vehemently opposed to the (supposed) destructive effects of fire on the indigenous vegetation.

At a special conference on veld burning convened in 1924, the renowned Cape botanist and champion of Table Mountain Rudolf Marloth argued that the 'glorious sight of whole hill slopes being sometimes covered with Watsonias and other tall flower spikes' was a 'sad spectacle' to the botanist who realised 'the tremendous slaughter that

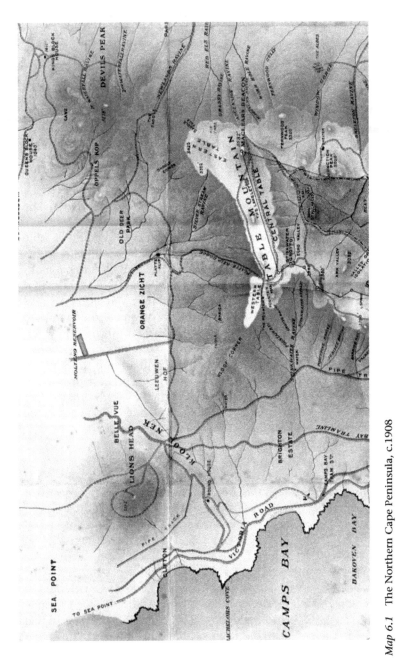

Map 6.1 The Northern Cape Peninsula, c.1908
Source: Detail from 'Map of Table Mountain', Cape Mountain Club, 1908.

ha[d] taken place to produce this display', with the ensuing 'certainty of the extinction of many species...of plants'.[19] Until the 1970s, all but a handful of botanists and foresters interpreted the response of fynbos to fire as a linear decline in floral diversity, rather than cyclical renewal through naturally recurring fires.

Veld fires and flower collecting

These were the two major issues the new Botanical Society of South Africa sought to address. In letters to the Director of Kew written in 1921, the well known Cape-based plant collector Thomas (T.P.) Stoekoe expressed his fear that the exceptionally rare *Orothamnus zeyheri*, 'only recently...rediscovered growing in the Klein River Mountains and the Hottentot Hollands [*sic*] mountains' was 'in danger of extinction through the persistent firing of the bush'. In another letter he also deplored the threat posed to this rare protea by:

> 'Wild Flower Shows', where the professional wild flower gatherers sweep the veldt and kloofs, and sell their harvest to the various competitors to the Exhibitions for the promotion of a love for the Native Flora. A case in point is the *Serruria florida* of French Hoek, reduced to a dozen specimens, just saved in the nick of time.

As described in Chapter 4, these two species were to prove pivotal in the shift to prescribed burning of fynbos catchment areas by the state forestry department in the 1960s.[20]

Kirstenbosch Garden's second director (from 1919 to 1953), Harold Compton, developed a particular antipathy to fire, expressed over more than 30 years of intemperate editorials in *The Journal of the Botanical Society of South Africa* (in practice a largely Cape-oriented journal). Typically, he argued that a large fire which had occurred above Kirstenbosch on 5–6 May 1923 had '[destroyed] practically all the scrub vegetation from Window Buttress to Constantia Nek'.[21]

In the 1920s, Compton and the Council of the Botanical Society successfully championed 'the establishment of local nature reserves in the Cape to protect the natural flora and fauna from the advance of urbanisation and agriculture, as well as by flower-picking, veld-burning, grazing...and other forms of vandalism...'. They joined with the Mountain Club and the Wild Flowers Protection Society to motivate for the establishment of a national park on the Peninsula (the Kruger National Park, South Africa's first, was proclaimed in 1926).[22]

In 1928 Sir James Rose-Innes (former Chief Justice of the Union) advocated the proclamation of Table Mountain as a nature reserve, noting the ease of access to many areas of the mountain and the limited surveillance over non-afforested areas by state foresters. Rose-Innes recommended employing a ranger 'for fire-protection [and] for combating the spread of exotics...'. Unfortunately, there would be no national park for another 70 years, but legislation was passed to protect wildflowers, as noted admiringly by speakers in favour of the Rural Amenities Bill in England in 1931.[23]

Sir Arthur Hill (1875–1941), Director of the Royal Botanical Gardens at Kew (1922–41), visited Cape Town in November 1929. He met prominent local botanists, explored the Peninsula and Stellenbosch, and spent several evenings with the Comptons, relaxing on their stoep [verandah] after dinner 'while [Compton] played Bach and Chopin and we enjoyed the music and the stars'. It is likely Hill was influenced by Compton's views on fire. In his private diary he noted the effects of a large bush fire which had burned above Kirstenbosch on 3–4 November 1929, just before his arrival. Hill recorded 'much damage by fires to the native scrub' on Table Mountain, and returning from Cape Point over Red Hill, remarked on the 'rather bare stony country spoilt by fires made to improve grazing to the sacrifice of indigenous mature plants'. Sir Arthur was also concerned over the invasion of the indigenous vegetation by introduced plants, notably pines and hakeas on Table Mountain, and described a 'long dull drive over the Cape Flats' where 'all vegetation [was] exotic Eucalypts, pines, wattle and venothera...'.[24]

Before leaving Cape Town, Sir Arthur gave an interview to the *Cape Times* in which he 'dilated on Kirstenbosch and planting of pines and wattles' and called for the 'reservation of Table Mountain...for native flora'.[25] In a report to the Union Government, he motivated that 'Table Mountain should be kept sacred as a Reserve'. Exotic (introduced) plants should be removed, and 'constant fires' prevented by outlawing all 'intentional fires'. Sir Arthur was particularly exercised by the threat to the 'wonderful silver trees' (*Leucadendron argenteum*).[26] These giant proteas are endemic to the Peninsula, and being serotinous are killed by fire, subsequently dropping their seeds. Too-frequent fires can prevent trees maturing and producing seed. On the other hand, total fire exclusion inhibits reproduction and results in senescence, which also threatens the species.[27]

Professional botanists weren't the only ones determined to protect the Peninsula's flora. Throughout the 1920s a Mr Julius Jeppe campaigned to obtain municipal legislation to protect the flora on the slopes above the City from fires (this may have been Julius Otto Jeppe, snuff dealer and honourary consul for Japan in Cape Town from 1910 to 1918). Finally in October 1933, a regulation promulgated in the Provincial Gazette extended the Municipal boundaries to include the catchment area on top of Table Mountain and Devil's Peak, and 'Paradys' at Papenboom Estate. City Council regulations 'prohibiting the lighting of fires in catchment areas' covered these extensions. This regulation was described in the *Cape Times* as 'the climax to 12 years of effort by Mr Jeppe... [aimed] at the preservation of the indigenous trees and flora'. Echoing Rose-Innes, Jeppe recommended recruiting a force of 12 rangers to patrol the mountain in plain clothes, especially on weekends and holidays from December to March.[28]

(Whose) 'mountain beauty ruined'?

Concerns about the mountain flora expressed in the 1920s and 1930s were brought into sharp focus by the Christmas Day fire on Devil's Peak in 1935. The spectacle of the fire deeply impressed onlookers, and the way it spread through plantations and up into cliffs invaded by cluster pines had made it very difficult to suppress. The press and public responses to the fires were unprecedented, with an outpouring of articles, editorials and comment from 26 December into the second week of January 1936. The first headlines were '£250,000 Havoc in Great Fire... blaze spreading to-day on suburban side... 700 men fighting the flames'. Aesthetic concerns loomed large in reports: 'Table Mountain, a scene of green and grey beauty when Christmas Day dawned, has been devastated by the most furious fire in its history'. Beneath the heading 'Mountain beauty ruined' it was averred that the material damage 'scarcely overshadows the aesthetic and sentimental loss caused to the city by the ruination of the mountain'.[29]

There would be numerous lamentations (throughout the century) about the disfiguring effects of fires on the mountain scenery, characterised by the use of words like 'mutilating', 'unsightly scars', 'black wilderness' and 'scarred hillsides'.[30] In the controversy that emerged over introduced trees after the big fire of Christmas Day 1935, it became obvious that there was a strong divergence in views over what constituted mountain beauty: unadulterated indigenous flora, or a hybrid landscape embracing indigenous flora augmented by introduced trees.

Clothing Table Mountain's 'bare cliffs'

Many correspondents to the newspapers following the Christmas 1935 fire focused on the destruction of the veld, and called for the removal of all inflammable introduced trees to reduce the fire risk.[31] However, some mourned the loss of trees planted for amenity purposes, and a minority even lamented the loss of plantations. In a rare letter focusing on the loss of plantations, 'Saddened' wrote:

> when I arrived from New Zealand 40 years ago, where our mountains mostly are forest clad right to the tops, Table Mountain looked very barren to me, and I have watched with joy its slopes being mantled in green ever higher and higher.... Alas, 40 years' work has gone up in smoke and flame, and I shall not be here to see it reach its glory again.[32]

This letter captures the difference between ecological (or 'mountain time') and human time. American historian and firefighter David J. Strohmaier reminds us that the famous nature conservationist Aldo Leopold died fighting a grass fire which threatened the pine trees on his Wisconsin farm. While Leopold knew that his pines could regenerate from seed, he also knew this would not happen in his lifetime. Strohmaier argues that 'one tree, or grove of trees, is not interchangeable with another. Places can be lost and never replaced.... Like a work of art, simply replacing it, even with the most exacting replica, will never fully replace the original, for part of an object's value derives from its history'. While this may seem a bit strong when applied to a pine plantation, his point holds that decisions about fire protection may come down to 'loss of place to fire versus ecological loss from the loss of fire'. These two dimensions, the personal and the ecological, may involve very different timescales. On an ecologically trivial scale, for someone enjoying their annual holiday, or a tourist enjoying a once-in-a-lifetime holiday to the Cape, their appreciation of the mountain may be marred by fire-blackened slopes.[33]

Others defended pine plantations for their aesthetic and practical functions in concealing and combating soil erosion. A *Cape Argus* correspondent maintained that it was 'the agitation in the Press of some 40 years ago about the ghastliness of the bare, eroded slopes of Devil's Peak' that had resulted in this good work of tree planting by foresters.[34] In an inflammatory editorial the *Cape Argus* decried the singling out of the pine tree 'as the enfant terrible, the public enemy No. 1 of the mountain':

It might almost be thought that the poor 'pinus insignis' was not only an eyesore but a self-igniting tree itching to burst into spontaneous combustion on public holidays, and that fires would be unknown if only the 'glorious native flora', which has no such suicidal tendencies, could be conjured back by a wave of the Forestry Department's wand to people slopes on which it never flourished.[35]

The editorial argued that 'mountain fires are not less but more prevalent when nature is left to look after itself, as witness the Caledon Mountains which are perpetually swept bare'.[36]

Worried by calls to remove their plantations from the Peninsula following the fire, Forestry Department officials utilised aesthetic arguments, arguing that the public would not approve of 'indiscriminate destruction of trees' nor the resulting 'bare cliffs and Gibraltar-like rocks' as a setting for Cape Town and Table Bay. It was argued that plantations could be laid out so as to flow with the landscape and improve the appearance of the environment (ignoring the fact that clear felling would result in a similar 'devastated' landscape to that following a fire).[37]

The Department did have a long history of strong aesthetic motivations for planting trees on Table Mountain. In bemoaning the loss of the planted area on the lower plateau of Table Mountain to Cape Town's second reservoir in 1899, Conservator of Forests David Hutchins had reported that:

> The flourishing young woods of Pine, blackwood, Oak and other trees, the Scotch Pine...are now being cut down to make way for the waterworks. The most beautiful spot of Cape Town...now ceases to exist. I understand that the actual money cost of these plantations has been refunded by the Town Council. But this does not replace the beauty of this lovely valley, with its charm of clouds and crag and the flowing streams fringed with the brilliant *Disa grandiflora* [now *uniflora*].[38]

In 1900 the Department was planting Turkey Oak and Kabul Ash below the waterfall above Rondebosch (presumably the one at the foot of First Waterfall Ravine, Devil's Peak). Hutchins commented that 'the view from this point is unsurpassed, comprising the whole of the southern suburbs of Cape Town, the best wooded portion of the Table Mountain ridge, and in the distance, False bay'. In 1901 Hutchins reported that 'The transformation of this bleak and forbidding spot [Peak's Plantation below the King's Blockhouse on Devil's Peak] is almost complete', and in November 1904 a marble tablet was built into the walls of the forester's cottage next to the King's Block House on Devil's Peak, reading:

> In memory of Forester Frank Jarman, who, from 1893 to 1902, had charge of the Forest work which covered this wind-swept mountain with trees. He left here for similar work at Elgin on the mountains opposite, and died as the result of an accident.... He found these barren stony slopes treeless: he left them covered with forests.[39]

The tablet stands there still on a remaining slab of the ruined cottage wall, like the inscription on the pedestal of Shelley's Ozymandias: all about now lie the 'barren stony slopes', cleared of the trees he worked so hard to establish during the campaign against introduced trees that was initiated in the mid-1990s and accelerated after the January 2000 fires (Photos 6.1 and 6.2).

Jarman and Cecil Rhodes shared, in a sense, a vision of the mountain's future. Rhodes, on whose estate and in the spirit of whose will Kirstenbosch was established, had no qualms about planting up the slopes with introduced trees and flowering plants. For this naturalised Englishman, his hybrid landscaping in no way detracted from the mountain's ennobling grandeur. Thus Jarman's modest monument perhaps echoes the rather more grandiose one to Rhodes (who also died in 1902) on the slopes below, though Rhodes' Memorial still retains a screen of oaks and pines between it and the nearly treeless fynbos that separates the two memorials.

Fencing Table Mountain

Sir Lionel Phillips, a Randlord who made a fortune on the goldmines of the Transvaal and retired to an estate near Cape Town, was instrumental in establishing both the National Botanical Gardens at Kirstenbosch and the Botanical Society. Following the big fire on Christmas day in 1935, his outspoken wife Lady Florence Phillips, 'noted for her efforts to preserve wild life, prevent soil erosion and save the natural beauties of the countryside', declared that 'Table Mountain should come under the control of Kirstenbosch... It is the gateway of South Africa, and everybody agrees that it is time it should be preserved.'[40] (In 1938, the *Cape Times* reported that '70 per cent of the overseas visitors to Union arrive through Cape Town', generating £2,000,000 a year for the city.)

Following the Christmas 1935 fire the idea of the Mountain as a commons was significantly challenged. The Deputy Mayor remarked

Photo 6.1 Jarman Memorial on Devil's Peak, with felled tree, and King's Blockhouse behind
Source: Simon Pooley.

that 'the Council would be reluctant to interfere with free access to the mountain, but if these outbreaks recur we shall be compelled to take drastic measures to protect the forests'. 'Saddened' wrote that it was time for the mountain be 'fenced in compound style, and that parties be allowed in at stated places only...all vagrants [should be] excluded'.[41]

It was through advocacy of a National Park that this exclusion of 'undesirables' and the enforcement of more draconian regulations was most insistently proposed. Lady Phillips recommended fencing and close supervision. A 'city man' maintained that 'under the National Parks Act of 1926, the mountain could be controlled as carefully as Alexander Bay diamond diggings' (it is an interesting analogy: great riches defended from thieves). Access would be controlled through fencing and a permit system. Cableway users and climbers would be closely supervised and 'vagrants and irresponsibles' excluded.[42]

Photo 6.2   Rhodes' Memorial, with stone pines
Source: Simon Pooley.

The less reactionary reasoning behind calls for a National Park recognised the problems posed by the fractured nature of landownership and
management. The *Cape Argus* recommended that private, municipal and
state lands 'be transferred to the nominal ownership and control of the
National Parks Board, provided that the rights of all parties were not
infringed'.[43] The final phrase sums up the major obstacle to the plan:
nothing came of calls for a national park for another 63 years, and
invasive plant control was desultory until the 1950s.

Fire, grass and war

The 1930s were a period of heightened concern about soil erosion and
drought in South Africa. A parliamentary motion urging the investigation of the effects of veld burning in river catchment areas was passed
in Cape Town in 1934, and in the same year the Department of Agriculture established a Pasture Research and Veld Management Section in the
Division of Plant Industry.[44]

In January 1938 the southwestern Cape including the Cape Peninsula was plagued by bushfires. In February, pasture researcher C.J.J. Van
Rensburg argued that suitable pastures would replace the need to burn
the Cape mountains to provide grazing for cattle. On the Peninsula,

Fernwood Estate was acquired by the Government 'as a grass nursery', apparently as a result of the keen interest of Prime Minister Barry
Hertzog. It was established primarily 'to replenish Rhodes Estate as the
game nearly starved last summer'. Approximately 60 acres (24ha) of
grass were planted on the lower slopes of Devil's Peak where 'it thrived
without irrigation'. Van Rensburg was of the opinion that:

> ...when all the lower slopes of our mountains of the Cape are cov-
> ered with grass, the veld flowers will have a better chance to survive,
> burning will decrease, and sufficient grazing will be available for all
> cattle. Farmers will then have no excuse for wilfully burning in order
> to obtain young grass.[45]

This scheme was controversial, and eventually abandoned. Following
the disastrous fire of Christmas day 1935, W. P. Steenkamp M.P. had
complained that 'the whole mountain is being covered with a grass of
very fine fibre, and when it catches fire it is like sulphur'. The grasses,
and a few zebra, remain on the lower slopes of Devil's Peak just north of
the University of Cape Town today, providing an incongruous tableau
of the grasslands and megafauna of the interior, set against stone pines
and the fynbos-clad slopes above.[46]

On 6 February 1938, a 'terrific fire' at Tokai swept up over the
Constantiaberg and down towards Hout Bay, consuming the little
nature reserve Dr Sydney Harold Skaife (1889–1976) had created in the
Baviaanskloof above Hout Bay. The *Cape Times* reported that:

> The place is a little paradise known to travellers from all over the
> world. Here were rare trees, plants and bushes collected by Dr Skaife
> from all over South Africa and beyond. They are, indeed, irreplace-
> able. Here, too, was the last patch of historic tierbos trees which once
> covered Hout Bay and gave the place its name.[47]

Skaife was Cape Inspector of Science and a well-known entomologist.
He was a passionate conservationist, first chairman of the (short-lived
first incarnation of the) Cape branch of the Wildlife Protection Society, and a key figure behind the Cape Divisional Council's purchase of
Smith's Farm near Cape Point for a nature reserve in 1939. In February
1938, Skaife told the *Cape Times* that 'this fire has broken my heart.
Eight years of irreplaceable work has been ruined. The place was a little
Eden'. A *Cape Times* editorial lamented that 'another few square miles
of assorted Peninsula scenery have gone up in smoke' and letter writers

decried the 'devastation of our mountains [and the] destruction of the beauty of the Cape Peninsula'.[48]

From the late 1930s into the 1940s, then, nature lovers on the Peninsula felt that the natural environment was under siege. In addition to the depredations of flower pickers and the effects of destructive veld fires, they felt that the way the Peninsula was being developed threatened its aesthetic and ecological integrity. The scenic roads planned in 1938 as a means of providing work for the unemployed, notably mooted roads from the lower cableway station to Devil's Peak, and from Lion's Head to Signal Hill (both were built), were opposed on the grounds that they would destroy the beauty and peace of mountain walks and increase the fire risk on the mountain (by 1944 fires were being reported 'above the new road' on Devil's Peak).[49] In November 1940 the *Cape Times* editorialised against quarrying on Table Mountain, calling on the City Council to: 'stop this aesthetic horror once and for all'.[50]

Veld burning as treason, 1939–46

Amid the ferment of World War II, which was deeply politically divisive in South Africa, emotions ran high over the management of the country's natural resources. Jostling with news stories about the latest antics of Oswald Pirow and his Nazi sympathisers and the saboteurs and 'dynamitards' of the extreme right-wing Afrikaner Ossewabrandwag movement, were stories about soil erosion and arson. Alongside news of a survey of the Peninsula's caves and kloofs for an evacuation plan in case of a bombing attack, the details of petrol rations and blackouts, experts fumed that 'soil abuse' and careless veld burning was tantamount to treason.[51]

The language and anxieties of war pervaded public discourse on most matters including the natural environment. 'War' was declared 'on soil erosion'. Editorialising on 'the enemy within', the *Cape Times* worried that 'there are far too many insects in the Cape Peninsula' including mosquitoes and Argentine ants [probably introduced with fodder for British cavalry horses during the South African War] which were 'undermining the foundations of our economy'. These invasive ants, it was later discovered, eat the fleshy 'food rewards' on fynbos seeds – evolved to encourage ants to bury seeds in their nests, safe from fire and rodents – without bothering to bury the seeds.[52]

In January 1943, the *Cape Times* reported that Smuts had appointed a committee to form a National Veld Trust, editorialising that: 'the Union is losing good agricultural and pastoral land at an alarming rate. Some err through ignorance . . . others err through indifference, and these are traitors to their country who ought not to be in possession of land at all'. The editorial continued:

> Another drastic need for a Union-wide trust has been grimly illus-trated on the mountainsides of the Western Province during the past few weeks. Every range within a day's run of Cape Town has seen conflagrations, some of them spreading over 100s of acres, which have aggravated the damage done by previous fires to our priceless mountain heritage.[53]

Amid rising fire incidence in the Cape, the Municipality's recently appointed (1939) first Chief Forest Officer Hugo Brunt's efforts at fire suppression on the Cape Peninsula were being heralded as exemplary at a time when the fire 'menace is a national one'. The *Cape Times* called for cooperation throughout the province, and suggested Brunt convene a conference to plan a zonal firefighting system for the Peninsula and the surrounding 30–40-mile area.[54] In December 1942, a massive four-day-long fire destroyed around 200 square miles ($518km^2$) of veld and mountain vegetation from the Nuweburg in Caledon District to the Olifants Hoek Mountains bordering the Franschhoek valley.[55] In the face of large mountain fires, amid the heightened emotions of wartime ('soil abuse' was again described as 'treason' in a *Cape Times* editorial[56]), public calls to extend a system of fire suppression being developed on the Cape Peninsula to the surrounding countryside carried much weight.

Agricultural and botanical experts added their voices to the clamour for a ban on burning. The Professor of Agronomy in the Stellenbosch-Elsenburg College of Agriculture, J.T.R. Sim, argued that 'intentional fires should be regarded as arson' and that 'no-one has the moral right to endanger the nation's water supply by burning the mountains'. Harold Compton endorsed Sim's sentiments in the *Journal of the Botanical Society of South Africa*. He feared the Cape's fynbos-clad mountains would be reduced to 'heaps of sterile rubble', as (he claimed) had occurred in the mountains of the Caledon District. He recorded nearly 60 fires in the Bishopscourt Estate adjoining Kirstenbosch over the summer of 1942/43, and two fires in Kirstenbosch, one of which 'destroyed' most of the vegetation of the Upper Kirstenbosch Nature Reserve and the buttresses above Newlands.[57]

These calls for fire suppression on the mountains of the Western Cape in 1942 and 1943, supported by the Forestry Department, motivated the retired forester John Spurgeon Henkel to write a memorandum to the

Royal Society of South Africa. He feared that attempts at fire suppression were doomed to failure, dangerous, and ecologically harmful. The Royal Society commissioned a survey of current threats to the fauna and flora of these mountains, chaired by the forester Christiaan Wicht.[58]

Wicht et al.'s 1945 report declared 'the flora of the Cape [to be] its chief beauty...unique...one of the richest, most varied and beautiful in the world'. This 'considerable asset' was, in the opinion of the commissioners, 'being lost, through the ravages of fire, browsing and erosion; the invasion of undesirable exotic species; the illegal gathering of flowers; and the indiscriminate conversion of veld to other uses'. Writing in the closing year of World War II, the commissioners emphasised the flora's 'amenity values', which 'touch the deepest sources of mental and spiritual refreshment, both conscious and unconscious'. Preserving the Cape's montane fynbos was a healing task for the nation requiring '[c]o-operation between State departments, universities, local public bodies, and the people' to sustain this 'part of our national heritage'.[59]

'Empty' landscapes such as the slopes of Table Mountain or the mountain flower reserves were devoid of inhabitants or cultural artefacts which could serve as reminders of the new country's long history of conflicts and dispossessions. As in the United States following the American Civil War, South Africa's relative paucity of written and cultural history facilitated the use of 'untamed nature' as a means to construct a national identity. These 'empty' landscapes could be viewed as timeless, and stripped of any connotations of ownership, to be preserved for 'the nation' and 'posterity'. Robert Adamson had presciently observed in his chapter in Wicht's 1945 report that most verifiable species extinctions on the Cape Peninsula had occurred on the Cape Flats as a result of construction and cultivation, not because of burning on Table Mountain or the (conveniently) distant mountains which were the focus of public and Royal Society concern.[60]

In Cape Town, it was in the spirit of preserving the mountain (and flora's) integrity that a suggestion in 1946 that a road be built to give cars access to the top of Table Mountain was almost universally condemned. Compton thundered that it would give bottle-smashers and litter-louts easy access to 'the special pride and glory of Cape Town...a majestic monument of nature as well as the abode of peace and loneliness...'. The coup de grâce was administered by General Smuts, who roundly condemned the scheme at the Mountain Club's Annual dinner.[61]

Burning Table Mountain

Wicht's Royal Society report of 1945 recognised the ecological role of fire and recommended controlled broadcast burning to manage Fynbos. In his book *Table Mountain* (1951), Carl Lückhoff (1914–60) described the resulting proposal that 'controlled burning should be applied to prevent the over-luxuriant development of scrub vegetation in order to lessen its inflammability' on Table Mountain. The idea was to burn at varying intervals ranging from four to seven years. However, Lückhoff, whose gravestone in Stellenbosch commemorates him as a 'protector of our mountains, defender of our heritage', maintained that 'controlled burning on any mountains...is to be unequivocally condemned' as 'morally unjustifiable'. He regarded prescribed burning as defeatist, a threat to insect and soil life, harmful to fynbos species that 'only seed profusely when old', and expensive.[62] With even committed and knowledgeable conservationists opposed to prescribed burning, the idea of block burning was abandoned.

Instead, the Cape Peninsula Fire Protection Area (FPA) was declared in February 1947, in terms of the Soil Conservation Act No.45 of 1946, to protect mountain catchments, valuable plantations and the symbolic importance and tourism value of Table Mountain and its flora. The strategy was to ensure inflammable material was cleared, firebreaks maintained, unregulated veld burning prohibited and firefighting improved.[63]

Motivated by the conjunction of a serious fire in January/February 1950, agitations in the press, and the Van Riebeeck tercentenary celebration scheduled for 1952, the Table Mountain Preservation Board was set up in 1952. Although the Board had no legal powers, it advised the National Monuments Board, and on 8 February 1957, Table Mountain was declared a national monument.[64]

In 1962 the Cape Provincial Administration's Department of Nature Conservation together with the Cape Peninsula Fire Protection Committee published an illustrated booklet to promote fire awareness entitled *Bokkie the Grysbuck*. The project was begun in the mid-1950s, along with a fire awareness poster. In his foreword the Acting Provincial Secretary G. A. van Oordt wrote that while fire was 'a good servant, it [was] a fearful master' which could be 'one of the greatest enemies of animal and plant life'.[65]

The book narrates the story of a young Grysbuck, a small antelope (*Raphicerus melanotis*) endemic to the Western Cape. When a veld fire

rages through the countryside, all the animals flee, but Bokkie is overwhelmed by the smoke. He awakens to a fire-devastated landscape.
Bokkie is rescued by valiant firebeaters who are angered by 'the happy-
go-lucky picnickers who had gone off leaving the ashes of their campfire
still smouldering dangerously'. These kindly men return with Bokkie
through the blackened bush, where lie 'the charred bodies of many lov-
able veld creatures', including Bokkie's mother. Nursed back to health
by children, Bokkie is released into a nature reserve, where 'all animals
and flowers are protected [from] their greatest enemy MAN', and also
from 'the fires which ravage the mountain and destroy so many animals'. Next, a heavy storm washes precious soil from the fire-denuded
mountain slopes, damaging farmers' land downstream. Nature's healing
powers mend the scarred landscape, but alas the 'hackea [sic] and pines,
which are strangers and intruders in South Africa, had crept stealthily
up the mountain slope'. After the fires, these 'greedy foreign' plants
threaten to overwhelm the mountainside. The book concludes with
inspiring scenes of children chopping out invasive plants and using
them to block gullies to prevent further erosion. They are exhorted
to visit Nature Reserves, where Bokkie's 'very look' will implore them
not to light fires in unsafe places. This 'look' is clearly evident on the
iconic 'prevent bush fires' poster based on the Bokkie character, showing a weeping, terrified fawn in a fire-devastated landscape.[66] The Bokkie
poster was soon deployed countrywide in South Africa, and has been
resurrected at the Cape following the January 2000 fires (Photo 6.3).

Environmentalists versus the 'engineering mentality'

In 1963 Douglas Hey, Director of the Cape's Department of Nature
Conservation and the Environment (from 1952 to 1979), wrote that
uncontrolled veld fires and the 'green cancer' of invasive introduced
vegetation posed grave threats to the indigenous flora. On the Cape
Peninsula, the combination of growing environmental awareness and
rapid urban development resulted in public debate over the environmental state of the Peninsula. In 1964, the Cape Town City Council
declared nature reserves on its land on Table Mountain, Lion's Head and
Signal Hill, and at Silvermine. Contemporary environmental concern
came to a head over the Kirstenbosch freeway controversy.[67]

In November 1971 it emerged, apparently by accident, that the
City Engineer Solly Morris was planning a ZAR13-million freeway link
between Union Avenue in Newlands and the Simon van der Stel freeway
to Muizenberg. The proposed route required two tunnels and a concrete

Photo 6.3   The iconic 'Bokkie' anti-wildfire awareness poster
Source: Simon Pooley.

viaduct in front of Kirstenbosch National Botanical Gardens, and would cut off a corner of the Gardens. Professor Brian Rycroft, Compton's successor as Director of Kirstenbosch (from 1954 to 1984) was outraged, especially as he had not been consulted about the plans. Local and national nature conservation organisations, botanical and horticultural societies rushed to support his campaign against the road.[68]

The *Cape Times* encapsulated the public mood in an editorial entitled 'Hands Off Kirstenbosch':

> so secret [were the plans] that neither the Kirstenbosch Trustees nor the Director of the Gardens himself were consulted during the planning of an operation that could disfigure one of the glories of South Africa. But for the vigilance of the local newspapers Professor Rycroft could have woken up one morning to find the City Council bulldozers at his front door.... [A]s the full disagreeable picture emerges, telephone calls telegrams and letter of protest are pouring into the Kirstenbosch headquarters. And rightly so, where there is a plan to turn those peaceful and internationally famous gardens into a noisy annexe to a six-lane freeway.[69]

The outpouring of outrage against the freeway plans were an outlet for wider fears over development in the region. Mr W.A. de Klerk, chairman of the Society for the Protection of the Environment, commented in the *Cape Times* that: 'Everyone is road-building mad these days. It seems to be a modern disease. The whole of the Western Cape is threatened by this type of thing'.[70] A Botanical Society of South Africa member wrote:

> freeway building is a modern disease, and those who plan these roads seem to choose the richest farmlands and the most glorious unspoilt areas for their concrete viaducts and massive embankments. Wynberg Park has been cut in two in the interests of progress and the lower Constantia Valley is an eyesore today.[71]

Morris attempted to defend the plan in the *Cape Times*, but Rycroft and the public were in no mood for compromise.[72] Rycroft pointed out that Kirstenbosch was part of a National Monument (Table Mountain) and lay outside of the municipal area. Letter writers bemoaned 'the devastation, all in the so-called name of progress', of development on the Peninsula, and a series of other environmental issues on the Peninsula were drawn into the debate. *Cape Times* columnist Bob Molloy fomented a 'Peninsula Campaign' against the (alleged) plan to 'straighten' Sandy Cove near Llandudno to improve travel times for heavy traffic, coining the acronym EPIC ('Enemies of Progress in the Cape').[73] Ken Godbold, chairman of the Hout Bay Civic Association, linked the Kirstenbosch freeway plan with the Council's plan to remove sand dunes and harden the beach at Hout Bay as examples of a destructive 'engineering mentality'.[74]

In the event, public sentiment backed by prominent individuals and numerous conservation-oriented organisations prevailed. The Kirstenbosch freeway plan, the Hout Bay dune levelling and the plan to 'straighten' Sandy Cove were all blocked. Of course extensive road building, industrial and urban development (including Marina da Gama on the wetlands near Muizenberg) continued on the Peninsula. Nevertheless, public and institutional support for environmental issues had been galvanised, and the City Council had been forced to take account of this in its planning.

Bush fire incidence on the Peninsula had been increasing steeply since 1967, and a typical newspaper report on a veld fire in the 1970s was: 'Fire sweeps Cape Point: a mile of rare flora destroyed.' The warden of the Cape of Good Hope Nature Reserve lamented that 'some of South Africa's most valuable vegetation has been destroyed...it will take us many years to get this right again'.[75] Addressing the Botanical Society in 1973 Philip van der Merwe of the Department of Nature Conservation warned that 800 plant species on the Peninsula were at risk from fires. Professor Rycroft confirmed that while most *Protea* species could survive fire intervals of eight to ten years, more frequent fires could destroy many species.[76] Following a fire above Hout Bay on 20 April 1973, Sydney Skaife argued that 'more and more of the indigenous plants were being destroyed', to be replaced by 'Australian exotics'.[77] In a response to this same 'mutilating' fire, a *Cape Times* columnist concluded that 'South Africans appear to be criminally contemptuous of their beautiful natural environment'.[78]

A master plan for the Peninsula

Despite heightened environmental awareness and fears for the indigenous fauna and flora of the Peninsula, conservation management remained fragmentary. The high fire incidence of the early 1970s, and concerns over human pressures on the environment, did however trigger research and planning which would ultimately result in a more comprehensive approach to conservation. Reports on the conservation status of Table Mountain by the Cape Town branch of the Mountain Club (1974), and botanists at the University of Cape Town led by Eugene Moll and Bruce Campbell (1976), raised public concern about the future of the mountain. The journal *African Wildlife* reported that 60 per cent of the Cape's fynbos had disappeared, and an article lamented that only 23 plants of the beautiful golden gladiolus (*Gladiolus aureus*) survived in the wild on the southern Cape Peninsula, hemmed in by 'a wall of Port Jackson wattle' (in 2014, only a handful of these plants survive).

In 1977 State President Nicolaas Diederichs appointed Douglas Hey to investigate the ecological state of Table Mountain and the southern Peninsula, and his report (1978) recommended that all of the Peninsula's mountains above the 152m (500ft) contour should be conserved, and managed by a single management board. The Commission's report was considered by Parliament in 1980, and laid the foundations for the establishment of the Cape Peninsula Protected Natural Environment (CPPNE).[79]

Following a big fire on Devil's Peak in March 1982, a master plan was drawn up to reconcile 'conflicting aspects of conservation and recreation' on the 'Western Table environment'. This involved closing barbecue areas, improving firebreaks and access roads. By this time, fire was regarded as a potentially useful management tool, rather than an outright danger to be suppressed in all circumstances. The Municipality's report on the 1982 fire observed that nature reserve managers in the region were introducing controlled block burning to reduce uncontrolled fires, and rejuvenate the indigenous vegetation. Burning to create a mosaic of vegetation ages was held to maximise species diversity, and fire intervals of 15–20 years were recommended.[80]

There were still serious obstacles to implementing prescribed burning. As late as 1978, botanists at the University of Cape Town argued that the natural fire interval on Table Mountain was 30–40 years, and supported the existing total ban on fires. With professional nature conservationists and academics speaking out about the alleged destructive effects of fire on fynbos, it is unsurprising that public opinion remained opposed to prescribed burning and none were implemented until 1984.[81]

A big fire that started on the north face of Table Mountain in December 1986 further stimulated public concern over environmental threats to the Table Mountain chain. Echoing the sentiments of December 1935, Minister of Environmental Affairs and Water John Wiley suggested that access to the mountain should be controlled at authorised points. Douglas Hey, chairman of the Cape Peninsula Nature Area Management Committee, called for the mountain to be closed to the public during the summer fire season (as he had suggested following the 1982 fire). However, the *Cape Times* launched a campaign to 'save the mountain for the public, with the help of the public'. In an editorial 'You can't lock the mountain', the *Cape Times* acknowledged Hey's justified 'sense of impotence in the face of public carelessness or, worse, deliberate intent'. However, they argued it would be impossible to enforce any closure of the mountain to the public, and doing so would alienate those whose co-operation was most needed. For residents, living in a city 'whose greater population is becoming denser by the day', the mountain provided 'sanctuary'. Further, the tourist industry would be badly affected by any such closure. Thus, 'more effective control, not total exclusion, should be the aim'.[82]

On 20 December 1986, the *Cape Times* launched 'an urgent campaign to preserve Table Mountain, one of South Africa's greatest natural assets'. Two days later, a Save Table Mountain Fund was initiated by the South African Nature Foundation, the Cape Town section of the Mountain Club of South Africa, and the Mayor of Cape Town. In 1989 the Cape Peninsula Protected Natural Environment (CPPNE) area was established in terms of the Environment Conservation Act 73 of 1989. This provided a degree of protection from development for around 60 per cent ($291km^2$) of the Peninsula, but did not resolve the problem of fragmented ownership and thus fragmented management.[83]

A major fire burned on Devil's Peak in February 1991. Afterwards, University of Cape Town Botany Professor Eugene Moll criticised the City Council for 'inept management of the ecology in the metropolitan area'. He said old vegetation should not have been allowed to accumulate close to the city, and recommended planting beds of low-growing vegetation around the urban fringe to act as fire barriers. A few months later, winter storms caused major flooding below the burned area.[84]

In response to the vigorous debate that followed, the Administrator of the Cape appointed the Attorney General Advocate Frank Kahn 'to negotiate and reach consensus on a plan for rationalising and consolidating the control, management and ownership of public land in the CPPNE'. Professors Richard Fuggle and Roy Siegfried, and Dr John Raimondo, led a University of Cape Town Environmental Evaluation Unit study to develop a 'policy for the multi-purpose use of the Cape Peninsula'. The Fuggle and Kahn reports (1994) recommended that the entire CPPNE be managed by a single statutory body. Fuggle et al. suggested South African National Parks (SANParks) should take over management until a new authority was set up under national legislation, while Kahn's group recommended the area be managed by the transitional metropolitan council until local government had been restructured. The CPNNE's management advisory committee had not been reconstituted after its statutory term expired in December 1993, and it was uncertain as to whether it was yet possible to implement the commissioned management recommendations. The country and region were in the throes of a major political transition. In 1995 the Minister of Environmental Affairs, Dawie de Villiers, appointed Professor Brian Huntley of the National Botanical Institute to invite submissions from those interested in managing the CPPNE and make a recommendation. His recommendation that SANParks be appointed was accepted in December 1995.

After sometimes fraught negotiations with the various national, provincial and municipal authorities, 200 private small landowners and NGOs, SANParks and key parties signed an agreement transferring management authority for approximately 16,000ha (39,540 acres) to SANParks on 29 April 1998. On 29 May 1998, then-president Nelson Mandela proclaimed the Cape Peninsula (now Table Mountain) National Park. The need to manage wildfire on the Peninsula had been a major motivation behind the realisation of this long-debated national park (Map 6.2).[85]

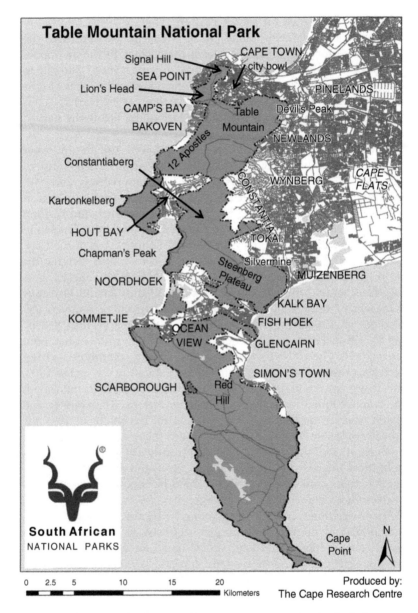

Map 6.2 The Peninsula showing Table Mountain National Park (shaded)
Source: Cape Research Centre, SANParks.

This was a revolutionary period in South African politics, with the release of Mandela in 1990, the first elections in 1994 and the enactment of the new constitution in 1996. Creating a new national park for the 'new South Africa' was a unifying act. Mandela referred to the multivalent cultural significance of the mountain, as well as its ecological and economic significance for the new nation:

> For centuries Table Mountain has been a symbol of our mother city, Cape Town. During the many years of incarceration on Robben Island we often looked across Table Bay at the magnificent silhouette of Table Mountain. It was in the shadow of this mountain that the first wars of resistance to colonial domination were waged... To the people of South Africa, the Table Mountain Range... is of immense ecological, cultural, religious and economic significance...

Creating the Park was also a means of allowing the country to rejoin the international community as a respected and responsible country. By now internationally recognised as a biodiversity hotspot, the Park was dedicated by Mandela as a 'Gift to the Earth'. Just as Table Mountain had offered a source of national pride for a deeply politically divided nation following Union in 1910, so in the late 1990s Mandela and the ANC looked to the country's natural heritage for sources of national unity.[86]

It had taken 70 years – and, perhaps, a unique historical moment with the transition to democracy in South Africa – to achieve the creation of a national park on the Cape Peninsula. However, this was a national park bordered by a fast-growing city, on a Peninsula still divided among numerous landowners, and there was still no overall consensus on fire management. Following the big fires of January 2000, the public and institutional response (despite the protestations of an informed minority) focused (once again) on the question of 'invasive alien species'. It is to the history of the introduction and spread of these species on the Peninsula that we now turn.

7
Afforestation, Plant Invasions and Fire

> The generations-old controversy over the planting of trees on Table Mountain was resurrected after the [Christmas Day 1935] fire – actually while it was still raging – and some claimed that the fire should be allowed to burn the whole mountain bare from Lion's Head to Cape Point.
>
> Annual Report for the Division of Forestry, 1936[1]

> There is a pleasure in strolling across a shady pine slope, but a sun-scorched, snake-ridden bush is just a slog for the hardy.
>
> Mr R. Pothier, 'For Pine Haters', *Cape Times*, 1 February 1950

Thomas Pakenham's beautiful book on his 'safari' to find great trees in southern Africa was controversial (for some) in that nearly a quarter of his chosen trees are introductions to the region. Pakenham feared that these trees were under threat as undesirable 'aliens', including Tokai Arboretum's world famous collection of Eucalypts and other introduced Australian trees:

> It turns out that when the boundaries of Table Mountain national Park were decided in the 1990s Tokai Arboretum was included within the park. And the authorities who manage this national park are seized by one simple idea supplied by the Talibans: restore the former glory of Table Mountain by removing all the trees... within the park which are not indigenous to the region.[2]

Pakenham's fears for the Tokai Arboretum capture the essence of a long debate over the natural environment at the Cape: how should we balance managing the cultural and ecological dimensions of the landscape? To what extent, and in what localities, should aesthetically, culturally and economically significant introduced plants be allowed to exist alongside the iconic indigenous fynbos and renosterveld? Trees are naturally scarce on this often hot and windy Peninsula, and they provide an important role in providing shelter for walkers and picnickers. They provide fuel and construction materials for the poor. Some of them have proved invasive of the indigenous vegetation, crowding out local species and increasing the risk of intense fires. This chapter shows that both public and expert opinion have long been divided on how (and whether) to find a balance in managing the introduced and indigenous flora of the Cape Peninsula.

Afforestation of the Cape Peninsula

The Cape Peninsula has never had much indigenous forest, and the first European settlers used accessible forest patches and woody shrubs for construction and firewood. The Dutch East India Company encouraged tree planting, including a wide range of introduced fruit trees, oaks and other species, but large-scale plantations were only begun in the mid-1800s. Australian wattles (*Acacia* species), hakeas and casuarinas were imported from the 1840s to stabilise driftsands on the Cape Flats. The first timber plantations were established on private farms and estates on the lower slopes of Table Mountain at mid-century, for instance Van Breda's pine forest on his property Oranjezicht, full grown by 1861. Count Médéric de Vasselot de Regné professionalised forestry on the Peninsula in the 1880s, establishing a nursery, and an arboretum at Tokai in 1885. The colony's foresters encouraged tree planting, selling tons of seeds, and hundreds of thousands of young trees at cost price to the public annually. In this period, German immigrants established wattle plantations on the Cape Flats.[3]

From the 1870s there had been significant pressure to afforest the country with fast-growing timber trees, which increased locally with the building boom that followed the end of the South African War in 1902. The population of Cape Town had nearly quadrupled between 1875 and 1904, and in addition to construction needs, sleepers were required for railways from the 1860s. The well-watered mountains of the Cape Peninsula were regarded as a key area for the development of timber plantations, which would also be conveniently close to where the timber would be needed. Pines in particular were useful for a wide range of timber products and fibre. They were much faster and easier to grow than indigenous tree species (seed is abundant and easily gathered).

Importantly, they along with wattles grew well in the poor soils of the region. Afforestation accelerated from the 1890s, and foresters aimed to entirely exclude fire from these valuable plantations until they were harvested (on 40-year rotations).[4]

State forestry

The first government plantations established on the Peninsula were Tokai, Cecilia (named after Cecil Rhodes) and Uitvlugt plantations (1884), followed by (Devil's) Peak (1893) and Rifle Range/Ottery (1897). The majority of trees planted were species of pines and eucalypts, with wattles (*Acacia*) at Uitvlugt and Ottery on the sandy Cape Flats. Of the 13,370ha earmarked for state forestry on the Peninsula c.1900, the once extensive area on top of Table Mountain was lost to the Cape Town and Wynberg municipalities for water works early in 1901. (Large planted areas survived as 'shelter belts' around the reservoirs until they were finally removed in the mid-1980s). In 1905, East End Plantation was promised to the 'people of Woodstock and Cape Town' as a park, becoming Trafalgar Park in 1909).[5]

The Peninsula's area of state plantations was dealt a blow by a timber famine during World War I from which it never fully recovered. The high price of importing timber as a result of disruptions to shipping put great pressure on local resources. Timber was required for construction, for special purposes such as box- and match-making, railway sleepers and for mining. In this period the gold mines on the Rand were using an estimated 250,000 tons of timber per annum, and large quantities of pine timber were sent up from the southwestern Cape. Inland from Cape Town, plantations on the Helderberg and at Worcester were disposed of outright. Tokai never again attained even half of its maximum pre-World War I extent (3407 acres; 1378ha). The smaller Cecilia Plantation, however, more than doubled in size to nearly 600 acres (242ha) in the mid-1970s.[6]

Uitvlugt, initially the largest plantation (3901 acres or 1578ha at its peak in 1916/17), comprised mostly cluster (or maritime) pines (*Pinus pinaster*) and some wattle. It was declared a failure by the end of World War I and 313ha (773 acres) were reassigned for development into the 'garden city' suburb of Pinelands in 1920. A new 'model' township (Langa) was created for Africans in the northeastern corner of Uitvlugt in 1927. Two plantations were developed in the mid-1920s to compensate for heavy utilisation during the War. The Hout Bay extension of Tokai had been granted in 1906, but was opposed by the public, notably in 1917/18, as it would replace indigenous vegetation.

The Groot Constantia plantation was developed on the slopes of the Vlakkenberg. Both plantations were dispensed with after one rotation: much of the Hout Bay plantation had been burned out, the final 179ha (442 acres) being removed in 1962/63; and Constantia was harvested in 1966/67.[7]

Heavy wartime demand for local timber during World War II, together with the impact of the first rotations (felling of trees for timber), meant that by the late 1940s the Peninsula's state plantations had been nearly halved again from their pre-World War I extent. Both Tokai and Cecilia plantations became important outdoor recreation areas after World War II, a function they retain today. In 1946 the Cape Town City Council decided to develop the Uitvlugt and Epping Forest Reserves (the latter comprising pines for railway sleepers) for industrial purposes, and together with Rifle Range/Ottery, all had disappeared by mid-century ('Plantation Road' is the only reminder of Rifle Range).[8]

There was some new state afforestation in the 1950s but by the late 1960s only Tokai, Cecilia and Devil's Peak survived. Devil's Peak extension around to the north face of Table Mountain had been returned to the City Council in 1962. The Department had reoriented its afforestation efforts to the eastern regions of the country. Most of the Devil's Peak extension was burned out or removed in the 1980s, and most introduced trees on Devil's Peak's eastern slopes were removed in the 1990s. This left only Cecilia and Tokai, and when state forestry was dismantled in the late 1980s, they were handed over to the South African Forestry Company Limited (SAFCOL). They were subsequently managed by SAFCOL's subsidiary, Mountains to Oceans Forestry (MTO), now known as Cape Pine.[9]

Figure 7.1 shows the trends in state afforested area (acres), for the four major plantations in the study area, 1905–83.[10] The data is for individual years listed only. Devil's Peak was phased out as a state forestry plantation from 1967 to 1971, but there were still 212 acres (85ha) of trees on the ground in 1975. Hout Bay (Tokai extension) had 738 acres (298ha) in 1910, but figures are missing until 1948. Thereafter it was 174–79 acres (70–72ha) until its disappearance in 1962/63.

Municipal forestry

The Cape Colony's foresters in Cape Town encouraged the various municipal authorities to develop plantations of exotic timber trees on their mountain lands. In 1891, the Cape Town City Council resolved to plant 'the barren slopes of Table Mountain and Lion's Hill' to 'not only beautify the Table Valley, but... secure our springs and water supply'.

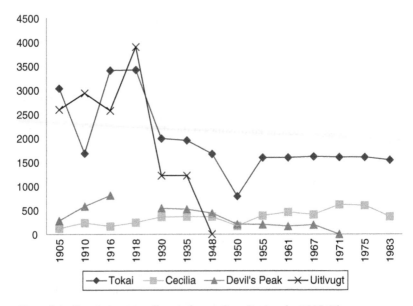

Figure 7.1   Trends in state afforested area, Cape Peninsula, 1905–83

The first plantation was established in 1894 at Kloof Nek, growing *P. pinaster* and *P. halapensis*. Frequent grass fires and browsing by live-stock and 'wild buck' presented challenges for the City Council's staff, but they persisted and plantations were maintained on the eastern slopes of Lion's Head until the 1960s.[11]

The Municipality's largest plantations were at Newlands Forest and around Kloof Nek below Lion's Head. They also planted pines and eucalypts on exposed areas of the mountainside to provide shade and shelter for walkers and picnickers, notably on the slopes of Lion's Head and Signal Hill, and along the pipe track below the western slopes of Table Mountain. Municipal plantations were heavily utilised to maintain City Council building programmes during World War II, when timber imports were once again interrupted. The Roeland Street plantation was used up at this time. The Department of Defence drew supplies from the Newlands Forests, using the timber for ammunition boxes, amongst other things. After the War, Newlands was reafforested with indigenous trees to create a recreational area not as 'eerie and uninviting to visitors or picnickers' as the large pine plantations had been.[12]

Following a large fire on Signal Hill in 1950, firebreaks of eucalypts were planted on Signal Hill, Kloof Nek and Orange Kloof, intended to screen out fires, suppress inflammable grasses and reduce maintenance costs. They ultimately proved unsuccessful and most were removed in the late 1970s. Following the proclamation of the Table Mountain Nature Reserve on Municipal lands in 1964, afforestation of new areas in the vicinity of Lion's Head and Signal Hill ceased. The municipality's afforested area peaked in 1971 at 3101ha (7662 acres). As a result of big fires in 1973 and 1974 and subsequent felling and clearing, by 1975 the municipal plantations had been reduced to 823ha (2033 acres), after which municipal commercial timber growing was phased out on the Peninsula. A notable area of pines remained on the slopes extending from Constantia Nek west to a screen of eucalypts protecting the indigenous forest in Orange Kloof.[13]

Plantations and fire

Until at least the 1930s, badly maintained private plantations on the slopes of Table Mountain remained a fire hazard. However, the fire protection measures concentrated around state and municipal plantations limited the spread of wildfires. Fires spreading from the Muizenberg and Constantia mountains towards the southern suburbs were cut off by the Tokai, Constantia and Cecilia plantations. Fires spreading eastwards around Devil's Peak or upslope from Groote Schuur Estate were blocked by measures at Peak Plantation.[14] However, many fires were started in plantations (some deliberately), spreading to the surrounding veld, for example a series of fires at Tokai, Peak and Cecilia plantations in February/March 1920, and a big arson fire in January 1934, that spread from Tokai plantations west over the Muizenberg Mountains towards Chapman's Peak.[15]

Of course, fire protection activities were not consistently applied, and even when they were, could not always keep out large fires from adjoining private plantations and estates. When fires did take hold in mature plantations, they developed into much bigger and more dangerous fires than could have developed in the fynbos with its lower fuel loads. Foresters were never able to achieve total fire suppression and fire incidence increased across the century. State and Municipal foresters were responsible for burning most firebreaks, and inevitably some controlled fires escaped. This is an insignificant category of ignitions until a period of (relatively) high incidence from 1963 to 1972, averaging 23 fires per annum – excluding an egregious high of 41 runaway fires in 1970. Across the country, there was a spike in fire protection measures in 1969/70, the year in which the Forestry Department took on protection of all mountain catchments, and prepared to implement a policy of prescribed burning. From 1982/83 to 1992/93, only 37 fires were attributed to escaped controlled burns. Two big fires were alleged – groundlessly – to have been the result of prescribed burns getting out of control: the Devil's Peak fires of 7 February 1974 and 8 February 1991. In the 1980s and 1990s land managers were increasingly inhibited in burning as a result of litigation, particularly on the Cape Peninsula where fires could easily approach residential areas expanding into the fynbos.[16]

Invasive introduced plants

The establishment of timber plantations and dune stabilisation work by state and municipal foresters had another important impact on the flora and fire regimes of the Cape Peninsula. In the long run, certain plant species spread well beyond these demarcated areas, invading the indigenous vegetation and transforming the fire ecology of parts of the Peninsula.

The history of plant invasions in the Cape complicates Alfred Crosby's idea of 'ecological imperialism'. Crosby suggested that in the wake of European explorations of the globe, the introduced diseases, plants and animals of the 'old world' (Eurasia) had overwhelmed regions with similar climates in the 'new world' (the Americas and the colonial 'neo-Europes', notably Australia and New Zealand), helping Europeans to dominate these regions. However, this biological 'imperialism' is perhaps better understood as a global biological interchange between regions with similar natural environments interconnected by political and economic circuits, as well as the personal trajectories of settlers, colonial officials, plant collectors and horticulturalists.

Local colonial experts played a vital role in developing the biological basis and technological expertise to enable biological transfers (or transplants) within and between colonial territories, and Europe and the colonies. Australian colonial foresters promoted and exported wattles and eucalypts worldwide. South African foresters pioneered a climate-informed approach to afforestation with introduced trees, conducting experiments with a wide array of species across a spectrum of local environmental conditions. The success of introduced plants also depended in important ways on social factors – they were not just spread through official imperial and colonial networks of exchange for the purposes of economic and utilitarian improvements of local landscapes. Colonial foresters vigorously promoted tree planting at the Cape, and middle class locals adopted species like Australian myrtle (*Leptospermum laevigatum*) as garden plants. Hedges of this plant still today provide a source of invasions of the surrounding fynbos. Poor people in rural areas and in shanty towns outside Cape Town have also long made use of invasive wattles (*Acacia*) for fuel and building materials.[17]

'Alien plants' is the local, pejorative, terminology for introduced or 'exotic' plants. I prefer 'introduced plants' because it is more value neutral, but acknowledge this is too bland for raising awareness or funding for control programmes. Introduced plants are plants introduced into local ecosystems by humans which could not have otherwise crossed biogeographical barriers to their arrival. Invasive introduced plants are a subset of introduced plants which are able to 'sustain self-replacing populations over several life cycles, often in large numbers and at considerable distances from the site of introduction'. The Cape's fynbos has been a crucible for research on invasive introduced plants, with South African researchers playing a key role in the development of the scientific subdiscipline of invasion biology (or ecology). There is also a growing body of literature on the cultural and historical dimensions of how biological (particularly plant) invasions have been understood and represented in the region.[18]

On the Cape Peninsula, most problematic (invasive) species come from a suite of fire-adapted Australian trees and shrubs. These were introduced for timber and tannin (eucalypts and wattles), driftsand stabilisation (Australian myrtle, hakeas and the wattles *Acacia saligna, A. longifolia, A. pycnantha* and *A. cyclops*), or planted in belts to act as fire screens around pine plantations (eucalypts and wattles). In the long run, the two most invasive trees have proven to be *P. pinaster* from Mediterranean Europe and *P. radiata* from California. All these invasive plants can be divided into two main groups. The pines, hakeas and Australian myrtle are serotinous, carrying winged seeds in fire-proof cones or follicles, which are released and spread by wind after fires kill the parent plants. The wattles (*Acacia*) drop numerous hard-coated seeds that form seedbanks, which are stimulated by fires to germinate. They are spread mainly along waterways. Many of the wattles can also sprout after being felled, and so require repeated interventions to eradicate. Little is known about how *Eucalyptus* and the Australian myrtle invade local ecosystems.[19]

In the early 1900s, foresters planted belts of *Eucalyptus diversicolor* and *E. corynocalyx*, acacias and casuarinas at Uitvlugt Plantation to stabilise driftsands and protect pines from wind and fire. 'Fire screens' were planted around the Peak plantation extension below the north face of Table Mountain (*E. resinifera, Acacia decurrens*), and along the bridle path from Kloof Nek to Devil's Peak (*E. pilularis*). Similar belts were grown where railways passed through or bordered plantations, for example at Uitvlugt and Wynberg Rifle Range (the latter had a station built in it, today named Ottery Station).[20]

The Cape botanists Peter MacOwan (in 1888) and Rudolf Marloth (in 1908) had expressed concern that introduced plants would replace the indigenous vegetation. However it was in the 1920s that such concerns attracted wider support. Local botanists publically urged the removal of invasive introduced plants, including *P. pinaster*, *Hakea* and wattles from the indigenous fynbos.[21] Travelling to a spring flower show in Tulbagh, well-known local author Dorothea Fairbridge complained of the replacement of the native flora with Australian wattles. The University of Cape Town botanist Robert Adamson's analysis of the plant communities of Table Mountain (1927) included 'plantations and alien plants', and he worried that introduced pines, eucalypts and wattles were '[eliminating] the native vegetation' on Table Mountain. *Pinus pinaster* in particular were spreading rapidly from seed and 'dominating the native flora in stations near planted or established trees'. Hakeas, notably *H. pectinata*, were spreading rapidly on the western slopes of Table Mountain, forming closed communities which excluded all other plants. Both *Hakea* and *P. pinaster* were 'competing with the native shrubs in areas recovering from fires', and their more rapid growth favoured their 'ultimate domination'. When Sir Arthur Hill, Director of Kew, visited the Cape in 1929, he complained of a 'long dull drive over the Cape Flats' where 'all vegetation [was] exotic Eucalypts, pines, wattle and venothera...'. Experts and informed laypersons were being primed to defend the local flora against invasions by introduced plants.[22]

A major public controversy over introduced species was sparked by the fire at Devil's Peak on Christmas Day 1935. The Forestry Department's annual report recorded that:

> The generations-old controversy over the planting of trees on Table Mountain was resurrected after the fire – actually while it was still raging – and some claimed that the fire should be allowed to burn the whole mountain bare from Lion's Head to Cape Point. The complaint made here against trees is mostly as a result of the fact that over the past years the 'cluster pine' has spread all along and even on top of the mountain in a manner that was certainly originally not anticipated and is undoubtedly now undesirable.[23]

The principle arguments for the removal of introduced species were that they were reducing water supplies, were a fire hazard, and were invading the natural vegetation.[24]

In a newspaper interview Benjamin Simmons, Conservator of Forests at the Cape, acknowledged that:

> There has been a great agitation against our work of planting pine trees of late. Coniferous trees are the only type of timber we can grow on account of the soil of the mountain... Whatever happens, we will not plant gums [eucalypts] again... they have a desiccating effect.

Simmons argued that whereas the prominent politician and avid botanist Jan Smuts wanted the Department to 'clear away every pine tree', there was no need to be 'so drastic', especially as: 'our work of planting pines has achieved its aims of checking erosion. Look at the lower slopes of Devil's Peak where there are not trees and you will see that'.[25]

Indeed, in the 1890s, Capetonians had complained about soil erosion occurring on the bare slopes of the mountain, notably on Devil's Peak.[26] Cluster Pines were planted to control, and hide, the soil erosion on the slopes. In the long run the plan was to use the pines to 'tame the soil' and make it suitable for the planting of a more diverse range of tree species – which never happened.[27] Instead pure stands were developed for timber.

Lady Florence Phillips came out as a prominent voice of opposition to introduced species following the 1935 fire. She asserted in a newspaper interview 'that the Forest Department should be deprived of the right to plant Table Mountain with pines and gums. They are nothing but a source of danger'. The Minister of Agriculture and Forestry, Deneys Reitz, commented on the alleged desiccating effects of pines and eucalypts, concluding: 'the question might arise whether it would be advisable to afforest Table Mountain with a type of timber not so liable to fire'.[28]

In response to this controversy the Division of Forestry resolved to 'destroy the Cluster Pine, which has invaded the veld in places where it threatens to distract the eye from the outstanding landmarks of the mountain or suppress desirable areas of natural vegetation'. However, the Division wanted to avoid any precedent for the wholesale destruction of 'forests', arguing that a great portion of the public would not approve of 'bare cliffs and Gibraltar-like rocks'. The Division hoped to 'retain single or clumps of trees at picnic or camping places', and 'leave rows of trees around the reservoirs for the protection they offer against flooding and erosion and as windbreaks to reduce evaporation'.[29] In all of this, the Department's views were to prevail.

Little came of these resolutions regarding invasive plant control. The keen amateur botanist and plant collector Carl Lückhoff wrote that: 'in 1936 desultory attempts were made and numerous groups of pine were felled'. The first mention of control measures appears in Municipal reports in 1939, and the focus was on Hakeas. In 1940, the City Forest Officer reported that self-sown pines were removed from 16ha (39 acres) atop Table Mountain, a few Hakea from above Muizenberg, and *Acacia longifolia* from Newlands. It was resolved that 'next year a vigorous programme is to be instituted on the slopes of Table Mountain', but the War intervened. There is some evidence that the clearance of 'noxious weeds' notably *Hakea* and *Acacia* species was resumed after the War, but no major eradication campaign was attempted.

Certainly, the lack of control measures contributed to the intensity of big fires in invaded areas. When in January 1942 six square miles (15.5km^2) of bush burned near Glencairn, 'Simonstown CPS firefighters had a risky job as Port Jackson bush there was over 20 feet high'. In December 1943 a fire between Glencairn Quarry and the Fish Hoek municipal border was 'confined...to several hundred acres of Port Jackson, rooikrans [both introduced wattles] and indigenous bush'. On Table Mountain, voluntary parties of Mountain Club members 'attacked' the pines scattered across the upper plateau.[30]

The state Forestry Department likewise did little to clear introduced trees and shrubs invading the surrounding fynbos on the Cape Peninsula, despite their promises in 1935, and despite Wicht's warning in 1945 that: 'one of the greatest, if not the greatest, threats to which the Cape vegetation is exposed, is suppression through the spread of vigorous exotic plant species'.[31] Carl Lückhoff relates that: 'the first concerted attempts by the Forestry Department to clear the mountain were only begun following the very large fire of [January] 1950' above the City Bowl.[32] This fire reignited the controversy about growing pines on the mountains. Gwen Edwards wrote to the *Cape Times*, citing Harold Pearson, first director of Kirstenbosch, to the effect that pines (being northern hemisphere plants) lose water through transpiration most heavily in the Cape dry season. She argued that indigenous vegetation is more effective at protecting the soil and uses less water than pines. A Mr R. Pothier of Burg Street contributed 'For Pine Haters'. He noted that the indigenous flora burns very well, claimed fires are easier to fight in plantations, and expressed his preference for strolling through shady pines rather than 'sun-scorched, snake-ridden bush'. Removing pines because they are foreign, he pointed out, would mean removing oaks, gums, poplars and many other attractive introduced trees.[33]

According to Lückhoff, in 1950 'exotic' flora posed a serious threat to the indigenous flora of the Peninsula: 'the relentless advance of [Cluster pines] on the lower slopes, summit plateaus [of Table Mountain] and [especially the] slopes and high ledges overlooking Newlands' was overpowering the fynbos. The pines had spread from the plantations around the Woodhead and Hely-Hutchinson reservoirs, and plantations above Newlands. Hakea species were abundant on the upper slopes of Kirstenbosch Estate, above Camps Bay, and on the Constantiaberg west of Constantia Nek. Wattles were 'running riot' on the lower slopes of Table Mountain from Devil's Peak to Lion's Head, on the slopes below the Twelve Apostles, the Constantiaberg and on Chapman's Peak.[34]

The City Council's decision in 1951 to create a Van Riebeeck Park provided the impetus to clear invasive plants from the Highlands Deer Park area of the City Bowl (117ha; 289 acres). The irony of clearing invasive plants to create a park to celebrate a human invasion heralding a major biological transformation was no doubt lost on the organisers. More pragmatically, the area had been prone to 'innumerable fires during the last twenty years' and had become overrun with weeds and wattles. Clearing vegetation and 'fire debris' considerably improved this situation.[35]

In 1953 Robert Adamson warned of the threat posed by the spread of introduced species aided by veld burning. He complained that 'when a visitor is taken for a drive in the neighbourhood of Cape Town, the plants most prominent are in nearly all cases not indigenous, oaks, pines and even more wattles predominate'. The coastal plains north and east of Table Bay were infested with wattles, as were the hills between Fish Hoek and Kommetjie on the southern Peninsula.[36]

In 1959/60, University of Cape Town botanist Anthony Hall undertook a survey of plant invasions from the Muizenberg Mountains north to Table Mountain. Hall recorded indigenous and invasive plant cover around 87 precisely recorded and marked sample points. He found *P. pinaster* around 82 of these points, with the next most widely distributed arborescent invasive species being *Hakea gibbosa* (at 22 sites) and *Acacia cyclops* (20). These arborescent or overstorey plants shaded out fynbos plants beneath them. He recorded three invasive pine species, three *Hakea* and four *Acacia*.[37]

Although the problem of plant invasions had been well publicised in Cape Town since the 1920s, and major public debates followed big fires in 1935 and 1950, it was only in the late 1950s that the problem was formally tackled by a conservation body. In 1958 the Kirstenbosch Botanical Gardens' Wild Flower Protection Society Committee set up a

Control of Alien Vegetation Committee, which published a book entitled *The Green Cancers of South Africa: The Menace of Alien Vegetation*
(1959). Anthony Hall's quantification of the problem was also influential in stimulating action. In 1964, the Cape Town Municipality's
programme of clearing and reclearing introduced invasive species finally
became a 'vigorous campaign'.[38]

The state Forestry Department had begun invasive alien plant control in fynbos catchment areas in the 1950s, first recognising this as a
discrete management goal in 1960 when the annual report commented
that 'Hakea, wattle and self-sown *P. pinaster* are causing trouble in the
catchment areas in the Western Cape Region'. In 1961 Forestry Department representatives attended a conference on 'the Hakea problem'
convened in Stellenbosch by the Department of Agricultural Technical
Services. Knowledge was exchanged on control measures and 'intensive
research' was called for. In 1964 the Forestry Department acknowledged
'great pressure from outside ... being brought to bear on the Department
to combat ... so-called invader plants'.[39]

Unfortunately, by 1973 (the year of the OPEC oil crisis) funds were
lacking for this kind of work. The Forestry Department acknowledged
that invader control programmes were being scaled back as 'whenever
funds have to be cut back these operations are rated a lower priority than
others'.[40] The City Council was concentrating its efforts at clearing invasive plants on the catchment areas of its Wemmershoek and Steenbras
dams.[41]

In 1976 D. McLachlan, Eugene Moll and Anthony Hall resurveyed
Hall's sample points. They found a reduction in the distribution of
P. pinaster and *Acacia cyclops*, but increased distributions of *Hakea
gibbosa*, *H. sericea* and *Acacia saligna*.[42] It is unclear whether they
included areas cleared by the City Council in that year – an unusually large area was cleared for security reasons in response to political unrest following the Soweto Riots. Either way, there was clearly
still cause for concern. The policy was to clear all invasive introduced plants from above the 300m (984ft) contour line, leaving non-
invasive introduced species below this to provide shade for 'high
usage nature-orientated recreation'.[43] However, this was not a budgetary
priority.[44]

Following a major fire on Devil's Peak in March 1982, the Municipality phased out the system of *Eucalyptus* fire-belts on Table Mountain.
They were judged effective as spark arrestors to protect pines, but of little value in indigenous vegetation.[45] The Municipal report on the fire
revealed that fire was being used to control invasive plants:

after a dense alien stand has been felled, the area is usually burnt... to break down... remaining trunks and stumps, kill off much of the remaining seed-store, [and] stimulate mass germination of exotic seedlings so they can be easily removed.[46]

Unfortunately, budget cuts and public opposition to burning seriously hampered control measures.

In 1989/90, Moll and Trinder-Smith resurveyed Hall's survey points and analysed 30 years' of invasive plant control on the Peninsula. The distribution of *P. pinaster* had been reduced but it remained the most widely distributed invasive species, with density reduced on Table Mountain and at Muizenberg but increased on the Constantiaberg. *Acacia cyclops* was the next most widespread invader. *Acacia saligna* had been controlled in some areas, but had spread elsewhere, forming dense stands. *Hakea* species had been successfully contained post-1976, though not eradicated. The Constantiaberg had proved a nexus for invasions, being downwind of southeasterlies blowing seeds across from the Tokai pine plantations, and upslope of poorly controlled *Hakea*- and *Acacia*-invaded slopes above Hout Bay.[47]

The control efforts of the 1950s and 1960s were also relatively ineffective as a result of ignorance of the reproductive strategies of the invading plants. An improved ecological understanding developed during the 1970s by Anthony Hall, Hugh Taylor, Eugene Moll and others, resulted in more effective control measures. Managers realised that invaded areas require repeat treatments, and new invasions must be quickly checked before dense stands can form. A systematic invasive plant management plan was adopted in the Cape of Good Hope Nature Reserve in 1981.[48]

Scientific research into the mechanisms of plant invasions took off in South Africa in the 1980s, following a discussion at the 1980 MEDECOS meeting in Stellenbosch between Fred Kruger, Hal Mooney and others. They were intrigued that undisturbed fynbos vegetation was being invaded by introduced pines and hakeas (received wisdom was that only disturbed vegetation should be vulnerable to invasion.) In 1981 David Richardson was hired to work at Jonkershoek on plant invasions in catchment areas (he is now Director of the Centre for Invasion Biology in Stellenbosch). In 1982 the Scientific Committee on Problems of the Environment (SCOPE) identified biological invasions as a problem of global concern and initiated an international project to research this theme. The South African report (1986) provided a scientific basis for management in the country.[49]

Unfortunately, limited funds made implementing invasive introduced plants management plans increasingly difficult. Further, the fractured nature of landownership and management meant well-controlled areas could be easily invaded by seeds spread from ill-managed neighbouring areas. Finally, prescribed burning for invasive plant control (or any purpose) was opposed by the public.[50] Even after attempts were made to undertake prescribed burning on the Peninsula in the early 1980s, the very short burning season approved as ecologically safe by fynbos ecologists, and fears about fires spreading to built-up areas, meant many burns necessary to control invasive plant seedlings after felling could not be implemented.[51]

By the late 1980s the Apartheid state was crumbling and State Forestry collapsed. Serious funding shortfalls left invasive plant (and fire) control operations in a parlous state. An example of the combined problem this could cause followed the big Devil's Peak fire of February 1991. On 25 and 26 May 1991, winter storms sent $400m^3$ ($14,126ft^3$) of fire debris, rocks and mud hurtling down Tin Mine Ravine, overwhelming debris traps and stormwater drains, to flood streets as far downhill as Darling Street. The cost of post-fire mitigation was estimated at ZAR4.5–6.5 million. Researchers at Forestek, formerly the Forestry Research Institute and now a division of CSIR, concluded that the flood was the result of high soil repellency, which greatly increased runoff. This repellency was caused by the hardening of the soil resulting from the presence of introduced plants intensifying the fire. Invaded sites have as much as eight times more fuel (10–25 tonnes per hectare) than typical fynbos shrublands (three to five tonnes per hectare).[52]

In 1994, South Africa became a democracy led by Nelson Mandela. The Western Cape's new Ministry for the Environment, Nature Conservation and Tourism requested submissions from the City Planner for the President's 100 days Reconstruction and Development Programme. The resulting Clean and Green Campaign focused on creating job opportunities and improving peoples' quality of life through environmental actions. One of the approved projects was the removal of invasive vegetation from Devil's Peak, which had proliferated since the large fire of 1991. The Botanical Society of South Africa supervised the project, which commenced in April 1995 and employed 27 people to clear an area of 99ha (245 acres).[53]

In 1995, Guy Preston (with others) persuaded the Minister of the new Department of Water Affairs and Forestry (DWAF) Prof. Kader Asmal, to help fund what would become Working for Water (WfW). This programme allied environmental management with poverty relief. The 'poorest of the poor' were recruited and trained to manually control invasive introduced ('alien' in the language of WfW) plants, focusing on clearing catchment areas and waterways.[54]

Preston had studied environmental and geographic sciences at the University of Cape Town, and was interested in conservation and economics. He believes that that 'politics is about the allocation of resources, and conservation is about resources too', and was acutely aware that the transition to democracy in South Africa in 1994 provided an opportunity for enormous positive change. Of equal concern at this time were the future of research and conservation funding under the new dispensation, and the consequences of the collapse of catchment areas management. These concerns were aired at the Fynbos Forum, an influential annual gathering of a wide range of stakeholders including researchers and managers, which had developed to extend the cooperative approach of the recently terminated Fynbos Biome Project.

State funding for fire and invasive plants research dried up in the 1990s, and Brian van Wilgen and others turned to ecosystem services arguments to motivate for funding for environmental research and management. They acknowledged that: 'this economic justification is important where strong competition exists for the public funding needed for conservation management'. They focused on the negative impacts of alien plant invasions of fynbos on the water resources of the Western Cape's mountain catchments, concluding that 'existence values [value conferred regardless of actual or potential use] are becoming increasingly difficult to quantify and to defend, particularly in developing countries, where basic human needs and economic growth are the overriding concerns'. They aimed to link conservation with basic human needs, and put a financial value on ecosystem services (and the cost of their deterioration).[55]

A 'road show' was put together to communicate these views to policy makers in the new government. At a presentation attended by DWAF Minister Kader Asmal, Van Wilgen presented findings of a study led by fellow CSIR researcher David Le Maitre that modelled the consequences of lack of invasive plant management in Western Cape catchments. Their study suggested that uncontrolled invasions would result in average annual water losses equivalent to 30 per cent of Cape Town's annual water supply. Preston presented a paper critiquing the DWAF's focus on dam building, arguing that it was preferable to prioritise conserving the water supply, which involved controlling invasive plants in catchment areas. After hearing his paper, and meeting with Preston, Asmal invited him to become his advisor. With the aid of World Wide Fund for Nature (WWF) South Africa, for whom Preston had developed a 'user pays' scheme for power and water in nature reserves, they initiated a national water conservation campaign, and one of the first foci was invasive plants.

Invasive plant control was labour-intensive and environmentally positive – an ideal project for the African National Congress (ANC)-led government's new Reconstruction and Development Programme (RDP). Preston secured funding from the WWF, and together with several local academics, CSIR ecologists, CapeNature managers and other Fynbos Forum members, developed the WfW programme. Asmal subsequently secured substantial funding from government and a somewhat reluctant forestry industry. In their first year the WfW programme was the only RDP project to spend their budget.[56]

The considerable success of the WfW programme meant it grew exponentially, and with inevitable teething troubles. An important instance was the chopping down and stacking up of woody invasive species, which were then left uncollected on hillsides. As they dried out, these became in effect tinderboxes. Of the area burned in the January 2000 fires, 90 per cent had been invaded by introduced plants, and 26 per cent of the area had been cleared of invasive plants in 1998 and 1999, with the stacks of cut wood providing hotspots during the fires.[57] On the whole, however, the programme offered a good solution in a time of political revolution to the existing vacuum in environmental management.

Introduced versus indigenous plants

In the long run, advocates of afforesting the Peninsula with introduced trees have lost out to those in favour of conserving the indigenous vegetation. Although plantations remain at Tokai and Cecilia, and a few stands remain at Newlands, the Peninsula's many other pine and eucalypt plantations have been harvested, or burnt, and not replaced. For many, Table Mountain National Park protects the most iconic landscape supporting South Africa's unique fynbos vegetation, and exotic trees and shrubs are not welcome within its borders. Small stands of introduced trees persist in pockets around Table Mountain, appropriately enough near Rhodes Memorial and in Van Riebeeck Park above the City Bowl, and also near Constantia Nek, but elsewhere most have been chopped out. Even the stand of trees that provided shelter for picnickers atop

Silvermine was removed following the fires of January 2000. It is Park policy to remove all pine plantations from the Peninsula and hand them over to SANParks to rehabilitate sites for indigenous vegetation.

Some feared that the historic eucalypts of Tokai Arboretum were similarly under threat, and this led to a sharp exchange between visiting author Thomas Pakenham and local environmentalists in 2007. Pakenham's book *In Search of Remarkable Trees: On Safari in Southern Africa* includes several species introduced to the Peninsula, namely the 300-year-old camphor trees at Cellars Hotel in Constantia, the immense Moreton Bay fig tree in Arderne Gardens, Claremont, the stone pines and oaks around the Rhodes Memorial, and Australian apple gums in Tokai Arboretum. In the introduction to his book Pakenham castigated the South African authorities for their 'excessive zeal...towards the culling of "alien" trees'. He accused 'a small number of hard-line ecologists' of being 'white "eco-fascists"', seeking to ethnically cleanse the country of its 'alien' flora. He coined the term 'tree taliban', which was gleefully seized upon by the local media when the book was launched in Cape Town in 2007.[58]

In fact, campaigns to rid the Peninsula of introduced plants have focused on invasive species and were implemented by poor African and coloured people under an ANC-government-supported programme. They should not be misinterpreted as antipathy to all introduced trees per se. In addition to all the introduced flowers so beloved of the city's gardeners (to the vexation of local botanists), the many fruit and ornamental trees introduced by European settlers since the seventeenth century have become deeply ingrained in local residents' 'sense of place'. Oak trees are particularly associated with the few remaining colonial Dutch estates, and of course grapevines are both a scenic and economic feature of the landscape. The Long Street palms have been a noted feature of Cape Town for decades, as have the Camps Bay palms. In 1943 the falling of a 'venerable' gum tree (eucalypt) in the Company Gardens was noted with regret in the *Cape Times*. Among the invasive species, even the dreaded invasive wattles have been celebrated for their cheerful golden blossoms.[59] For Stephen Watson (1954–2011), the city's greatest poet of place, stone pines were an integral part of the Peninsula's imagined spaces. The cover of his collection *Presence of the Earth* features an etching of stone pines in silhouette by well known Cape Town artist Pippa Skotnes, from Watson and Skotnes' collaboration *Cape Town Days*.[60] These non-invasive pines have been retained at Rhodes Memorial, and at Kloof Nek and The Glen below Lion's Head. Together with the screen of large eucalypts separating the western slopes of Signal

Hill from the suburbs below, they have been preserved as 'vegetation of stature'.[61]

Originally introduced to combat driftsands, and used to try and keep fires out of pine plantations, invasive Australian Acacias and Hakeas remain a threat to the indigenous fynbos on the Peninsula. So do the cluster pines. However, it is also important to remember that the vegetation maps in many books, reports and publications about the flora and fauna of the Peninsula are really informed imaginings of what might have once grown there, or what might potentially grow there given ecologically sensitive management interventions. In 1996, Cowling et al. produced the first detailed and comprehensive vegetation map of the Peninsula. Rainfall and soil data were used to reconstruct the pre-colonial vegetation distribution for areas where introduced plants and urbanisation had transformed the indigenous ecosystems. Twenty years later, the SANBI vegetation map (2005) indicates expanses of Cape Flats Sand Fynbos where there is now almost exclusively suburban development and shanty towns, and most of the area labelled Peninsula Shale Renosterveld is buried under the Cape Town city bowl (see Map 5.2).[62]

In reality, the Cape Peninsula has been a hybrid landscape for almost 350 years, and especially so since the plant introductions and afforestation which began in earnest nearly 150 years ago. The distributions of some of the Peninsula's indigenous plant species, and the compositions of its plant communities, were modified by landscape gardening throughout the twentieth century. Showy indigenous plants were planted along roadways and in picnic areas. Sugar bushes (probably *Protea repens*) were selectively removed from Signal Hill to create a more aesthetically pleasing effect, and silver trees were planted and protected with strips of pines. Throughout the 1920s and 1930s the municipality's Parks and Gardens department sowed 'seeds of various indigenous trees and shrubs... including 12 different species of protea... to improve our native flora'.[63] In December 1943, *Cape Times* columnist Alan Nash announced:

> ... the restocking of the Cape Peninsula's common lands and mountains with their original wild flowers. Gangs of men [will be put to] work on Table Mountain... to plant disas, Jersey lilies, and the hundred other wild flowers that have been savaged from the kloofs and the plateaux in the last 40 years.[64]

This was to be ecological restoration on a grand scale – but I have not found evidence that this project was implemented. Nevertheless, the

City Lands and Forests Branch continued this indigenous gardening work in the newly established Van Riebeeck Park, and in 1953 recorded that thousands of proteas and indigenous trees had been planted and sown. This work was continued at Silvermine, on Table Mountain, Newlands and Orange Kloof through into the 1980s under the rubric of 'scenic and recreational forestry'.[65]

In 1998 South African National Parks (SANParks) took over the management of the Peninsula's protected areas. Subsequent attempts to eradicate introduced trees from the Table Mountain National Park have met with controversy and fierce resistance from some residents. SANParks have been obliged to take account of both ecological and cultural heritage considerations, as well as recreational needs, in developing their management plan for the Park.

Conclusion

Trees were introduced to the region primarily to provide local timber supplies. Some Acacia and Hakea were introduced to stabilise dunes. Nineteenth-century botanists' and foresters' claims that afforestation improved water supplies and local climatic conditions were questioned from the 1890s, and in effect disregarded on the Peninsula in the twentieth century. By the 1930s it was widely asserted that introduced trees actually desiccated the landscape.

Complaints about invasions of the fynbos, notably by pines, hakeas and wattles, gained momentum in the 1920s. After the big Christmas Day 1935 fire, state foresters acknowledged the problem of pines invading the surrounding fynbos, and claimed they would not afforest with eucalypts because of their desiccating effect. Arguments for introduced trees remained primarily economic, but additional arguments included aesthetic justifications (clothing the barren mountain slopes) and to combat soil erosion.

The most voluble opposition to plantations of introduced trees arose following the spectacular fires that resulted when mature trees did catch fire, notably in 1935, 1938 and 1950. The intensification of bush fires in invaded veld was noted from the early 1940s. Opposition followed a boom and bust cycle, however, with flare ups after major fires followed by swift forgetting and disregarded promises to tackle the problem.

By the mid-1960s, clearing invasive species had become policy for both state and municipal forestry, primarily to prevent big bush fires but also for water and nature conservation (see Chapter 4). Nature reserves were declared and further afforestation halted, to be completely abandoned by the late 1970s. By the 1980s only three significant planted areas remained.

The first really effective attempts to clear invasive plants were initiated in the late 1970s, following breakthroughs in ecological understanding of how these plants reproduced and invaded indigenous vegetation. Unfortunately the combination of fractured landownership and management on the Peninsula (allowing reseeding from adjacent untreated land), the collapse of state forestry and dwindling funds for control programmes as political sanctions took their toll, meant control efforts declined from the mid-1980s into the mid-1990s. Prescribed burning, a key tool for 'recleaning' land cleared of invasive plants, was publically opposed and little implemented.

In post-democratic South Africa, invasive plants once again became the focus of eradication programmes under the WfW programme. In a sense this was reviving the arguments of the 1930s about the hydrological impacts of introduced trees in catchment areas (arguments concerning impacts on ecosystem services are nothing new in South Africa). However, the impact on water supplies wasn't a major factor on the Peninsula because Cape Town's major water supplies now lay elsewhere. Fire prevention remains the key motivation for eradication programmes, for the safety of humans, property and the preservation of the indigenous flora and fauna. Eucalypts, the least favoured trees in the first half-century because of their desiccating effect, have been replaced by pines as prime targets for eradication because they are far more invasive of the indigenous vegetation.

Attempts to link the high-profile campaign to eradicate 'invasive alien plants' after the January 2000 fires with xenophobia in the 'new South Africa' overlook the long history of debates over invasive introduced plants in South Africa, conducted alongside attempts to link the indigenous flora and fauna with regional and national identities. As discussed in Chapter 6, floral nativism is nothing new on the Cape Peninsula, with Nelson Mandela's statement in 1998 that the Table Mountain chain is of 'immense ecological, cultural, religious and economic significance not only to the Western Cape Region, but also to the rest of the country' echoing the sentiments of Harold Pearson in 1910 and Jan Smuts in 1923.[66]

It seems that while nature conservation managers should control the harmful effects of invasive species, and continue to use fire to do so where appropriate, they should perhaps let go of the idea that either

Photo 7.1   A hybrid landscape at Constantia, with vineyards, oak trees (right, middle), Tokai plantation (pines) on the lower slopes of the Constantiaberg and fynbos above
Source: Simon Pooley.

these plants or the fire regimes they modify can be eradicated from the Peninsula. Managing the Peninsula with the culturally and ecologically hybrid landscapes that exist in mind, rather than lost landscapes from the precolonial past, should enable effective, culturally acceptable and sustainable interventions.[67]

8
Socio-Economic Causes of Fire: Population, Utilisation and Recreation

> In the neighbourhood of Cape Town bushfires are usually started deliberately, either by the farmers... or by the wood gatherers who eventually remove the dry brushwood thus provided.
>
> Margaret Levyns, 1924[1]

In 2008, CSIR researchers Greg Forsyth and Brian van Wilgen showed there had been an increase in short-interval (more frequent) fires in nature reserves on the Cape Peninsula over the period 1970–2007. They attributed this to the increasing human population and expanding wildland–urban interface (WUI) in this period, drawing on research from California. In particular, they noted a population boom in the 1980s, linked to the end of Apartheid-era influx control regulations designed to prevent Africans from settling permanently in urban areas (it follows that most new immigrants were Africans).[2] However, this politically influenced demographic change complicates mapping the American example onto a South African context.

The City of Cape Town has a complicated history of racial segregation which was formally instituted and implemented under Apartheid in the 1960s, and despite some blurring of boundaries, still largely determines the demographics of the city today. The city has also developed in such a way that it little resembles a California-style wildland–urban intermix: rather, the interface between sprawling suburbs (and a few satellite settlements) has increased considerably across the century, as outlined in Chapter 5. To consider the long-term impacts of the increasing population of the City on the fire regimes of the Peninsula requires a more nuanced investigation of the socio-economic causes of fires. An apparently strong historical relationship between fire incidence and overall population growth on the Cape Peninsula raises more complex questions about access to and usage of undeveloped land on the Cape Peninsula.

Population growth

Aside from short-lived expansions during the South African War and World War I (the latter including an influx of 'poor whites' fleeing drought and the economic depressions of the later 1910s), the population of the City grew steadily to 1941. After that it experienced its first significant influx of African migrant labourers to service the expanding wartime economy (and replace enlisted Europeans), along with an influx of soldiers and their families. By 1943, Capetonians were concerned about 'the native influx', and the accommodation situation was so chronic for Europeans that non-Union nationals were banned from holidaying in a congestion zone including Muizenberg, St James and Kalk Bay from 1 January to 30 April 1944. This was in order to 'prioritise for soldiers' wives, essential war workers and families, as well as Parliamentarians and officials'.[3]

The population surged from 331,756 in 1942 to 593,533 in 1960, followed by a brief slump in the early 1960s due to Apartheid removals of 'non-whites' to the Cape Flats (see Figure 8.1). In 1962 the population of the entire Peninsula was estimated at 843,291. The appearance of three new shanty-towns on the Cape Flats in the early 1970s prefigured the failure of the State's attempts at influx control, and the decade saw another population boom. The Peninsula's population surpassed a million in 1986. The rescinding of some of the major influx laws controlling the movement of Africans to urban areas resulted in the creation of Khayelitsha Township and informal settlements on the Flats, together accommodating around 750,000 people by 1989.[4] Official census figures undercounted Africans and Coloureds in greater Cape Town (i.e. including the Cape Flats), particularly 'illegal' Africans and those living in squatter settlements.[5]

For the urban poor, including the influx of African labourers during the early 1940s, cooking, heating and lighting were achieved by means of open flames (mainly paraffin, candles, firewood). That said, most of the study area, skirting the mountains, had been incorporated into the Cape Peninsula area of electricity supply by 1945.[6]

The long-term trend in population growth does show an apparent relationship with wildfire incidence as recorded by the Fire Brigade – accelerating in the 1940s, and showing high growth/incidence in the

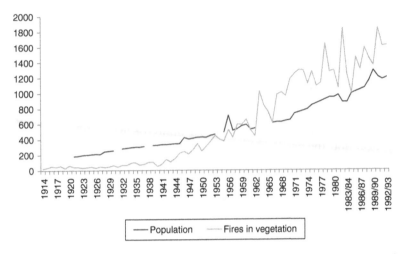

Figure 8.1 Growth in population (Cape Town Municipality) and incidence of fires in vegetation attended by the municipal fire brigade, 1914–1992/93. Population in 1000s, rounded down. The reporting period changed in 1982 from annual, to June–July.

1970s and from the mid-1980s. The brief time lag between the initial acceleration in fire incidence and in population growth may simply reflect inaccurate population records. Attempts to correlate population growth with fire incidence must remain speculative given the high level of uncertainty over the accuracy of population records.

Caution should be exercised when quoting census figures for the Greater Cape Town area, as conservationists sometimes do when linking population growth with escalating fire incidence in the area now encompassed in the Table Mountain National Park. Population should be considered spatially. During the latter half of the century, Apartheid laws and the patterns they established meant that the bulk of the growing population lived some distance from the mountain. Many Africans (in particular) and Cape Coloureds felt excluded from the Table Mountain range, seeing it as a space of privilege, encircled as so much of it was (and is) by affluent white suburbs.[7]

In 2001, census results showed that the M5 highway south to Claremont and then the M3 to Lakeside still delineated mostly white areas close to the mountains from Cape Coloured and then African settlements stretching east across Flats. Even in the uncertain interregnum era beginning in mid-1980s, only three pint-sized informal settlements were established on the Peninsula itself (excluding the Cape Flats). These were Imizamo Yethu in Hout Bay (8063 people in 2001, with 48 per cent of households using open flames for light), Masiphumelele between Noordhoek and Kommetjie (population 5518 in 1996, with 38 per cent using open flames; 8242 in 2001, with 10 per cent using open flames) and Red Hill above Simon's Town (population 406 in 1996, with 100 per cent using open flames; 654 in 2001, with 92 per cent using open flames).[8] It is likely that one of the major fires of January 2000 was started by someone from Red Hill settlement. Electrification was provided in 2005, partly to reduce the risk of wildfire.[9] However, the other big January 2000 fire was started near a picnic site at Silvermine little used by Africans.

Thus while there clearly is a relationship between urban population growth and wildfire incidence in the study area, both need to be understood in spatial and socio-economic context. Most of the population growth over the past four decades has occurred on the Cape Flats, away from the undeveloped parts of the study area, and with the exception of a few small informal settlements, since World War II most of the population has enjoyed electric power and has not had to rely on open flames. The growth in population and fire incidence can be more plausibly linked, I will argue, to increased outdoor recreation on the Peninsula. There are also other important socio-economic factors contributing to the increase in fire incidence from the 1940s.

Socio-economic causes (attributed)

'Socio-economic' is used here in a very limited sense as an umbrella term for activities practised by poorer, mostly Coloured and African people on the Peninsula. Tourism and recreation are discussed separately. It is notable that the attribution of fires to socio-economic causes so-defined is concentrated in the earlier decades of the century, before Apartheid removals of non-Europeans to the Cape Flats.

Flower sellers

Flowers have been sold in street markets in Cape Town since the 1880s, and by the late nineteenth century the Adderley Street flower sellers regularly featured on postcards. In the 1890s flowers were being exported wholesale by steamship to Europe and by rail to Johannesburg, and the Forestry Department was moved to put restrictions on flower gathering on Table Mountain. At the foundational meeting of the South

African National Society, intended primarily to protect historical arte-facts deemed culturally significant, in 1905, legislation to restrict flower gathering in the wider region was discussed. This resulted in a string of Wild Flower Protection Acts whose regulations impacted primarily on the collection of wild flora on 'commons' land, which Coloured flower collectors relied upon to turn a profit.[10]

These Acts were admired as progressive. When the Rural Amenities Bill was read for the second time in the British Parliament in London in 1931, a Mr Lovat-Fraser noted that:

> In Cape Colony there is a law providing that the flowers which are regarded as best worth preserving are to be scheduled, and the law forbids the gathering, uprooting, selling or exporting of them with-out an approved licence. What South Africa does we may very well do, and I should like to see a provision in the Bill to preserve our wild flowers.[11]

Revised regulations to Cape Ordinances on wild flowers recom-mended by the Wild Flower Protection Society in 1922 were however frustrated by legal disputation over definitions of 'wild' and 'cultivated'. In Caledon, a magistrate ruled that 'flowers which are protected from fire, pruned and weeded, and generally protected by farmers on whose property they grow, may be regarded as cultivated'. As Harold Compton complained, 'How are the police to know the cultivated from the uncul-tivated wild flowers? How can they know from what area they have been gathered?'[12]

From the 1920s into the 1940s, professional flower pickers were blamed for starting fires on Peninsula's mountains. In 1930, Compton explained that flower pickers burned 'in order to stimulate temporary flowering'.[13] Following the big fire of Christmas Day 1935, Minister of Agriculture and Forestry Deneys Reitz claimed that most wildfires on Table Mountain were started by flower pickers.[14] The Wild Flowers Pro-tection Ordinance No.15 of 1937 aimed to prohibit the sale of wild flowers within the municipal area from 1 January 1938, except in spe-cially nominated places set apart by the Council. A hurried scramble to decide these places ensured flower selling could continue, on a section of pavement kerb on Adderley Street in central Cape Town, on Albert Road in Salt River and the corner of Grove Avenue and Main Road in the suburb of Claremont.[15]

This conservation ordinance had cultural repercussions for the city. Local writer Sally Starke bemoaned 'the diminution of the old riotous, haphazard flower market of Cape Town'. She maintained that 'Cape Town is more flower-conscious than any other town in South Africa. Artistic arrangement of flowers is almost a sacred rite in Cape homes.' She also feared the impact on the Coloured flower sellers: 'The face of Cape Town changes, history and laws are made, but so far the little group of coloured people has remained beneath the shadow of the Post Office shouting their wares'. Never was a threat to 'local colour' so literal. Starke makes the striking argument that it was by selling local flowers to passengers from ships stopping at Table Bay that this 'little group of coloured people' had 'spread the name of Cape flowers across the world'. This provides an interesting corrective to the notion that academic botanists had generated the popular renown of the Cape flora beyond South African shores.[16]

Cape Town's flower sellers, and regional flower shows, were supplied by what became a small industry of collectors and growers. On the Cape Flats, farmers pumped up 'mulch' from the bottom of vleis (seasonally water-filled depressions) to fertilise the surrounding sands, to grow flowers for sale in Cape Town. Flowers came from as far afield as Somerset West and the Stellenbosch District, bulbs being delivered to Salt River bridge where they were sold to Coloured flower sellers. There were limits to white Capetonians whimsical appreciation of (their image of) 'colourful' Coloured flower-sellers, however. Coloured sellers were required to keep special flower sheds 'where no coloured people live or sleep, so that there is no danger of blooms carrying infection from the coloured dwellings where there is sickness to the homes of the purchasers'.[17]

One month after the flower sellers' reprieve, a huge fire burned from Tokai over the mountains to Hout Bay, destroying the small reserve filled with rare indigenous flora grown by Sydney Skaife. An irate letter-writer blamed flower pickers, but maintained that because he owned a valuable forest area he declined to prosecute them, fearing his forest would be burned out in revenge. The late Sir Drummond Chaplin, this correspondent (signed 'Constantia') continued, had taken a stand against flower pickers and suffered serious intimidation as a consequence.[18]

In the long run, flower selling was never outlawed: rather, flower sellers were licensed and flower collection was regulated. The *Cape Times* opined in 1946 that the flower sellers were 'as much part of the city as the flat-topped mountain under whose shadow they were born and raised'. Increasingly, commercial flower farmers provided them with their bulbs and blooms, although concerns persisted about unregulated flower picking, especially in more remote mountain areas.

The Apartheid removals of the 1960s, notably from Constantia where Coloured residents tended gardens and even farmed on a small scale, and flower collecting and selling were well established, further reduced flower collection on Table Mountain. In the second half of the twentieth century burning to stimulate flowers for collection disappears as an attributed cause of fires.[19]

Wood collectors

At a Symposium on Veld Burning convened in 1924, University of Cape Town botanist Margaret Levyns claimed that on the Cape Peninsula bushfires were usually started by farmers for grazing, or wood gatherers creating dry brushwood to sell. Neville Pillans, a well known botanist made the same claim.[20] The Christmas 1935 fire was attributed by some to:

> Coloured wood collectors. After a fire they are permitted to gather partly-burnt wood free, for the sake of clearing the burnt stretches. And when fire-wood is scarce, the wood-sellers are suspected of deliberately setting fire to plantations.[21]

The collection of fire-killed wood was later forbidden, and the wood was instead collected in heaps and burned 'in huge bonfires'.[22] Burning for wood collection was again cited as a cause of fires on the Peninsula following the big fire of February 1938.[23] In January 1944, a coloured woodcutter by the name of G. van Staaden, of Elsies River, was found guilty of lighting a fire near the municipal water works in Dido Valley and fined £20 or two months' hard labour with additional hard labour suspended for three years. This followed a spate of 12 fires between Simon's Town and Glencairn/Dido Valley. A *Cape Times* editorial celebrated the 'stiff and salutary' sentence which, it was hoped, would make burning for wood collecting unprofitable in the region. Electrification of the more prosperous suburbs of the city by 1945, and later Apartheid removals, reduced the demand for firewood in the immediate surrounds of the Table Mountain chain.[24]

Other causes of fire

Another supposed reason for fire lighting cited in the 1930s was unemployed men seeking payment for casual labour as fire beaters. After the Christmas 1935 fire, the Forestry Department decided to pay firefighters 'as little as possible' to avoid 'encouraging arson among men unemployed and wanting temporary work'.[25]

In the second half of the century the supposed fire-starting activities of flower pickers, wood collectors and would-be beaters are little mentioned. The most detailed figures on fire causes were kept by the Fire Brigade between 1930 and 1993 (see Appendix 2), and up to 80 per cent of causes were listed as unknown or attributed to 'dropping a light'. Of known causes, the fire brigade attributed most fires in vegetation to burning rubbish, children playing with fire, authorised burning to clear vegetation, steam locomotives and arson.

Minor causes throughout the century were fireworks, 'vagrants and bush sleepers'. In the late 1920s vagrants were cleared out of the Municipality's Roeland Street Plantation (defunct by the late 1940s) for lighting fires, and in 1949 the Parks and Gardens division complained that: 'a great deal of trouble was caused by vagrants erecting temporary shelters on the Green Point Common and setting light to the grass'. Fires attributed to vagrants peak from 1982/83 to 1992/93 (69 fires), which does overlap with increased immigration to the City as influx regulations broke down and were finally abandoned. Fires that required fire brigade intervention in this period no doubt included those lit purposefully or accidentally by the indigent *bergies* (*berg* meaning 'mountain' in Afrikaans) who shelter on the mountain slopes at night. It is unknown how many were lit by Africans newly arrived in the city sleeping rough on the mountain. None of these causes, however, is adequate to explain, or keeps up with, the steady increase in fire ignitions over the century.

Tourism and outdoor recreation

The carelessness of picnickers, hikers and campers (categorised separately from 'vagrants') was an important and persistent attributed cause of fires on the Peninsula. Examples include some of the century's biggest fires, namely those on:

- 21 January 1909, on Lion's Head, attributed to careless picnickers;[26]
- New Year 1933/34 (numerous), blamed on 'criminally careless walkers and picnickers;'[27]
- 21 April 1973, above Camp's Bay, attributed to picnickers in Kasteelspoort ravine;[28]
- 7 March 1982, on Devil's Peak, started when Johannes Strydom's portable barbecue was blown over (for which he was later charged[29]).

The use of the Peninsula for outdoor recreation by locals was significantly supplemented by tourists from as early as the 1880s. Railways to Johannesburg and Kimberley brought seasonal tourists, especially to Muizenberg. The Union and Castle steamship lines offered three-week cruises to Cape Town from England, reduced to only 15 days' journey time in the 1890s. Visits by famous travellers including Mark Twain and Anthony Trollope in the late nineteenth century helped establish Cape Town as a luxury long-haul destination. After the South African War ended in 1902, wealthy travellers were joined by the relatives of soldiers coming out to visit significant sites or war graves. The Cape Peninsula Publicity Association (CPPA) was formed in 1908 and relentlessly publicised the Peninsula as a welcoming 'tavern of the seas', a dramatically picturesque landscape, and 'the mother city of South Africa'. Cape Town was also marketed as 'the gateway to Africa': until the 1950s, the fastest way to reach South Africa from Europe remained the mail steamers.[30]

The cutting of paths and beautification of the mountain slopes above the City by the Cape Town Municipality extended sources of ignitions into the indigenous vegetation. Picnic spots were provided above Kloof Nek in 1912, and were apparently 'a great success … largely used by the public, especially on holidays'.[31] *The Summit*, published by the Cape Peninsula Publicity Association in 1913, listed 70 ways to climb Table Mountain. Chapmans' Peak Drive along the spectacular cliffs of the Atlantic coast of the Peninsula between Hout Bay and Noordhoek was opened in 1922, drawing comparisons with Italy's Amalfi Coast drives, and opening up the southern Peninsula to day-trippers (those few who had motorised transport).[32]

Increased recreational usage of the mountains, especially in a culture with a cherished tradition of barbecues ('braais'), brought an increased risk of wildfires. Galvanised by a large fire on the slopes above Kirstenbosch National Botanical Garden on 5–6 May 1923, the Director Harold Compton instituted fire control measures. Fireplaces were constructed in the Upper Kirstenbosch Nature Reserve 'with the object of localising picnic fires in safe spots'. Notice-boards urged the public to '[refrain] from flower-picking and promiscuous fire-making'.[33]

The cableway to the top of Table Mountain opened in 1929, making the mountaintop accessible to thousands of sightseers. Increasing ownership of private motor vehicles from the 1930s opened up undeveloped parts of the Peninsula to hikers, campers and picnickers. In 1936 the Cape Peninsula Publicity Association published *With Your Car at the Cape*, featuring a circular scenic route down the west coast to Cape Point and back via the east coast road. This increased road traffic provided complications for firefighters, blocking access and complicating firefighting manoeuvres.[34]

In 1924, Compton observed that many bush fires started on weekends, which he attributed to the carelessness of campers.[35] Discussing the big fire of Christmas 1935, Conservator of Forests at the Cape Benjamin Simmons said: 'these fires...always break out at times like Christmas, and on Sundays and public holidays'.[36] The high fire incidence of the mid-1930s was attributed to careless campers and picnickers, with fire lighting, cigarette smoking and discarding bottles (alleged to start fires by concentrating the sun's rays onto dry vegetation) given as the sources of ignitions.[37] Long-term fire brigade records show that while atmospheric conditions conducive to fires spreading last from November to March, fire incidence is concentrated in December/January, the period of the Christmas holidays (see Appendix 1, figure A1.1): this provides evidence of anthropogenic influence on the Peninsula's fire regimes.

In the face of public calls for the removal of plantations of introduced trees following the big fire of Christmas Day 1935, the Deputy Mayor, W.C. Foster, explained that picnickers sought them out for shade and shelter from the wind because there are no trees in fynbos. Foster advocated fencing off all plantations and (belatedly following the example of Kirstenbosch) providing 'recognised camps marked out where fires could be lit'. Over the next few years state and municipal authorities concurred that it was 'picnickers, roadside smokers, car smokers, mountaineers and hikers' who posed a serious fire risk to plantations and mountain veld.[38]

In 1947, the City Lands and Forests Branch were busy constructing picnic areas and bridle paths for hikers, under great public pressure to make all the Peninsula's forestry areas accessible for recreational purposes. Their 1949 report grumbled 'that public carelessness' [in setting fires] had 'left in its wake vast areas of desolation and blackened and scarred hillsides'. The author urged that 'increased efforts must be made to educate the public to a proper sense of responsibility in their behaviour when in the open, either hiking, picnicking or driving'.[39]

In the 1960s a series of scenic roads were built on the southern Peninsula, specifically intended to improve tourism by opening up the region to picnickers, walkers and other day-trippers. A nature reserve was opened at Silvermine and a new picnic area opened at Red Hill, en route to Cape Point. In 1960, the Cape of Good Hope Nature Reserve received more than 100,000 visitors (excluding luxury bus tours) for the first time.[40]

The era of environmental awareness ushered in by Earth Day in 1970, later encapsulated in the state Forestry Department's 'Our Green Heritage Campaign' launched in 1973, also augmented demand for facilities for outdoor recreation on the Peninsula.[41] The population pressure was intense over the Christmas period, which is also peak season for fires. Discussing the tourist influx, on 24 December 1971 the *Cape Times* announced that:

> by the end of this month about 400 extra trains will have been brought into service, starting from Johannesburg. Today 58 special trains will be leaving Johannesburg station – a record and there will also be seven extra South African Airways flights.[42]

This was a period of high fire incidence. Entry permit figures for Silvermine Nature Reserve on the Muizenberg Mountains, accessible via the Ou Kaapse Weg (Old Cape Road), show that by 1975 around 40,000 people were visiting annually, and more than 50,000 by the early 1980s. In 1976 an estimated 265,000 visited the top of Table Mountain, and more than two million people made use of Table Mountain, Lion's Head and Signal Hill for some form of outdoor recreation.[43]

Following a huge fire in April 1973, blamed on careless picnickers, the *Cape Times* published a statement from the City Engineer's Department 'outlining the rights and duties of campers and picnickers in Peninsula mountain areas'. Penalties for illegal fire lighting provided for in the Forest Act of 1968, it was noted, included 'a fine not exceeding R1000, imprisonment for two years, or both, or up to six strokes'. A reward 'of a quarter of a fine imposed' was offered to 'a person who materially assists in bringing an offender to justice'. The public was reminded only to make fires in demarcated areas, and to report fires immediately to the nearest police or fire station.[44]

From 1974 to the end of the decade, the making of fires in open areas (unless authorised) was forbidden on the Peninsula, and law enforcement officers prosecuted erring picnickers and overnight campers.[45] In the face of 'tremendous human pressure' in the early 1980s, particularly after the big fire of Sunday 7 March 1982, a master plan was drawn up to help reconcile 'conflicting aspects of conservation and recreation' on the 'Western Table environment'. It was resolved to remove barbecue sites from Tafelberg Road (below the north face of Table Mountain) and Camps Bay Drive. Perversely, 45 were installed at Silvermine.[46] In 1984/85, 54 barbecue areas were removed from below the north face of Table Mountain and Lion's Head and Signal Hill.[47] However, this policy was challenged by the *Cape Times* following the big fire of March 1986. The fire was started near a popular picnic site at which fires were prohibited and where there were therefore no barbecue facilities. The question, which remained unresolved, was: should more facilities be made available to allow safe barbecuing; or should barbecuing be outlawed entirely? (National Park policy at the time of writing is to limit barbecuing to properly appointed areas in sheltered, low-fire-risk zones and seasonal bans can be imposed.)[48]

The increase in outdoor recreation on the Peninsula caused concern over the impacts on natural ecosystems.[49] However, in Silvermine annual visitor numbers peaked in 1990/91 at 95,973, thereafter declining steeply to stabilise around 60,000 in the mid-1990s.[50] The fire that burned through the area in January 1992 may have had a short-term impact in reducing visits to the reserve. It may also be that some tourists and middle class visitors felt less secure here in this politically turbulent period.

Fire brigade records end in 1993, so it is not possible to assess whether there was a precise correlation between trends in outdoor recreation and decreasing fire return periods in the 1990s. We do know that from 1993 to 2000 there were big veld fires in every year except for 1996 and 1997, climaxing in the huge fires of January 2000.[51] Tourism grew appreciably in this period: with the advent of the 'new South Africa' in 1994, overseas tourism to the country surged, growing by 30 per cent in 1994, and then again by 52 per cent in 1995 (when the rugby world cup was held in South Africa), of whom more than 250,000 visited Cape Town. By 1999, a survey revealed that the new Cape Peninsula (now Table Mountain) National Park was receiving around 4.7 million visits per year, of which 61 per cent were by locals, and 39 per cent by visitors. Walking was cited by 48 per cent of overseas visitors and 46 per cent of local users as their main purpose for visiting the Park. Four per cent of overseas visitors and 13 per cent of locals visited to picnic (10 per cent of locals to barbecue). Most local visitors were from the middle- or high-income suburbs of the City (89 per cent), 72 per cent of whom lived within ten km of the Park, and 87 per cent travelled to the Park in private motorcars.[52]

Conclusion

Forsyth and Van Wilgen are correct that in the broadest sense, population increase has resulted in more bush fires on the Peninsula. My long-term comparison of population and fire incidence bears out the relationship. However, it is my argument that the spatial arrangement of the enduring population, and the dynamics of the transient population, must be taken into account. In fact because of apartheid-era legislation, the effects of which still underpin the demographic distribution of Greater Cape Town, much of the population growth – notably the surges of the 1970s and 1980s – occurred in areas far removed from the Peninsula's undeveloped mountain areas. Thus, activities likely to result in ignitions require closer attention. Among these, outdoor recreation is arguably the most significant, and this is tied on a local level to weekend and holiday activities, and more broadly, to major seasonal influxes of both domestic and international tourists.

Forsyth and Van Wilgen draw on Keeley et al.'s work on historical fire incidence in California brushlands (akin to Western Cape fynbos) to support their argument that increasing fire incidence on the Peninsula since 1970 is attributable to population growth. Keeley et al. (1999) found that fire extent in California brushlands was correlated with number of fires, which correlated with population density. However, they later concluded (2001) that the increase in destructive fires was not due to population density per se, but rather from the expansion of city boundaries and concomitant increase in the WUI.[53] In the case of Cape Town, the expansion of the city and its infrastructure, particularly into the southern Peninsula, has certainly increased the possibility of ignitions in the surrounding undeveloped areas. However, it is also important to look at the activities most likely to cause more ignitions. These have included the activities of the urban poor, dependant on the environmental commons for fuel, shelter and saleable natural products like firewood and wild flowers. It is my contention that outdoor recreation by locals and non-residents and the greater access offered to undeveloped areas by scenic roads, trails, the cableway and nature reserves with visitor facilities, have played a greater role in extending human ignitions into the undeveloped parts of the Cape Peninsula. There is some evidence for this in the clustering of fire incidence in the holiday months of December/January, within the 'natural' fire season of November–March.

9
Fire on the Cape Peninsula, 1900–2000

> After all the Table Mountain system is a scenic whole, and neither nature's architect nor fire – nor the tourist – knows anything of boundaries between local and Government authorities.
>
> *Cape Times*, 'Up in Smoke', Editorial, 9 February 1938

In January 2000, the largest fire event yet recorded on the Cape Peninsula burned for six days, threatening lives and property, despite a massive firefighting effort. While acknowledging the bravery and dedication of the firefighting efforts on the Peninsula, the official report into the fires concluded that 'Shortcomings in the state of veld management, including poor maintenance of firebreaks and the incidence of invasive plants, and weaknesses in the control and coordination of firefighting efforts contributed substantially to the magnitude of the disasters.' The authors emphasised the increasingly important role the general public and private landowners were required to play in preventing fires (in terms of the new National Veld and Forest Fire Act No.101 of 1998), while lamenting the lack of fire awareness among these individuals.[1]

It seems incredible that these shortcomings should prevail, after a century of intense debates and a long history of dangerous wildfires. A history of attempts to manage fire on the Peninsula reveals some explanations for these shortcomings, and the high fire risk, at the close of the twentieth century. It also should caution us against taking the 'unprecedented' fires of January 2000 as a ground-zero for fire management.

The social and biophysical template

As explored in Chapter 5 (and see Appendix 1), an examination of the natural template of the Peninsula shows that the areas most likely to burn are fairly predictable in space and in time, if vegetation cover and atmospheric conditions are taken into account. Most of the indigenous vegetation is inflammable but unlikely to burn again before three to four years after a fire. Some introduced plants can burn within a shorter period, and with longer fire intervals form dense stands with high fuel levels. Mature tree plantations don't burn easily but if fires take hold they are difficult to control. Fire risk is highest in dry conditions with high temperatures, low humidity and strong winds. This weather is very seasonal at the Cape, usually occurring in summer (November–March) (Figure 9.1).

Firefighting agencies on the Peninsula faced the challenges of rugged topography and lack of access paths and roads in the first half-century. High-speed and changeable winds complicate firefighting. Access to water for firefighting is limited higher up the mountain slopes and away from areas supplied with water mains and fire hydrants. On the upper slopes of the City Bowl, water pressure is poor. In the early decades of the century water had to be carted to the site of a fire by horses, donkeys and men, or later (where roads provided access), by motorised fire engines or tankers. There are few sources of standing water to draw water from on the mountain slopes: the larger reservoirs are located within the urban area in the City Bowl, on top of Table Mountain, at Newlands and atop Silvermine. It was only with the advent of helicopters with fire buckets in the late 1980s that water became a significant firefighting tool

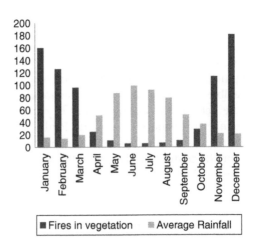

Figure 9.1 Comparison of distributions of monthly rainfall (average for 1899–1999, Cape Town Fire Station) and monthly vegetation fire incidence (fires attended by the fire brigade 1931–70), for the Cape Peninsula

for tackling large mountain fires. Before then, back-burning and beating were the main tools – dangerous and ineffective in the case of big fires in windy conditions.[2]

Fire causes: Anthropogenic

The major source of ignitions on the Peninsula in historic times has been humans, and the only long-term data on (attributed) human ignitions are in the records of the city's fire brigade. From the 1930s they recorded all fires attended, classified by location, including 'fires in bush, grass and rubbish', and recorded attributed causes of ignitions (see Appendix 2). Less systematic records of ignition causes can be recovered from forestry and newspaper reports.

In the first half-century on the Cape Peninsula, the major attributed causes of bush fires were deliberate ignitions either by arsonists or by children. The numerous minor causes included sparks from steam locomotives, vagrants' fires and burning to clear vegetation. In the 1960s and 1970s, Fire Brigade records show children, locomotives and controlled burns as the major sources of bush fires. In the final period for which Fire Brigade records exist (to 1992/93), the major attributed causes of fires were vagrants, arson and controlled burns.

The Cape Town Municipality's Parks and Forests Division (established in 1939) occasionally recorded causes of fires on its lands. In the 1940s, most fires were attributed to public carelessness, with vagrants, burning for grazing and flower picking given as minor causes. Prosecutions of flower pickers, especially at Silvermine, reduced burning for flower picking by the close of the 1940s. Public carelessness and incendiarism (arson) are most frequently cited as major causes of wildfires in the late 1950s into the early 1960s, and in the 1980s. Both periods were politically turbulent featuring mass protests against Apartheid, but insufficient evidence exists to determine the extent of politically motivated arson in connection with bush fires.

What most of these records really reveal is how little was known about the precise causes of fires in vegetation on the Peninsula. Some categories reveal social judgements, for instance sorting causes into 'children' and 'arson', or 'campers' and 'vagrants'.

Firefighting and control institutions

Responsibilities for fire control and firefighting in and around Cape Town in the first decades of the twentieth century were divided between several authorities, a problem which has dogged fire management ever since. The major relevant authorities (and landowners) on the Peninsula were the unified Cape Town municipality (from 1913), the Cape Divisional Council and the colonial (later state) Forestry Department. The South African Defence Forces controlled land, including the Navy on the eastern slopes below Silvermine, and around Simon's Town from 1955 (it was previously a British Royal Navy base). There were also numerous private owners of land, including farms and vineyards, and large estates for example Cecil Rhodes's Groote Schuur Estate, which by 1900 had merged several earlier estates below the eastern slopes of Table Mountain, and several smaller estates including the Van Breda, Rocklands and Highlands estates in the Cape Town City Bowl.

The Cape Town Municipality and State Forestry department were the two major fire management and firefighting agencies in the region in the twentieth century. As a result of the remit of these institutions, my focus on bush fires, and the need to maintain relatively uniform coverage across the study period, this fire history focuses on a particular area of the Peninsula, excluding the Cape Flats (see Map 6.2 for the study area, and place names). The periodisation has been guided by the only consistent record of wildfire incidence in the period, that of Cape Town's fire brigade – triangulated with other municipal departments, state forestry and newspaper records. The fire brigade only recorded fires in vegetation that it was required to attend, in order to defend the city and its suburbs. The overall increase in fires attended across the century thus results partly from increased fire risk as the city expanded. In practice, the Fire Brigade attended most of the large fires that broke out on the mountain slopes above the City and its suburbs – reflecting the extent of the wildland–urban interface and the social-ecological nature of fire on the Peninsula. In the absence of other data, the brigade's records are here used as a proxy for fire incidence trends in the study area.

Five broad periods of fire incidence and fire control emerge: 1900–38, 1939–59, 1960–79 and 1980–93, with 1994–2000 making up the years to the January 2000 fires.

Cape Town municipality

Cape Town's municipal fire brigade, established in 1845, was responsible for fighting structural fires and defending the city and its suburbs against bush fires. The municipality was also responsible for numerous parks, commons and catchment areas, and had its own timber plantations on the Peninsula. These were all looked after by a Parks and Gardens division, until in 1939 the catchments and plantations were handed over to a new branch run by a City Forest Officer. In practice all divisions of the municipality were deployed in firefighting as and when necessary. In extreme cases, the Defence Force, which had barracks in the Castle of Good Hope in the City Bowl, and in the southern suburb of Wynberg, would be called in. African and coloured causal labourers would be rounded up (for a nominal daily wage).[3]

Cape Town fire brigade

Following large losses to fire in Cape Town in 1903 and 1904, William Frost, Chief Officer of the fire brigade of the City of Sheffield in England, was engaged to modernise the fire brigade.[4] He was succeeded in 1930 by Stanley W. Thorpe, formerly Chief Officer of the city of Brighton's fire brigade. By 1933 Thorpe had transformed fire protection in built-up areas on the Peninsula. Fire Brigade headquarters was moved to the smart new Art Deco station in Roeland Street (opened in 1932). The Peninsula was split in two for the purposes of firefighting. The Northern Division – extending from Bakoven on the Atlantic coast clockwise to a line extending northeast from the King's Blockhouse (on a buttress of Devil's Peak) to Pinelands Railway Crossing – was served by stations at Roeland Street (HQ), Sea Point and Salt River. The Southern Division extended from east of the line, south to the Silvermine River Municipal Boundary, at Kalk Bay.[5] It was served by fire stations at Wynberg (HQ), Claremont (closed in 1933) and Muizenberg (moved to Lakeside in 1959).[6] Epping was added in 1959 and Mitchell's Plain in 1990, by which time the brigade was beginning to take on responsibility for townships and informal settlements built on the Cape Flats. (In a disgraceful episode in 1986, 40 shacks burned down in Crossroads township while local authorities passed the buck as to who should attend the fire. The army eventually stepped in.[7])

Until 1931, Cape Town's fire brigade had to deal with less than 50 bush and grass fires per year, except for a period of high incidence from 1916 to 1920. From 1932 to 1939, this increased to 50–100 fires per year (see Figure 9.2). Throughout this period the most common location for fires was 'dwellings', followed in most years by 'bush fires'. Calls to fires increased across the period, notably from the mid-1930s (false alarms wasting ever more of the Department's time and energy), but Thorpe's reforms brought down fire incidence in structures. From 1951, the fire brigade found themselves fighting more fires in vegetation than in structures.

Parks, gardens and tree planting

From the turn of the century until 1927, the Municipality's Parks and Tree Planting branch was in charge of fire prevention and firefighting on

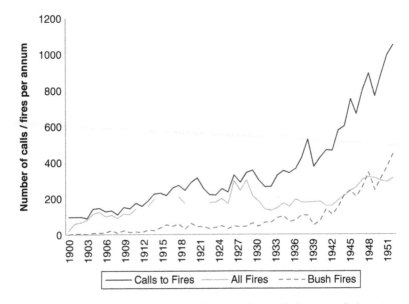

Figure 9.2 Comparison of calls to fires, total actual fires in all locations (excluding bush fires), and bush fires, from Fire Brigade records, 1901–53

undeveloped City Council lands. The branch became the Department of Parks and Gardens in 1928. The beautification and maintenance of green areas was the focus of this branch's work. The Branch was entirely opposed to fire, regarding it as a threat to their plantations, to the indigenous vegetation and in particular the endemic silver trees (*Leucadendron argenteum*), and to the beauty of the natural environment, decrying the resulting 'ugly black patches on the slopes'.[8]

Throughout the period, the municipality undertook fire prevention measures including cleaning fire paths, thinning pine plantations, clearing undergrowth and pine needles, and cutting and burning 'strong-growing native bush' above the city when they thought it was becoming a fire hazard (see Photo 9.1). The City Lands and Forests Branch was formed by the Cape Town Municipality in 1939, and for the first time the municipal plantations and undeveloped lands were run by trained foresters.[9]

In addition to their plantations (discussed in Chapter 7), the municipality were responsible for fire control on large catchment and other mountain areas, atop Table Mountain, above Muizenberg (the most extensive area) and at Orange Kloof, and also small areas of indigenous

Photo 9.1 City Council plantation and firebreaks on Lion's Head, photographed from east of the cableway on Table Mountain, c.1936
Source: C. Carlyle-Gall (ed.), 1937, *Travel in South Africa* (Johannesburg, South Africa: South African Railways and Harbours), p.23.

forest at Orange Kloof and Newlands. Until the late 1940s they had to contend with grass burning for grazing on Rondebosch Common, Signal Hill and at Silvermine above Muizenberg.

State forestry

The policies and management of fire by state foresters have been extensively reviewed. As discussed in Chapter 7, state forestry plantations were heavily utilised during World War I. After the War, afforestation was continued at Constantia and Hout Bay, but the Peninsula's plantations would again be heavily utilised during World War II. Some efforts were made to reafforest areas after the War, but by then the Department was expanding into the eastern regions of South Africa, and the Cape Peninsula's plantations became commercially unimportant. Afforestation had ceased by 1975, and the remaining plantations at Tokai and Cecilia were handed over to SAFCOL in the late 1980s.

Fire management: 1900–38

Fire management on the Peninsula in this period was fragmentary and reactive. It was a period of significant institutional change, with

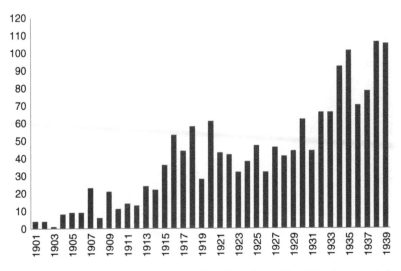

Figure 9.3 No. of fires in vegetation the Cape Town Fire Brigade attended, 1901–39

the political unification of the country in 1910, and the unification of the city in 1913. With municipal unification, the fire brigade became responsible for fires all the way around from the City Bowl to Muizenberg, and the brigade was fully mechanised. In the face of public protest following big fires in the early 1900s, the afforestation of Table Mountain with introduced timber trees accelerated up until World War I.

The first attempts to coordinate veld fire prevention and suppression around the City Bowl followed a spate of serious fires in January/ February 1909, notably big blazes on Lion's Head above Camp's Bay which ripped through cinder-dry pines and pine needles and damaged silvertrees. A big fire on 19 February below Platteklip showered the suburb of Orangezicht with sparks in the heavy southeaster, igniting the thatch roof of the old Dutch residence of Nooitgedacht, which was burned to the ground.[10] In response, bush fires east of Kloof Road were assigned to the municipality's Water Works Department, and fires around Lion's Head and Signal Hill to the Superintendent of Tree Planting. Twelve men were put to work clearing underbrush, and made available for firefighting, to assist the rangers employed on Lion's Hill and Signal Hill as fire lookouts and firefighters.[11]

During World War I there was a population boom in Cape Town and fire incidence was high from 1916 to 1920. There was a big fire at Kloof

Nek in 1915, started by careless soldiers, which burned an 'immense' area of veld and valuable stone pines along the Woodhead Pipe Track.[12] On 28 February 1920 a fire broke out on Steenberg farm, and was spread by a violent southeaster into Tokai plantation, traversing 139ha (343 acres) and causing damage valued at £11,431. The next day a fire at Devil's Peak destroyed 56ha (138 acres) of plantation valued at £8877. Letter writers and the *Cape Times* blamed the criminally careless, arsonists and flower gatherers. A *Cape Times* editorial called for urgent joint action by the City, Government and Divisional authorities.[13] The damage to the state forestry plantations resulted in a government inquiry, as a result of which telephone systems and a lookout station were installed at Tokai.[14] In 1922 the City Council adopted draft regulations on fire, most notably: 'no person shall set fire to any grass, bush or tree, garden or other rubbish, in any private or public place, within the limits of the Municipality, to the danger of life or property'. The penalty was a maximum of £20 for each offence, payment of all expenses incurred, and in default of payment, imprisonment for three months 'with or without hard labour'.[15]

A large fire above Kirstenbosch National Botanic Gardens on 5–6 May 1923 led the director Harold Compton to institute fire control measures including a safe barbecuing area and warning notices. By the late 1920s influential conservationists and botanists were calling for a nature reserve on Table Mountain to protect the flora from flower pickers and fires, and another fire threatened Kirstenbosch on 3 November 1929.[16]

In the early 1930s, as a result of 'the extensive damage caused by bush fires on private property', attempts were made through municipal and provincial legislation to make owners of private property responsible for fighting fires on their own properties, and to enable municipalities to recover the costs of suppressing such bush fires. Municipal boundaries were extended to cover catchment areas on Table Mountain, and municipal regulations prohibited the lighting of fires in catchment areas on the Peninsula. This came amid a prolonged dry period (1921–36), and at the same time the economic strictures of the Depression were, in the opinion of Inspector of Forests J.J. Boocock, dangerously hampering fire prevention measures.[17]

Increased private car ownership significantly extended the penetration of the Peninsula by sightseers, picnickers and hikers and foresters complained of irresponsible fire lighting on public holidays and weekends. The conjunction of all of these factors contributed to very high fire incidence in 1934 and 1935. Traffic congestion as onlookers drove to the scene of one of the numerous Lion's Head fires resulted in collisions and hampered firefighters. These fires were blamed on flower-sellers, wood collectors and careless picnickers, and Julius Jeppe called publically for a force of 12 rangers to patrol the mountains.[18]

At around 11 a.m. on 25 December 1935, a fire broke out on the north face of Table Mountain. Early responses were ineffective and it spread to the Peak Plantation (Photo 9.2). Being Christmas day, it was difficult to mobilise enough firefighters, and it took two days of firefighting by nearly 700 men to fully extinguish it. They faced pine trees flaming up 'like 50-foot torches' and tumbling down the slopes, red-hot boulders bouncing through their ranks and dry pine needles 'blazing like celluloid'. The city and suburbs were 'dusted with fine particles of charred wood' and on 26 December the white sands of Milnerton beach were streaked with ashes. The fire had destroyed 121ha (300 acres) of Peak Plantation, causing £4000 worth of damage. Following this fire, there were strident calls for a national park to restrict access to Table Mountain, and to remove inflammable plantations of introduced trees from the Peninsula. Neither call resulted in significant management action. Instead, in 1937/38 fire prevention was focused on the urban poor, with efforts to control flower-sellers and firewood collectors.[19]

There was very high fire incidence in 1938 and 1939, particularly on weekends and public holidays. On Sunday 6 February 1938, a 'terrific fire' swept through the Peninsula's mountains, the *Cape Times* reporting that 'an enormous area [had been] devastated by the flame'. First noticed at 11:30 a.m., the fire had been spread quickly by a strong southeaster, sweeping over the Constantiaberg and on to threaten the Chapman's Peak plantations. Firefighting was difficult due to the lack

Photo 9.2   The Christmas Day 1935 fire on Devil's Peak
Source: Carl Lückhoff, Courtesy of A.A. Balkema Publishers.

of standing water and more than half a mile of hose was run out to a vlei (marshy pool) below the main road in Hout Bay. Chief Forest Officer Hugo Brunt reported that: 'Flames spread to inaccessible crags on the mountain [Steenberg], and from there burning pine cones and red-hot stones fell down to start new fires....Loud explosions were heard as great rocks, heated by the flames, split in two'. By 17:30 more than 500 men were fighting the fire, including the army and navy, Africans from Langa location, and inmates of Tokai Reformatory. On 8 February the headlines read: 'great fire sweeps down into Hout Bay: residents flee in terror'. Baviaanskloof above Hout Bay had been 'reduced to a blackened waste', including well-known local naturalist Sydney Skaife's cottage and private nature garden. Several other houses and gardens were also damaged or destroyed before the fire was finally extinguished on 8 February, after the southeaster died down.[20]

The Tokai fire had started simultaneously in different places, and it was believed to have been ignited deliberately. There had also been two fires above Camps Bay and one above St James on that same weekend. The Camps Bay fires had spread quickly through cut bush which had been cleared two months previously, but still awaited removal, providing ready fuel (as would be the case in the January 2000 fires). Four letter writers to the *Cape Times* noted that although these bushfires were routinely blamed on coloured wood gatherers and flower pickers, it was noteworthy that several bush fires in the course of the month had started alongside roads, on weekends and public holidays, and instead blamed cigarette-smoking motorists and pedestrians. Reg de Smidt of Rosebank complained that 'after the disastrous fire on Table Mountain in January 1936 [presumably he meant Christmas 1935], the Mayor convened a meeting of citizens to discuss future preventive measures. [However, this was] postponed owing to the death of King George V, and was never held afterwards.' He urged that a lorry and squad of men should be available to fight fires in or around each suburb, 'or the total destruction of our veld and flora is merely a matter of a few more summers'.[21]

Sidney Skaife was of the opinion that the 700 men who had converged on the Hout Bay fire on the night of 8 February 'were almost useless, because they were untrained and unorganised'. The *Cape Times* quoted from a *Reader's Digest* article on firefighting in the United States, where the Civilian Conservation Corps had been formed in 1933. The Corps provided work for unemployed youth, who were given skills and employment for a maximum of two years, and Skaife recommended such a system for the Cape. A civilian corps was subsequently organised during the War years, but was disbanded in 1945. The idea would then lie dormant until it was resurrected in the form of the poverty relief programme Working on Fire in 2003.[22]

Three days after the fire, the *Cape Times* editorialised that while 'post mortems are almost as thick as the smoke on Sunday and Monday', under the prevailing conditions, 'it was almost impossible to expect any authority to quench the resulting conflagration'. A wet winter and spring had resulted in lush vegetation growth on the mountains' lower slopes, and the dry summer and severe southeaster had left the litter tinder-dry. The problem lay rather with the lack of co-ordinated fire prevention measures, 'for after all the Table Mountain system is a scenic whole, and neither nature's architect nor fire – nor the tourist – knows anything of boundaries between local and Government authorities'. Fires were almost inevitable, whether caused by 'fools' who drop burning cigarette ends, flower pickers seeking to 'promote the growth of bulbs, poor households who want to make certain of a plentiful supply of dry wood, and unemployed ambitious of earning 1s.6d. an hour as beaters'. What was needed on the Peninsula was unified mountain fire control.[23]

Cumulatively, the high fire incidence and public debates over fire of the late 1930s did lead to the first attempt to formally coordinate fire management and promote fire awareness on the Peninsula as a whole. In 1938 the Minister of Agriculture appointed a (voluntary) non-statutory Committee for the Prevention and Control of Forest and Veld Fires on the Cape Peninsula. This committee zoned the Peninsula into 11 fire control zones, and a map was prepared for issue to all Forest Fire Stations. Methods for closer cooperation in firefighting were drawn up between the Government Forest Department, the Cape Divisional Council, the municipalities of Simon's Town, Fish Hoek and Cape Town and Private Estate owners. These bodies voted a sum of £175 'for propaganda purposes' to educate the public 'as regards fire dangers', to which the City Council contributed £25. The Forestry Department contributed £75 in 1939/40, and kept up payments until 1949.[24]

Fire management: 1939–59

The fires of 1938 and 1939 significantly influenced public and expert concern over wildfires. The building of scenic roads, and quarrying on the Peninsula's mountains, contributed to a general sense that the natural environment was under threat. At the same time the country was entering a politically explosive period as right-wing Afrikaners opposed

South Africa's entry into World War II on the side of the Allied Forces. Environmental issues became politically loaded, with soil erosion and veld burning characterised as unpatriotic, even treasonous. Fire incidence rose sharply from 1942, with 140 significant bush fires in the year, the highest annual incidence to date. In January 1942, a fire burned 283ha (700 acres) of mountain veld from Silvermine towards Chapman's Peak, and there were two big bush fires above Clifton in mid-February. All of these were judged to have been started deliberately.[25]

It is unsurprising that 'the fire menace' became a major public issue, or that a defensive approach focusing on fire suppression was favoured. Hugo Brunt, Chief Forest Officer of the Cape Town Municipality's City Lands and Forest Branch was praised for the apparent success of his fire suppression campaign on municipal lands. After a fire burned for five days and destroyed four houses in Gordon's Bay settlement 50km (30 miles) east of Cape Town, there were calls to extend Brunt's 'scientific' fire suppression approach to all of the Cape Province's mountain areas. This seemed even more urgent following a massive fire in December 1942 which destroyed around 200 square miles (520km^2) of veld in the Cape mountains before burning out near Franschhoek [65km (40 miles) east of Cape Town].[26]

On the Peninsula, the propaganda fund set up in 1939 to raise awareness about the dangers of wildfire was used to create educational films, posters, radio programmes and competitions. During the 1940s, firefighting competitions were held in schools, and the film 'This Heritage', about fire and fire control on the Peninsula (with a commentary by Hugo Brunt) was widely screened. In his report for 1940, Brunt reported that: 'much useful propaganda work was carried out, and the spirit of co-operation between the various [firefighting] bodies is excellent.' Members of the Voluntary Fire-Fighting Corps had joined the CPS [the Civilian Protection Service] Fire-Fighting Squads, cooperating on fighting forest fires. The City Lands and Forests Branch drew up detailed fire plans detailing 'all field control, staff, labour, duties, costs, and methods of firefighting. Plans showing the subdivision of the forests and mountains into units of control are included.'[27]

The War years proved challenging for this new coordinated approach to fire management. Budgets were squeezed and all organisations lost staff to the armed forces. The volunteers of the CPS, a civil defence organisation formed during World War II, provided valuable assistance with firefighting.[28] They required training, which was undertaken from 1940 by the already hard-pressed Fire Brigade. The City Lands and Forests Branch had to focus on supplying timber for home consumption as a result of wartime disruption of imports, rather than fire prevention.[29] The population influx during the war years, notably from 1942, no doubt also contributed to the high incidence of ignitions on the Peninsula.[30]

The fire season of December 1943/January 1944 was severe, with numerous fires including two at Oudekraal, fires at Lakeside, Fish Hoek, Clifton, and two fires on Devil's Peak. There were 12 separate fires in the hills between Simon's Town and Dido Valley over the Christmas/New Year period, and during a large fire above Simon's Town searchlights from warships were turned on the mountain slopes to help beaters fight the flames. In a rerun of the late 1930s, the principal result was that a coloured wood cutter was given a severe fine for lighting fires, and flower selling was regulated. This had little impact on fire incidence, which continued to increase steeply, exacerbated by a very dry period from 1946 to 1949. Fire prevention efforts and firefighting forces had been much reduced during the War years, and the end of the war made matters worse: the CPS was disbanded. Hugo Brunt, mastermind of the Peninsula's defensive fire management approach, died in April 1945 and was replaced by Colin Grohl. In these circumstances it is unsurprising that the recommendation of the Royal Society of South Africa's 1945 report into the preservation of the montane vegetation of the Western Cape – that controlled rotational burning be used to manage mountain catchments – was ignored on the Peninsula.[31]

In February 1947, the Cape Peninsula Fire Protection Area (FPA) was declared in terms of the Soil Conservation Act No.45 of 1946, covering 40,993ha (101,296 acres) on the northern Peninsula. The Soil Conservation Scheme called for urgent action in light of a worsening fire situation – despite considerable anti-fire propaganda and the expenditure of £50,000 per annum by the Government and local bodies to prevent and extinguish unauthorised veld and forest fires. It was a defensive strategy: inflammable material was to be cleared, firebreaks maintained, unregulated veld burning prohibited, and firefighting improved.[32]

On 17 February 1947, King George VI, Queen consort Elizabeth and the princesses Elizabeth and Margaret arrived in Table Bay on HMS *Vanguard* to begin their Royal Tour of the Union of South Africa. They were presumably pleased with the fine, hot weather as they descended the gangplank from the forbidding grey battleship dressed in light summer clothes, having left a bitterly cold England on 31 January. Cheering crowds lined the streets of their route to the City Hall, bringing the city to a standstill. Unfortunately, conditions were also ideal for bush fires, with temperatures in excess of 40°C (104°F) in the shade measured inland at Paarl. A huge fire developed on the Hottentots Holland Mountains, drawing firefighting forces from the Peninsula. When another fire started on the slopes of the Bakkeman's Kloof above Bishopsford in the Hout Bay valley, it was difficult to muster firefighters as some were fighting the Hottentots Holland fire, and others had been given the day off to witness the arrival of the Royal Family.[33]

On 18 February the front page of the *Cape Times* featured the arrival of the Royal family alongside accounts of the fires, headlining breathlessly: 'National road cut by fire' (the pass over the Hottentots Holland mountains) and 'Hout Bay homes threatened'. Beaters fought the flames 'in terrific heat' but failed to stop them crossing the Vlakkenberg plateau and entering the upper section of the Tokai plantations, where it burned 25 acres (10ha) of trees. The flames loosened the soil and great boulders tumbled down among the beaters, as a heavy pall of smoke unfurled across the Cape Flats. By midnight more than a hundred men were trying desperately to prevent the fire from crossing the Hout Bay road and entering the Orange Kloof and Cecilia plantations. Another fork of the fire was burning west past the old magnesium mine to threaten the unfortunate Sydney Skaife's property Tierbos, and the houses and two hotels below. In the end it took 500 City Council, Cape Divisional Council and Government Forestry firefighters more than three days to control the two fires. The *Cape Times* praised the efforts of the Committee for the Prevention and Control of Veld and Forest Fires for arranging a 'very well planned co-operative effort between the various authorities and interests concerned' on the Peninsula.[34]

In 1947 the acute staffing problems of the City Lands and Forests branch were resolved with the appointment of a Regional Forester and three Senior Forest Rangers, all trained at the Government forestry school. The Branch resumed 'both pre-suppression and fire-fighting operations in the forests and catchment area, and on other open city lands bordering the mountains, and also on private property'. This proved necessary during a sequence of years experiencing three- to four-month dry periods (1945–1952/53). It was also advisable for safety reasons because from 1947 the branch was busy constructing picnic areas and bridle paths for hikers, in the face of public pressure to make the Peninsula's forestry areas accessible to the public.[35]

Despite all these dramatic fires on the Peninsula, in 1948 the City Lands and Forests Branch report expressed frustration that 'public carelessness should...leave in its wake vast areas of desolation and blackened and scarred hillsides'. Alas 1949 proved to be worse, with 343 bush fires attended by the fire brigade, and 100 by the City Lands

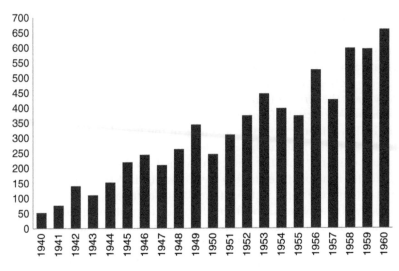

Figure 9.4   Fires in vegetation attended by the Cape Town Fire Brigade, 1940–60

and Forest Branch. In response, the *Cape Times* and the Cape Peninsula Fire Protection Committee announced a competition for the best slogan 'which will rouse residents and visitors to the menace of bush fires in the Peninsula'. Mr H.C.D. Steel of Sea Point won the first prize of £30 for: 'Don't let your mountain go to blazes'. The runner-up was 'Wonderland or Cinderland' by Mr Sheasby of Mowbray (£15) (Figure 9.4).[36]

In 1949 the non-statutory committee of 1938 was established as a statutory committee by Government Notice No.979. This Cape Peninsula Fire Protection Committee was empowered to enforce the provisions of the Forest Act of 1941. The City of Cape Town Municipal Regulation No.1949 provided for the clearing of fire hazards on private land within the Municipality, a measure emulated by Simon's Town and Fish Hoek municipalities and taken up by the Soil Conservation Scheme, Section 13. This latter empowered the Fire Protection Committee to give notice to landowners to clear inflammable matter or make fire breaks where necessary. All veld burning was prohibited, except under licence, between 15 October and 15 May of the following year, and from 10:00–14:00 at any time of year.[37]

The Committee also set about establishing a force of 500 Honorary Fire Wardens, volunteers who would be trained to assist either in policing the mountain and enforcing anti-fire laws (Passive Group), or active firefighting (Active Group). The Active group comprised mostly mountaineers, fit and familiar with the terrain. Volunteer firefighters had not previously been encouraged because (a) they were unruly and got in the way of hazardous and difficult tactics such as counter-firing (setting a fire to burn towards an approaching wild fire), and (b) they were not insured. Now, trained Active group volunteers would be insured against injury.[38]

The major hindrance to firefighting at this time was identified as lack of access, notably to areas like Orange Kloof, Silvermine and the Hout Bay mountains. It was acknowledged that firebelts were not in them-selves effective barriers to fires, but rather provided a secure line from which to combat fires through beating or counterfiring. Although it was possible to make firebelts wider in non-afforested areas, fires had jumped belts several hundred yards wide on the Peninsula. Making belts any wider than this would defeat the point of making them, that is, to '[preserve] the scenically attractive mountainside with vegetative cover to prevent erosion'. Notices were served on 250 property owners 'to clear their properties of excessive bush and inflammable undergrowth'. The State Forestry Department installed a radio station at Tokai, issuing walkie-talkie portable sets for use at the scene of major fires.[39]

The worsening fire situation, despite all measures to the contrary, came to a head on 26 January 1950, when a fire broke out on the north-ern slopes of Table Mountain and burned out of control for four days. The difficult topography and weather conditions, exacerbated by the work of arsonists, made the fire nearly uncontrollable from the start. In another big bush fire at Simon's Town, two navy personnel were trapped by the flames and died of burn wounds. As a result the Mayor of Cape Town called a public meeting to gather support and cooperation for the prevention of mountain fires. The Minister of Agriculture autho-rised special financial support to these fire prevention efforts, granting £15,300 to the Cape Peninsula Fire Protection Committee for 1950/51. It was channelled through the Soil Conservation Scheme to provide for the construction of firebelts, access roads, static water supplies, and the coordination of firefighting on the Peninsula.[40]

This was the pattern for fire prevention on the Peninsula in the 1950s. In his report for 1951, Colin Grohl wrote:

fire-control work comprised the major part of the operations of the City Forests Branch during the year. Extensive pre-suppression work was undertaken, and over 30 miles of firebelts were constructed and maintained.... There is no doubt that the elaborate system of fire-breaks established with the aid of substantial financial assistance

from the Government through the Statutory Fire Protection Committee has assisted very materially in the control of veld and forest
fires.[41]

For the remainder of the decade the municipality's emphasis was on
pre-suppression, including clearing of 'abnormal fire hazards' (notably
above the city bowl for the new Van Riebeeck Park created in 1952),
and the serving of fire hazard notices under Regulation 1949.[42] Grohl
proved a pro-active and imaginative City Forest Officer, collaborating
with Lückhoff on his Table Mountain book and coming up with the idea
for the educational booklet *Bokkie the Grysbuck* with its anti-fire message.
From 1951 fire brigade records show the brigade was fighting more fires
in vegetation than structural fires (see Figure 9.2) and this continued,
with 55.8 per cent of all actual fires attended occurring in vegetation
from 1950 to 1959, 61.8 per cent for the period 1960–69, 62.5 per cent
for the period 1970–79 and 59 per cent for the period 1980–1989/90
(reporting periods changed).

At the close of 1952, city parks and gardens were handed over to
the City Forest Officer's branch, and from 1956 to 1959 some 1958km
(1216 miles) of tracer belts were skoffeled (cleared and hoed), 283km
(175 miles) of firebelts were burned and 74km (46 miles) of access roads
were constructed. Between 1955 and 1959, 65 people were prosecuted
for 'negligently allowing fires to spread'. The municipality also began to
try to control flammable introduced plants on the Peninsula.[43]

State foresters similarly adhered to a defensive fire exclusion strategy
in catchment areas, despite reports from 1952/53 of flowers disappear-
ing from flower reserves where fire had been excluded.[44] However, in
September 1952 the behaviour of a large fire at Jessivale Plantation near
Pretoria led the department to conclude that:

Although the measures adopted to protect the forest estate from fire
are generally speaking effective, conditions arise from time to time
when the relative humidity is so low, the temperatures so high and
the wind so strong that it is practically impossible to check a fire
when once it has taken hold.[45]

In the long, hot, dry summer of 1958/59, 80 acres (32ha) of the
state forestry plantation at Tokai were burnt out and R4000-worth of
extraction equipment was destroyed.[46] By the end of the decade the
department was questioning whether the expense of these defensive fire
protection measures was justified, or whether less intensive measures

might not suffice, at least where stands were sufficiently developed to allow much of the burnt timber to be salvaged.[47]

Fire management: 1960–79

The Peninsula experienced a spate of fires in early 1960, notably blazes that threatened Hout Bay village on New Year's Day, and 21 February, and the big Tokai Plantation fire of 22–26 January which destroyed 27ha (66.7 acres) of plantation and the whole remaining area of the Hout Bay plantation. In November, following a record-breaking 36 call-outs for the Fire Brigade on Guy Fawkes' night, two houses were burnt down by a bush fire at Kommetjie.[48] In 1961 a fire burnt 24ha (59 acres) of plantation and 97ha (239 acres) of abandoned plantation at Devil's Peak.[49] A big veld fire on Red Hill above Simon's Town in early January swept down through a naval gun battery and threatened houses and a marine oil factory, consuming a large area of fynbos and Port Jackson (*Acacia saligna*). The next day the *Cape Times* lamented '100 square miles of "black Cape carpets": 80 fires in 3 months devastate country'.[50]

These fires occurred during a period of apparent management complacency. The *Cape Times* had reported in the dry, hot December of 1960 that 'it is highly unlikely that any really appreciable tract of country will be burnt out in the Cape Town municipal area'. This was, explained by municipal forester P. J. Bruwer, because the extensive firebelts on municipal lands on the southern Peninsula and the approximately 40 miles (64km) of mountain access roads provided fast access for the municipal fire truck. A lookout was on duty 24 hours a day from a hut atop the Noordhoek Mountains in the fire season, with a telephone link to municipal forestry headquarters at Kloof Nek. Foresters were in radio contact with headquarters at all times. Municipal lands were divided into five areas, run by 27 supervisors, with 60–70 beaters on duty at all times.[51]

Nevertheless, Bruwer was frustrated with the public for making fires in the bush despite the provision of more than 60 official fireplaces on municipal land on the southern Peninsula. He expressed concerned over hunters firing the mountainsides to attract antelope (including grysbok). A series of new scenic roads approved in 1960 improved access to the southern Peninsula and in January 1961 the *Cape Times* recorded that more than 10,000 people (excluding bus tours) had visited the Cape of Good Hope Nature Reserve between Christmas Day and 4 January. The Divisional Council's new picnic site (for 'whites' only) at Red Hill was also deemed a success, with all barbecue places occupied at peak times.[52] Silvermine Nature Reserve would be opened in 1964, with paths and picnic sites opening up this inflammable mountainside to new sources of ignitions.

The opening up of the southern Peninsula made it necessary to raise public awareness of the risks and consequences of fires. In 1962 the provincial Department of Nature Conservation together with the Cape Peninsula Fire Protection Committee published an illustrated booklet promoting awareness of environmental issues entitled *Bokkie the Grysbuck*. This illustrated book etched in readers' minds the message that fire was destructive to the indigenous fauna and flora, caused soil erosion and encouraged invasive introduced plants.[53]

Despite a sudden spike in fire incidence in 1963 and 1964 (very dry years after a very wet 1962), and a fire which burned 24ha (59 acres) of pines and eucalypts at Tokai Plantation in 1964, bush fire control measures began to fall away. Maintenance of municipal firebreaks declined steadily from 217km (135 miles) in 1963 to only 80km (50 miles) in 1969. In part, improved firefighting capabilities were judged sufficient to control the inevitable fires which humans ignited on the Peninsula. Protective measures were further reduced because the state Forestry Department's Groot Constantia and Hout Bay plantations were discontinued (much of the latter had been burned out). Following a huge wildfire in the Outeniqua Mountains in 1962, the Department had decided that fire prevention measures alone would never prevent large fires for long, and removed firebreaks, often planted with eucalypts, from within its plantations. Revised fire protection plans were completed for Cecilia, Tokai and Devil's Peak plantations in 1965/66.[54]

The short-term response to the apparent anomaly of high fire incidence in 1963 and 1964 (see Figure 9.5) seems to have been principally propaganda and further legislation: the Forestry and Agriculture departments made two fire prevention films, 'The Land of Smoke and Fire' and 'Operation Wildfire' (1966). The Forest Act of 1968 was amended to make landowners or occupiers responsible for clearing firebelts to restrict fires on their properties, with severe penalties for negligence or failure to comply, and rewards offered to anyone bringing offenders to justice.[55]

In 1970 the Department of Forestry took over the administration of Fire Protection Committees in South Africa, and developed a policy of controlled rotation burning for catchments in the Cape. Unfortunately this coincided with intense development pressure on the Cape Peninsula. After 30 years' of anti-fire propaganda on the Peninsula, public opinion was not receptive to prescribed burning, and the official

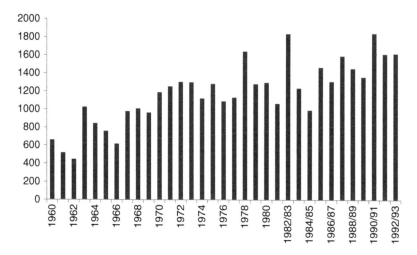

Figure 9.5 Fires in vegetation attended by the Cape Town Fire Brigade, 1960–92/93

Note: Reporting periods changed from calendar year to June–July. The transition year 1982/83 is an 18-month period so actual annual incidence for 1982 was likely c.1200–1250.

response to increasing fire incidence was once again to attempt to exclude all fires. Press stories and conservationists routinely reported the 'destruction' of fynbos by fires on the Peninsula, despite a new wave of scientific research confirming that fynbos requires fires (within certain ecological parameters) to regenerate.

From 1969/70 to 1975/76 there were unusually long periods (of three to four consecutive months) of very dry weather on the Peninsula, with very low annual rainfall from 1971 to 1973, and the early 1970s were characterised by high fire incidence. The most dramatic effect was the loss of 1876ha (4635 acres) of municipal plantations to fire from 1972 to 1975. There was also a major reduction in state afforested land, notably the clear-felling of pines on the slopes of Devil's Peak. The significant reduction in state and municipal plantations and management further reduced protection measures on the mountains. At the same time, population pressure was intense over holiday periods, particularly Christmas and New Year. In December 1971, 400 extra trains and seven extra South African Airways planes were laid on to bring tourists from Johannesburg to Cape Town.[56]

In early April 1973, the City Engineer Solomon Morris warned that the fire hazard on the Peninsula was unprecedentedly high. More than 200 fires had broken out over the summer of 1972/73, and Parks and Forests rangers had spent 19,376 hours fighting 98 of these. Morris maintained that dry conditions, population growth, car use and increased accessibility to mountain areas meant greater attention would have to be paid to reducing fire hazards and making mountain areas accessible to firefighting crews. The air force agreed to supply helicopters to aid the transport of firefighters in the event of a large fire, but argued that aerial spraying wasn't feasible because of the rugged terrain.[57]

On 19 April 1973 (Thursday before the Easter Weekend), a fire broke out in Kasteelspoort Ravine above Camps Bay at about 23:00 (later attributed to careless picnickers). Gusting, gale-force northeasterly winds (helicopters were not deployed) spread the fire in three directions. One prong of the fire burned southwards for 6km (3.7 miles) along the Twelve Apostles. The second moved downslope towards the settlement of Camps Bay, but was stopped short by City Council firemen. The third prong moved upwards and eastwards over Table Mountain in the direction of Kirstenbosch and Newlands Forest. A picnicker, and four climbers over-nighting on the mountain, were badly burned. This fire was headed off by a counterfire burned along an existing fire belt on top of Table Mountain. Nearly 200 firefighters drawn from the City Council, Department of Forestry and Divisional Council, and volunteers, managed to bring the fire under control by the night of 20 April. During the afternoon of 20 April, another fire broke out on the slopes of the Constantiaberg above Hout Bay. Local resident Sidney Skaife described it as 'worse than the 1938 blaze that devastated the same area'.[58]

After the fires had been extinguished, David Bloomberg, Deputy Mayor of Cape Town, said fire incidence had reached 'crisis proportions', despite a 'plethora of legislation'. He recommended a focus on fire prevention rather than firefighting, revealing that the Council spent only R150,000 on fire protection for the mountains annually, and R1.2 million on the fire brigade. He called for the integration of the Parks and Forest Branch's mountain firefighting forces with the City Engineer Department's municipal fire brigade. Camping on the mountain should be prohibited, and all vagrants removed. President-elect of the False Bay Conservation Society, Mr E. J. Smith, called for 'vicious measures' including the shutting of all picnic and viewing spots along roads. In contrast to Bloomberg, John Wiley, MP for Simon's Town, recommended modernising firefighting equipment and methods. The *Cape Times* agreed that more effective firefighting was the answer, but opposed 'restricting the public unreasonably'.[59]

One outcome of the fires was the formation of a subcommittee to investigate firefighting in the Cape Peninsula catchment area. It grew out of an initiative set up in 1972 by the Minister of Defence (from 1966 to 1980) Pieter Willem (P.W.) Botha to investigate fire services in the country and South West Africa. Botha consulted the South African Navy, Department of Forestry, Cape Divisional Council and municipalities of Fish Hoek, Cape Town and Simon's Town. The committee concluded that fixed-wing waterbombers, much used in the United State and Canada, were unsuitable for firefighting on the Peninsula. Similarly, and also because of the challenges of the terrain and prevailing weather conditions in the fire season, air force helicopters would only be deployed for transportation of equipment and personnel to inaccessible areas. These decisions emerged in response to calls for air bombers to fight fires spearheaded by Simon's Town MPC Chris Visser, following a big fire in February 1974.[60]

On the night of 9 February 1974, a large fire burning on Devil's Peak was sent raging downhill towards the residential area of Vredehoek and Highlands Estate by a strong southeaster, chasing about 20 homeless 'bergies' off the mountain. Residents in Clifford Avenue began evacuating their homes, and 50 children were moved out of Herzlia School hostel in Clairwood Avenue. Fortunately, municipal firefighters managed to stop the fire short of any structures, while Parks and Forests men fought the bush fire from Tafelberg Road. The most important response to these fires was that the Forest Act, 1968, was deployed to forbid the making of fires in all open areas (unless authorised) on the Peninsula. This was repeated from 1975 to 1977, and in 1979 and 1980.[61] The ban was not applied in 1978, and there was a spike in fire incidence in that year (1641 bush fires attended by the Fire Brigade).[62]

There was a brief increase in firebreak maintenance from 1972 to 1974, but from the mid-1970s economic pressures began to seriously curtail all fire management activities. Although gold prices soared, a global economic slump and the appreciation of the Rand impacted negatively on the economy and inflation ran high. The Forestry Department report for 1975/76 recorded a 30 per cent decline in the demand for mining timber, and noted that 'measures taken by Government to limit expenditure by the public sector resulted in insufficient funds being available for the Department to carry out its planned programme of activities'.[63] The brutal suppression of the student uprising in Soweto in June 1976 would result in a net outflow of capital from the country which lasted until 1980, putting further pressure on the budgets of state departments.[64] In 1976, the City Engineer likewise complained of 'severe cutbacks'.[65] At the same time, plantation forestry and accompanying fire protection measures were drastically reduced on the Peninsula, and outdoor recreation was putting increasing pressure on the natural environment. By 1976 more than 40,000 people visited Silvermine annually, and more than two million utilised Lion's Head, Signal Hill and Table Mountain for outdoor recreation.[66]

Following the big fires in 1973 and 1974, high fire incidence and rising usage of outdoor recreation areas, in 1975 the Municipality finally employed five law enforcement officers to patrol its undeveloped land. They enforced the bans on fires in open areas, resulting in a reported drop in wildfires until 1979.[67] Regulation, when enforced, can be effective in controlling fire incidence (this does not necessarily limit the area burned.) Overall, however, fire prevention measures were steadily reduced on the Peninsula for the remainder of the 1970s.

Fire management: 1980–93

On Sunday 7 March 1982 a fire broke out just below Tafelberg Road on the western slopes of Devil's Peak when a portable barbecue was blown over by a gust of wind. The scattered coals immediately ignited the dry vegetation, and fanned by a near gale-force southeaster, the fire spread rapidly. The Kloof Nek Forest Station gave the alarm at 14:33 and within five minutes the first Parks and Forest branch firefighters were on their way. The fire was already too big for them to control and the Newlands and Orange Kloof crews were scrambled too. A pall of smoke blocked out the sun over the City Bowl as the fire spread eastwards towards Devil's Peak, and traffic officers had to close De Waal Drive, a highway linking the city centre to the southern suburbs. Municipal firemen ran out hoses and at first managed to keep the fire away from houses in the neighbourhood of Vredehoek on the lower slopes of Devil's Peak. Between 00:30 and 00:45, three houses caught fire, Lionel Hardenberg and his family escaping in their pajamas as their home went up in flames. Residents of the three cylindrical Disa Park towers were evacuated, with sparks and embers burning holes in their clothing as they fled. Ultimately all the Municipality's Peninsula forest fire teams except those stationed at Silvermine were required to control the fire, which they achieved by Monday evening. The fire had burned 397ha (981 acres), including 60ha (148 acres) of Forestry Department plantation, and three houses. Newspaper reports described it as the worst fire 'in almost four decades', and 'the city's worst mountain fire in 30 years'.[68] Both estimates reveal the fickleness of public memory, enhanced by an urge to dramatise current events. In fact, it had been less than a decade since the last major mountain fires, in 1973 and 1974.

The Administrator of the Cape, Gene Louw, 'called for a review of the competence and extent of fire-fighting methods'.[69] Correspondents complained in the *Cape Times* that following the big 1974 blaze the City Council had promised 'drastic steps' to prevent the recurrence of another huge mountain fire. Peter Geldenhuys, a city businessman and veteran of several major mountain fires, observed that 'the small fire-fighting teams were no better equipped than they were last time'.[70] Nicky Bräsler wrote that as a New Zealander she was 'appalled at the rather primitive methods of fire control employed', with water being transported by trucks and a team of only 200 firefighters (there were actually less). She argued that in New Zealand such a fire would have been quenched 'almost immediately ... by several helicopters equipped with massive "monsoon buckets" filled with water from a nearby ocean, lake or dam'.[71]

Although City Councillors passed a unanimous vote of thanks to council firefighters, and the *Cape Times* praised their 'orderly and disciplined' efforts, clearly the Parks and Forests Branch were stung by the criticisms of their efforts and methods. Their Annual Report commented that firefighting was 'a thankless task'.[72] Stanley Evans, regional chief of Civil Defence, pointed out the difficulties the 133 firefighters had faced, including a lack of water pressure in Vredehoek resulting from the heavy usage of garden hoses by home owners trying to protect their premises. Volunteer firefighters had not been engaged because their lack of experience would hamper operations and put them at risk (the active volunteer system seems to have fallen away). It emerged that in 1974 the laying of high-pressure mains around the mountain perimeter had been costed, but an appeal to the Cape Peninsula Fire Protection Committee and the Department of Forestry had not secured the necessary R3million.[73] This kind of money was certainly not forthcoming in 1982 as the gold price collapsed, and the Rand fell below parity with the US dollar for the first time.[74]

On Wednesday 18 December 1986 a fire broke out on the north face of Table Mountain, the day after two bush fires in the city's southern suburbs. The high winds that had been buffeting the Peninsula swung to a southeasterly, and just after 13:00 smoke was reported high up in Platteklip Gorge on the north face of Table Mountain. The fire crew at Kloof Nek Forest Station were mobilised, but the fire was difficult to access and the wind quickly fanned it into a 'huge blaze'. All Kloof Nek gangs and auxiliary firefighting forces were scrambled, but they couldn't contain the fire, which by 13:45 had leaped Tafelberg Road and was racing downslope towards the neighbourhood of Oranjezicht. It engulfed the pine forest above Oranjezicht, before a fortuitous change in wind direction deflected it from the houses. The fire now spread westward along the slopes above Tafelberg Road towards Kloof Nek, where a counterfire was lit along the vertical firebelt at Kloof Corner. Unfortunately the fire spread upwards into the cliffs above this firebelt and over onto the western slopes of Table Mountain above Camps Bay. Firefighters had to retreat from the dangerous conditions on the cliffs, and by 19:40 the situation was judged critical, with continuous spot-firing from the upper cliffs onto the slopes below. Meanwhile, on the north face of Table Mountain, the fire was spreading eastwards towards Devil's Peak (Photo 9.3).

In Camps Bay the high wind drove the fire downslope towards Rontree Estate residential area. A counterfire was started along the Pipe Track contour path, but it failed to halt the fire, which jumped first the firebelt and then Camps Bay Drive, advancing 'with tremendous speed' on houses in upper Camps bay. Chaos ensued as residents evacuated their homes. A roadblock was put across Kloof Nek because the hundreds of onlookers congregating on Camps Bay Drive were hampering firefighting efforts. At around midnight a counterfire was instigated above houses at Woodhead Glen, which helped to contain the fire.

Photo 9.3   The December 1986 fire above Cape Town city bowl
Source: Courtesy Independent Newspapers, Independent Newspapers Archive, UCT Libraries Visual Archive.

The fire in the pines above Oranjezicht was contained on Wednesday night, but east of Platteklip Gorge the fire was spreading rapidly. Soon after midnight on Thursday, gale force winds swept this fire downslope towards Van Riebeeck Park. At 01:19 the City Engineer asked the City Council's civil defence authorities to begin evacuating upper Vredehoek and Van Riebeeck Park, as the high winds made firefighting impossible. However, the wind dropped markedly at around 04:05, and good progress was made in firefighting on all flanks. Evacuation measures were suspended. Only a fire burning high up on Knife Edge, a ridge linking Table Mountain to The Saddle and Devil's Peak, continued burning out of control. By 08:00 the wind had picked up again, and this fire spread towards the Newlands side of The Saddle, requiring another two hours of firefighting to contain. By 15:13 all flanks of the fire were finally under control.[75]

Over the two days of firefighting the Parks and Forests branch used its entire complement of 1000 men, supplemented by personnel from State Forestry, the Cape Divisional Council, the Fire Brigade, South African Defence Forces and the Police, Kirstenbosch staff, the Mountain Club and a private company, Court Helicopters. In all, 29 firefighting vehicles, 16 water tankers and three helicopters were deployed. A local radio ham network assisted with communications.[76] For the first time on the Peninsula, helicopters were used in command-and-control and firefighting capacities. Court Helicopters waterbombed the Camps Bay fire using a 950-litre fire (or 'monsoon') bucket. South African Air Force spokesman Lieutenant Heinrich Nagel said the steep cliffs and high winds had made it impossible to use this tactic on the City side of the mountain.[77]

Inevitably, alongside the Council's resolution expressing appreciation for the firefighters' 'excellent and heroic service' came criticism. Ex-mayor Bloomberg demanded an investigation into the reaction time to the reporting of the fire, and what immediate action had been taken.[78] A resident of Oranjezicht, Tom Moore, claimed he had noticed the fire around 11:15, whereas the official investigation concluded it had started between 13:00 and 13:15.[79] The *Cape Times* opined that access to the fires had been hampered by 'hundreds of gapers', and in future access to threatened areas should be more strictly controlled.[80]

The Municipality responded (in an annual report) by arguing that occasional fires were inevitable because mature fynbos is highly inflammable and 'Table Mountain is surrounded by a metropolis and is used by millions of people every year'. The mountain firefighting force was judged 'without equal in the Republic of South Africa', and 'the introduction of other fire-fighting techniques, or purchase of additional equipment' unnecessary. (A degree of pride, and defensiveness, is evident.) Rather, better policing, restricting access to the mountain, education and controlled burning were recommended. Standard suppression measures were being carried out.[81] However, elsewhere in the same report, the City Forest Officer admitted that 'the reduction of funds and resultant cutback in labour [more than 50 per cent since 1976] once again adversely affected the branch's work in the Nature Reserves under its control'.[82]

Forsyth and Van Wilgen analysed fire incidence on the Cape Peninsula's protected areas from 1970 to 1990, and it is striking that most areas on the northern Peninsula experiencing short-interval ($\leq$6 years) fires are grouped around popular walking and picnic areas. Hotspots in the Silvermine Nature Reserve were all located along established walking trails, and include popular picnic areas in Echo Valley and at Silvermine. Hotspots on the northern slopes of Signal Hill, immediately south of Kloof Nek, along the Tafelberg Road below the northern cliffs of Table Mountain, and below the old tin mine on Devil's Peak all feature popular walking routes. On the southern Peninsula there was a hotspot around Redhill (with its popular picnic site).[83]

On Friday 8 February 1991, a fire broke out high on Devil's Peak. Firefighters airlifted by helicopter prevented it from spreading eastwards, but fanned by a southeasterly wind it raced westwards towards the City, jumping De Waal Drive to threaten buildings in District Six, upper Vredehoek and Deer Park. Choking smoke and soot drove residents from their homes and De Waal Drive was closed during peak traffic flow. So much smoke entered the air-conditioning system of the Houses of Parliament that it was switched off, and temperatures soared in the debating chamber. Members asked permission to remove their jackets, but the reliably contrarian Conservative Party objected, and MPs had to keep them on. An estimated 420ha (1037 acres) were burned, around 100ha (247 acres) of it plantation land, but around 500 firefighters supported by 22 fire engines kept the blaze out of any properties.[84] While firemen battled the Devil's Peak fire, three further fires broke out, above Llandudno, on the Hangberg (or Sentinel) above Hout Bay harbour, and between Kommetjie and Ocean View on the southern Peninsula, requiring further strenuous firefighting.[85]

After the fires, the *Cape Times* praised the efforts of firefighters, but supported Mayor Gordon Oliver's call for 'an urgent reassessment of mountain fire control procedures'. The Director of Parks and Forests, Peter Rist, was criticised for prioritising using helicopters for transporting firefighters over waterbombing. The City Council's subsequent investigation found that the decisions and actions of its staff were entirely satisfactory. They defended Rist's decision not to ask Court Helicopters to waterbomb the fire, because of 'steep slopes . . . dangerous wind changes, poor visibility due to smoke and the absence of ground-based crews'. The 50 firefighters ferried up to the fire had prevented it from spreading to Newlands. In 1994, David Daitz (who replaced Peter Rist as Director of Parks and Forests in March 1993) would convince the Council to hire an Mi-8 helicopter and crew to be on standby at Newlands Forest Station for fire control purposes (commencing January 1995). This arrangement quickly proved its worth when the helicopter saved the historic *kramat* atop Signal Hill and airlifted two German tourists to safety, during a big fire on Lion's Head on 23 March 1995. A helicopter has been retained since then (Photo 9.4).[86]

None of this addressed the larger criticism made by University of Cape Town botany professor Eugene Moll following the 1991 Devil's Peak fire: that the City Council was not properly managing the ecology of the metropolitan area. Under pressure from financial sanctions

Photo 9.4   Firefighters struggle to save the *kramat* on Lion's Head, 23 March 1995
Source: Doug Pithey, Courtesy Independent Newspapers, Independent Newspapers Archive, UCT Libraries Visual Archives.

and following the collapse of state forestry, from the 1980s into the early 1990s fire management on the Peninsula was essentially reduced to firefighting. By the late 1980s invasive plant clearing operations had been drastically reduced. The city was booming, informal settlements had sprung up on the Peninsula and the municipality was preoccupied with catering for previously disadvantaged communities on the Cape Flats. Detailed Fire Brigade records cease after 1992/93. The period 1994–2000 (with the exception of 1996) were notably dry. The high fire incidence of the 1990s was entirely predictable under the circumstances.[87]

The Peninsula experienced two major mountain fires in January 1992. The largest broke out above Hout Bay on 15 January, spreading eastwards towards Tokai State Forest, burning an estimated 1500ha (3706 acres) of trees and fynbos. The second fire broke out on 17 January on Noordhoek Farm and spread eastwards, threatening houses from Kalk Bay around to Muizenberg. An estimated 1000ha (2471 acres) of fynbos and 'alien vegetation' were 'destroyed'.[88]

Fire management: c.1994–2000

Detailed fire brigade records on veld fires were discontinued in this period, so other sources must serve. A database of *Cape Times* newspaper reports on fires from 1993 to 2000 shows there were big veld fires in every year except 1996 (very wet) and 1997.[89] Forsyth and Van Wilgen found a significant increase in the areas that burned frequently (≤6 years) on the Peninsula, in the period 1985–2007, compared to 1970–90. They mapped notable new localities for short interval fires for 1985–2007 on the northern Peninsula, including Lion's Head and west of the reservoirs on top of Table Mountain. New patches developed on the Constantiaberg and Vlakkenberg above Hout Bay (a hotspot for invasive plants), between Noordhoek Peak and Chapman's Peak, and in the Silvermine Valley above Noordhoek. It is significant that following the advent of the 'new South Africa' in 1994, overseas tourism to the country surged, boosted by the 1995 rugby world cup. By 1999 the recently proclaimed Cape Peninsula National Park was receiving around 4.7 million visits per year.[90]

The most dramatic increase in areas experiencing short interval fires was on the southern Peninsula. These areas included Slangkop above Kommetjie, and from Ocean View eastwards along the Brakkloofrant towards Glencairn Heights, also extending south to the Klaas Jagersberg. The area inland from Glencairn Bay was developed in the 1980s, and

Capri west of the M6 in the 1990s. Several new small satellite settlements here created an urban–wildland intermix in fynbos heavily invaded by introduced plants originally introduced to stabilise driftsands. When the Red Hill fire burned great swathes of veld in this area in January 2000, 19 per cent of the area burned had been cleared of introduced plants by Working for Water workers in 1998 and 1999. Stacks of cut wood were left standing on the mountainsides.[91]

In 1998, the proclamation of the Cape Peninsula National Park finally achieved unified management for the major protected areas on the Peninsula. This did not however include fire management in areas outside the Park, and by December 1999 the Park had only completed a draft management plan. An unforeseen consequence of the hand-over to SANParks was the loss of staff previously employed by the Forestry Department or the Municipality. Many left, taking with them decades of experience of working on fire on the Peninsula, with its complex terrain, unique flora and rapidly changeable weather conditions.[92]

Fire climax: January 2000

When two fires broke out at Silvermine and Red Hill on the Cape Peninsula on Sunday 16 January 2000 – in conditions of strong winds, high temperatures and low humidity highly favourable to the rapid spread of fires – the environment had been well primed by humans over many years for a so-called natural disaster. Two smaller fires were already burning on the Peninsula, one at Groote Schuur Estate below Devil's Peak, and one above the navy base on the lower slopes at Silvermine.[93]

The Silvermine fire was started shortly before 15:20 on the roadside near the junction of the Ou Kaapse Weg road and the access road to the picnic site in the Silvermine Nature Reserve. The fire spread fast through land heavily invaded by introduced trees and shrubs, driven by strong southeasterly winds gusting up to 50 kilometres-per-hour (30 miles-per-hour). Within two hours the fire covered 3.5km (two miles) to reach Chapman's Peak Drive above Noordhoek, flames flaring 10m (32ft) into the air above dense patches of invasive vegetation.

The fire was declared a Code Red fire – a recognition that it had spread beyond the control of the South Peninsula Municipality, led by Carl Theunissen and Jolyon Schmidt of the Parks and Recreation Department. The Fire Mutual Aid Agreement was invoked, with other fire services called on for help. SAFCOL and Simon's Town firefighters were already fighting a flank of the fire threatening vineyards and plantations at Steenberg and Tokai. Cape Peninsula National Park staff, led by Cas Theron and Philip Prins, were released from the Groote Schuur fire to join operations at Silvermine. Having nearly killed Theunissen and Schmidt when it encircled and engulfed the old forester's cottage on Chapman's Peak, the fire spread northwards, reaching the outskirts of Hout Bay by the early hours of Monday 17 January.

The Red Hill fire broke out near the informal settlement of that name above Simon's Town, and was reported to the South Peninsula Municipality at 22:30 on 16 January. Gale force winds sent it rushing through piles of slashed exotic trees, and then racing through stands of mature fynbos. By Monday, the 17th, 3 helicopters and 40 firefighters were battling to save the settlements of Misty Cliffs and Scarborough on the west coast of the peninsula. Blocked by this action, the fire spread east into the Cape Point Nature Reserve, and west towards the settlement of Kommetjie.

By Monday morning, helicopters were waterbombing the fires above Hout Bay, flying in a racecourse pattern (to avoid colliding in the dense smoke) from reservoirs atop Table Mountain then back to Hout Bay. By late afternoon the informal settlement Imizamo Yethu was being showered with sparks. On Wednesday Hout Bay was saved when the wind swung around to a northwester, sending the fire roaring towards Constantia Nek. In a rerun of February 1947, firefighters were now engaged in the 'battle of Constantia Nek'. If the fire jumped across the Nek (mountain pass) to Cecilia Forest and Orange Kloof, the whole of the southern slopes of Table Mountain would have burned, potentially spreading the fire to the city of Cape Town.

This change in wind direction also resulted in flare ups again at Red Hill, which spread east towards the harbour town of Simon's Town. Twenty houses and several naval buildings were destroyed there, and private and industrial properties were damaged in nearby residential areas. This fire then spread southward into the Cape Point Nature Reserve.

At this point the fires were declared a disaster at a provincial scale, mobilising provincial and national resources, and a Joint Operating Centre was (belatedly) established in Wale Street in Cape Town to coordinate the many firefighting forces now in action. This included four municipalities, the South African Air Force and Navy, SAFCOL, Heyns Helicopters, Cape Peninsula National Park staff and volunteers.

The change in wind direction blew flames into Price and Belair drives, Constantia, and five houses were burnt. However, the line was held, and on Thursday the wind switched to a mild southwester, bringing in moister marine air and lowering temperatures. By Thursday evening, the concentrated firefighting efforts of naval personnel, South African

Air Force helicopters and local volunteers, had stabilised the situation on the southern Peninsula, and the Silvermine fire was classified Code Green on Friday.[94]

In total, 8370ha (20,682 acres) were burnt by these two fires, by far the most serious fire event yet recorded on the Cape Peninsula. Eight structures were destroyed and 51 damaged, at an estimated cost of R10.4 million. The insurance industry reported claims of R40 million (much less than the R3 billion claimed in newspaper reports). Damage to tree plantations was estimated at around R1 million, and there was some damage to vineyards. The costs of controlling the fire were in excess of R2.5 million. The fires were traumatic to many people, but fortunately injuries were not serious and there was no loss of life. Ecologically, the only serious harm was intense soil damage in areas where there had been dense patches of invasive plants, though rehabilitation costs (notably to prevent soil erosion) were in excess of R2.5 million.[95] A combination of quick responses, courage, community spirit, good equipment and luck with the weather after Wednesday 19, all contributed to the containment of what could have been a much more serious fire disaster on the Peninsula. However, in light of the long history of fires surveyed here, what is also striking is the continuity in factors contributing to the spread of the fires, as well as the shortfalls in the responses to them.

Conclusion

Humans have influenced the fire regimes of the Western Cape for 1.5 million years, but there is too little evidence to assess the impacts of precolonial fire use. There is no evidence of long-term impacts, though it is inadvisable to assume it was always benign. The Khoikhoi would have had to develop their transhumance patterns and burning practices through a long (undocumented) period of trial and error and in the context of climatic fluctuations. The first evidence for environmental degradation in the region was recorded in the late 1720s, though again, this was localised and probably short term.[1] Throughout the period of continuous Dutch colonial occupation at the Cape (1652–1795), Resolutions of Council were passed to regulate the use of fire in urban and rural areas. Sometimes draconian proscriptions clearly failed to stop veld burning and the authorities attempted to regulate rather than prohibit burning.

During the period of British colonial rule (continuous from 1806–1910) two official botanists were appointed, both of whom fiercely opposed veld burning. Pappe and Croumbie Brown argued that burning was desiccating the region and impoverishing the indigenous flora, particularly valuable timber species. Influenced by the effects of recent droughts they reinforced a narrative of linear environmental degradation which they attributed to poor land management practices. The Forest and Herbage and Preservation Act No.18 of 1859 included the first regulations on fire aimed at protecting the indigenous vegetation of the Cape Colony. A series of large fires in the 1860s, and the development of mining and related infrastructure requiring high volumes of timber, along with the Royal Navy's unfavourable assessment of indigenous timber, led to concerted afforestation with introduced trees from the 1880s. Elaborate fire protection measures were implemented and the

Forest Act No.28 of 1888 empowered foresters to enforce restrictions on burning practices.

Colonial botanists and foresters introduced the idea that the region had been extensively deforested by anthropogenic cutting and burning. The fact that introduced trees flourished in the nutrient-poor soils, which they found covered only in treeless fynbos and renosterveld, seemed to confirm this. They misinterpreted small forest patches inhabiting moist, fire-free refuges in ravines and on scree slopes as remnants of larger forests. Certainly, existing patches of forest on the Peninsula had been heavily utilised in colonial times, but forest had never been extensive.

From the 1880s, foresters claimed afforestation would improve water supplies and stabilise soils. As early as the 1890s, others argued that introduced trees sapped water supplies and were a fire hazard. Until the 1950s it seemed inarguable that the Peninsula should grow its own timber, but despite this, large fires repeatedly sparked controversies over the planting of inflammable pines and eucalypts. This controversy was also played out in sectorial conflicts over catchment management in the country as a whole and came to a head at the fourth British Empire Forestry Conference in South Africa in 1935. The Conference recommended that state foresters investigate the hydrological effects of introduced trees and of veld burning. This was implemented at the Jonkershoek Forestry Research Centre.

By the late 1940s, groundbreaking hydrological research led by Christiaan Wicht at Jonkershoek suggested that controlled burning of fynbos might actually reduce water loss in catchment areas. However, until then, state forestry policy on fire was based on agricultural research on grasslands, influenced by Clementsian succession ecology. Fire was held to reverse the 'natural succession' of vegetation towards a more complex and stable state, to impoverish the flora and soils and cause soil erosion and water loss. This linear approach was misapplied to fynbos, and it was only in 1948 that Wicht and the reconstituted Forestry Department felt secure enough in their scientific evidence to propose controlled burning in fynbos catchments.

Unfortunately, this proposal came at a time of high fire incidence amid a period dominated by a national narrative of environmental collapse through soil erosion, fuelled by irresponsible veld burning. The expansion of the state Forestry Department's afforestation programme into the inflammable grasslands of eastern South Africa meant a national-scale experience of increasing fire incidence on forestry lands influenced policy in the Western Cape. Renewed hydrological controversies between farmers and foresters resulting from this expansion put further pressure on the Department and controlled burning was shelved.

The sectorial and institutional politics which shaped fire policy at a national scale were played out on a smaller scale on the Peninsula. Botanists aimed to preserve indigenous vegetation in its natural state, foresters aimed to afforest the mountains with introduced trees and municipal authorities planted introduced species to supply shelter for recreation and both introduced and indigenous species to beautify the landscape. For some, the fynbos-clad slopes were exposed and forbidding, whereas forests and plantations offered shelter, shade and beauty. For Jan Smuts and others, embracing the indigenous flora and fauna was a key to building a united nation. One of the few points of agreement was an antipathy to fire. The recommendation of Wicht's 1945 Royal Society report that prescribed burning should be applied in fynbos was rejected in favour of a high-profile campaign to prevent, and suppress, fires.

In the 1960s, wildlife conservation became a central objective of the Forestry Department's catchment management programme. In 1965, research was initiated to discover an ecologically sound method of burning fynbos, spurred on by evidence that the charismatic fynbos species, the Marsh Rose and the Blushing Bride, had dwindled to near extinction – in the absence of fire. These developments, alongside a series of very large fires that convinced the Department that total fire exclusion was impossible, resulted in the implementation of prescribed burning in the 1970s. This was feasible because the management of the country's catchment areas had been consolidated under the Department of Forestry, which was empowered to formulate, implement and enforce a scientifically informed fire management approach by the Mountain Catchment Areas Act 63 of 1970.

Ecological research on the effects of fire on fynbos became a priority once prescribed burning was introduced. The 1970s and 1980s saw a renaissance in fynbos ecology in South Africa, stimulated in part by contact with the international MEDECOS network for studying the ecosystems of Mediterranean-type climate regions. Exciting new ideas and approaches shaped the innovative Fynbos Biome Project of 1977–89. Important findings were made concerning the fire-adapted reproductive strategies of fynbos plants and recommendations developed regarding when and how to burn fynbos. The community of researchers, planners and managers involved in this project played an important role in coordinating research, management and policy on fire in fynbos in this period.

Despite these breakthroughs in fynbos fire ecology, management authorities on the Peninsula sustained the high-profile anti-fire campaigns developed in the late 1930s, through to the late 1980s. Thus public opinion, and that of local government, was firmly opposed to prescribed burning despite the best attempts of some local experts and managers to point out its ecological necessity and the dangers of long-term fire exclusion. A steady increase in fire incidence appeared to make the introduction of more fire to the Peninsula unnecessary and unwise.

By the mid-1970s commercial forestry was being phased out on the Peninsula, and the government began to dismantle state forestry in 1986. Where forestry managers may have been able to implement ecologically sound burning practices in the face of public opposition, this was not an option for municipal employees answerable to locally elected councillors. The collapse of funding for environmental projects, as international sanctions, a global economic slump and high inflation hit the country, meant that fire prevention and invasive plant control went into a steep decline from the mid-1970s.

Conservation-minded researchers shorn of institutional support and funding in the 1990s turned to arguments for ecosystems services, focusing on the impacts of invasive introduced species on water supplies. Fire was neglected, or seen largely as a problematic influence encouraging invasive species. This was embodied in the major environmental initiative of the post-democratic era, Working for Water. The overwhelming response to the January 2000 fires was that invasive plants must be cleared.

It is disheartening, given the long history of fire on the Peninsula, that the commissioners investigating the January 2000 fires identified an urgent need to 'cultivate a well-informed public' who understood fire danger ratings and risks and could 'understand sustainable veldfire management'. The report on the January 2000 fires also concluded that 'there is little to suggest that the fire regime has changed much over the past 50 years or more'.[2]

On the contrary, this history suggests that on a local scale humans have influenced nearly all dimensions of the 'natural fire regime'. Progressive urbanisation and infrastructure development has both limited fire ignitions and fire spread in some places and introduced ignitions into others. The dynamics of afforestation over time have influenced fuel loads and availability, and thus fire spread, intensity and severity. So have plant invasions and the varying efforts to control these. The development of transport infrastructures and facilities for outdoor recreation, along with urban sprawl, has influenced the distribution of areas with short interval fires. The fire season remains similar, but anthropogenic global climate change may be increasing the frequency of extreme fire weather conditions. The concentration of fires on weekends and holidays reveals a human influence on the temporal clustering of ignitions.

It may help to broaden the fire regime concept to incorporate both social and ecological elements. Understanding the fire situation on the Cape Peninsula requires an integrated, historically informed approach to understanding the interrelations of the human and natural dimensions of fire in the Peninsula's hybrid landscapes. This book has thus investigated key individuals, organisations and programmes involved with fire on the Peninsula. The key human actors had very diverse geographical and intellectual points of origin. They worked within a wide range of sectorial contexts, some experiencing notable evolutions in their understanding of fire, born of experience in the unique ecosystems and landscapes of region. Institutionally, fire policy and management have been shaped by changes in governments and their bureaucracies, divergent sectorial goals, fluctuations in economic circumstances and the social goals, recreational needs, aesthetic and environmental values and opinions of the public. Non-human actors shaping local fire regimes range from the underlying geology to the vagaries of atmospheric conditions and the adaptations and interactions of myriad species of indigenous and introduced plants and animals and fire.

This book identifies some important networks of exchanges of materials, expertise, values and ideas which have impacted on fire incidence and control on the Cape Peninsula. However, none of these offers a simple explanatory hierarchy, as all have operated in the context of a very diverse set of interests and influences. I have instead shown how interactions between, and conjunctions of, social and ecological cycles and events have driven changes in fire policy and management.

The complexity of interactions this reveals, and past failures or missed opportunities in fire research and management, should not lead us to despair. There is much to be inspired by in this history – namely the passion, ingenuity, intelligence, perseverance, bravery and dedication of generations of men and women who have wrestled with the challenges of understanding and managing the charismatic, beautiful and dangerous beast we call fire. Table Mountain, an ark of biodiversity in a sea of people, has been a crucible for attempts to integrate the social and ecological dimensions of fire. The iconic landscape, unique fire-prone vegetation and the long history of human occupation on the Peninsula make it an inspirational focus for a more integrated, more honest attempt to understand and live with wildfire. We need to better understand, admit and, where necessary and possible, mitigate our influence on the Peninsula's fire regimes. We cannot eradicate fire as we did the Cape's lions, and to do so would in any case impoverish the Peninsula's astonishing biodiversity. We should rather strive to retain the wildness of fire within the bounds of the Table Mountain National Park and ensure we share the responsibility of living with fire equitably.

Appendix 1: Cape Peninsula Vegetation, Climate, Weather and Fire

Botanists now estimate that fynbos and renosterveld naturally burn every 8–12 years and every 3–5 years respectively. Too-frequent burning (≤6 years) favours grasses and disadvantages some fynbos species (notably some large protea shrubs).[1]

Introduced pine trees are particularly vulnerable to fire in the early years of growth. Mature plantations are less inflammable, but once alight, they fuel intense, uncontrollable fires. Some introduced trees and shrubs invade the indigenous vegetation and regenerate quickly from fires to form dense stands. Although less inflammable than fynbos, they burn more intensely.

Climate and weather influence fire season, the likelihood of ignitions spreading and the area burned (see Chapter 5). Winter is often cool and wet. The hot and windy summer and early autumn (November–March) are conducive to big, fast-moving fires.

Rainfall and moisture distribution in space

Rainfall varies locally, as a result of the complicated mountainous terrain and prevailing winds. North- and west-facing slopes are drier and the upper south- and east-facing slopes moister. The latter support most of the Peninsula's natural forests and plantations. However, prevailing southeasterly winds can quickly dry out the fynbos on east-facing slopes in summer. These variations contribute (along with vegetation type and accessibility to human ignitions) to high fire incidence on Table Mountain's drier northern slopes, on Lion's Head and Signal Hill (Figure A1.1).

Winds

In most major fires high winds were recorded – usually southeasters – for example, in 1902, 1909, 1920, 1934, 1938, 1982, 1986, 1991 and 2000. The southeaster blows from spring to late summer. Its desiccating effect primes the vegetation for fire and it fans and quickly spreads ignitions. Firefighters have to contend with the movement of the winds around the mountains and big swings in wind direction.

Rainfall distribution in time

The period 1921–49, but especially 1922–36, represents the longest sustained period of below average annual rainfall experienced in the period. If sustained

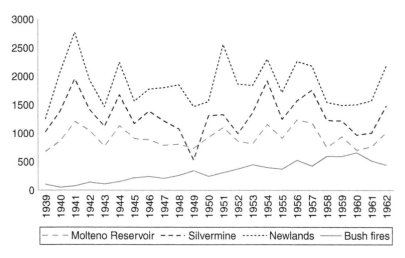

Figure A1.1 Annual rainfall (mm) for three Cape Peninsula stations and the number of bush fires 1939–62: Molteno Reservoir (city bowl); Silvermine (southern Peninsula); Newlands (eastern slopes, Table Mountain).

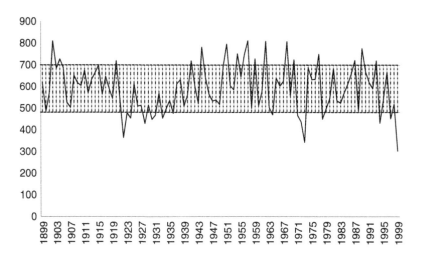

Figure A1.2 Annual rainfall (mm) for Cape Town, 1899–1999, with band representing one std deviation above and below the 100-year average (590mm)
Source: S. D. Lynch, 2003, *Development of a Raster Database of Annual, Monthly and Daily Rainfall for Southern Africa*, WRC Report 1156/1/03 (Pretoria, South Africa: Water Research Commission).

drought is conducive to wildfires this should have been a period of unprecedented fire incidence. However, this is not borne out by fire department figures, or the incidence of large fires, until 1934/1935 (Figures A1.1. and A1.2).

The overall increase in annual fire incidence in vegetation (1900–93) is weakly correlated with annual rainfall, which fluctuates throughout. The first notable acceleration in fire incidence occurred during the middle-third of the long dry phase from 1921–49. There is a more plausible link between years of high fire incidence and periods of unusual inter-annual rainfall patterns. The sequences of dry spells of three to four consecutive months in the periods 1942–53, 1959/60, 1963, the early 1970s, 1991/92–1992/93 and 1997/98–2000 are notable. Seasonally, most big fires occurred in January and February.

Appendix 2: Fire Causes

Fire Brigade records categorise fires by location, including fires in bush, grass and rubbish. Fires in trees, reeds and hedges are also listed (very minor sources) and included (in my analysis) with bush and grass fires as 'fires in vegetation'. Fires in rubbish are excluded. Information on causes is extracted from fire brigade records for 'fires in vegetation' only. Categories of causes were altered across time and records were not kept throughout the period. The proportion of unknown or vaguely attributed causes ('dropping a light') is high, around 77 per cent for 1930–53, and not specified thereafter. More often than not, no witnesses came forward to report causes of ignitions in fires requiring fire brigade attendance.

There are 34 categories of causes to which bush fires are attributed in Cape Town Fire Brigade reports, ranging from persistent, major causes, to one-offs like 'servant burning feathers off goose and igniting hedge'.[1] 'Spontaneous combustion' is listed as a cause – it is the second-largest attributed cause (19) in 1975. Overall, rubbish fires account for 9–13 per cent of fires in the category 'fires in bush, grass and rubbish'.

Major causes listed in fire brigade records for 1930–53 (ignoring rubbish fires) are 'children playing with fire' (116) and arson (77). Key minor causes were locomotives, vagrants, fireworks (Guy Fawkes Day falls within peak burning season) and burning to clear vegetation. Only 17 per cent of fires were ascribed specific causes.

Major attributed fire causes in state plantations and forests in South Africa from 1900–20 were unknown (51 per cent), entering forest/plantation by crossing exterior fire lines (43 per cent) and intentional firing (23 per cent). Figures are available for 1953–61, showing major causes as campers and smokers (31 per cent), unknown (24 per cent), arson (21 per cent), steam locomotives (11.8 per cent) and lightning or falling rocks (11.5 per cent).

Cape Town fire brigade records for 1962–77 show major causes as children (677), locomotives (256) and controlled burns (262). National state forestry records for the same period (3802 fires) show major causes as arson (744 fires), lightning and falling rocks (429), smokers, campers and picnickers (379), honey hunters (247) and departmental (246).

Cape Town Fire Brigade records attribute major causes during 1982/83–1992/93 to vagrants (69), arson (40, with 33 in 1985/86) and controlled burns (37).

Detailed state forestry records cease in 1984/85, with some categories recorded to 1987/88. Main causes for 1977/78–1987/88 were arson (526), lightning, and falling rocks (413), smokers, campers and picnickers (272) and honey hunters (212).

Notes

Introduction

1. *Cape Times*, 'Wind-Driven Fires Ravage Western Cape,' 17 January 2000, front page; *Cape Times*, 'Fire-Storm Sunday,' Editorial, 18 January 2000, 10; C.N. de Ronde, 1999, '1998: A Year of Destructive Wildfires in South Africa,' *International Forest Fire News*, 20, 73–8, 73; L. Aupiais and I. Glenn (eds.), 2000, *The Cape of Flames: The Great Fire of January 2000* (Cape Town, South Africa: Inyati Publishing); F.J. Kruger, P. Reid, M. Mayet, W. Alberts, J.G. Goldammer, K. Tolhurst, 2000, 'A Review of the Veld Fires in the Western Cape during 15 to 25 January 2000,' *Department of Water Affairs and Forestry Report*.
2. S. Watson, 1995, *Presence of the Earth: New Poems* (Cape Town, South Africa: David Philip).
3. For a review, see W. Beinart, 2000, 'African History and Environmental History,' *African Affairs*, 99, 269–302. See J.A. Tropp, 2006, *Nature's of Colonial Change: Environmental Relations in the Making of the Transkei* (Athens, OH: Ohio University Press); T. Kepe, 2005, 'Grasslands Ablaze: Vegetation Burning by Rural People in Pondoland, South Africa,' *South African Geographical Journal*, 87, 10–17.
4. D. Reitz, 1975 [1929], *Commando: A Boer Journal of the Boer War* (London: Faber and Faber), pp.147–9, 178; T. Pakenham, 1992, *Scramble for Africa* (London: Abacus), pp.576–8.
5. J.C. Scott, 1999, *Seeing Like a State: How Certain Schemes to Improve the Human Condition Have Failed* (New Haven, CT: Yale University Press), pp.45–6.
6. Key texts include R.H. Grove, 1995, *Green Imperialism: Colonial Expansion, Tropical Island Edens and the Origins of Environmentalism, 1600–1860* (Cambridge: Cambridge University Press); J. Hodge, 2007, *Triumph of the Expert: Agrarian Doctrines of Development and the Legacies of British Colonialism* (Athens, OH: Ohio University Press); W. Beinart, K. Brown, and D. Gilfoyle, 2009, 'Experts and Expertise in Africa Revisited,' *African Affairs*, 108, 413–33; J. Hodge and B.M. Bennett (eds.), 2012, *Science and Empire: Knowledge and Networks of Science in the British Empire 1800–1970* (London: Palgrave Macmillan).
7. S. Dubow, 2006, *A Commonwealth of Knowledge: Science, Sensibility, and White South Africa 1820–2000* (Oxford: Oxford University Press).
8. Grove, *Green Imperialism*; P. Anker, 2001, *Imperial Ecology: Environmental Order in the British Empire, 1895–1945* (Cambridge, MA: Harvard University Press); W. Beinart, 1984, 'Soil Erosion, Conservationism and Ideas about Development: A Southern African Exploration, 1900–1960,' *Journal of Southern African Studies*, 11, 52–83; J. Carruthers, 2011, 'Trouble in the Garden: South African Botanical Politics ca.1870–1950,' *South African Journal of Botany*

(henceforth *SAJB*), 77, 258–67; B.M. Bennett, 2011, 'Naturalising Australian Trees in South Africa: Climate, Exotics and Experimentation,' *Journal of Southern African Studies*, 27, 265–80; B.M. Bennett and F.J. Kruger, 2013, 'Ecology, Forestry and the Debate over Exotic Trees in South Africa,' *Journal of Historical Geography*, 42, 100–9; R. McLeod, 2000, 'Introduction' to 'Nature and Empire: Science and the Colonial Enterprise,' *Osiris*, 15, 1–13. On networks, see D. Wade-Chambers and R. Gillespie, 2000, 'Locality in the History of Colonial Science,' *Osiris*, 15, 221–40.

9. Notably: J. Fairhead and M. Leach, 1996, *Misreading the African Landscape: Society and Ecology in a Forest-Savanna Mosaic* (Cambridge: Cambridge University); C.A. Kull, 2004, *Isle of Fire: The Political Ecology of Landscape Burning in Madagascar* (Chicago, IL: Chicago University Press).

10. W.J. Bond and B.W. van Wilgen, 1996, *Fire and Plants* (London: Springer), vii–viii.

11. J.E. Keeley, W.J. Bond, R.A. Bradstock, J.G. Pausas, and P.W. Rundel, 2012, *Fire in Mediterranean Ecosystems: Ecology, Evolution and Management* (Cambridge: Cambridge University Press), p.3.

12. D.M.J.S. Bowman et al., 2011, 'The Human Dimension of Fire Regimes on Earth,' *Journal of Biogeography*, 38, 2223–36.

13. Ibid., 2225.

14. On 'novel ecosystems,' see E. Marris, 2009, 'Ragamuffin Earth,' *Nature*, 460, 450–3. On the WUI, see S.J. Pyne, 2008, 'Spark and Sprawl,' *Forest History Today*, Fall, 4–10. For urban ecology on the Cape Peninsula, see A.H. Neely, 2010, ' "Blame It on the Weeds": Politics, Poverty, and Ecology in the New South Africa,' *Journal of Southern African Studies*, 36, 869–87; A.G. Rebelo, P.M. Holmes, C. Dorse and J. Wood, 2011, 'Impacts of Urbanization in a Biodiversity Hotspot: Conservation Challenges in Metropolitan Cape Town,' *SAJB*, 77, 20–35; the Special Feature on 'Urban Ecological and Social-Ecological Research in the City of Cape Town,' *Ecology and Society*, 2012, 17, 3.

15. A. Shlisky, R. Meyer, J. Waugh, K. Blankenship, 2008, 'Fire, Nature, and Humans: Global Challenges for Conservation,' *Fire Management Today*, 68, 36–42, see 38–9.

1 Fire at the Cape: From Prehistory to 1795

1. H.B. Thom (ed.), 1958, *Journal of Jan van Riebeeck*, Vol.3 (1659–62), trans. C.K. Johnman and A. Ravenscroft (Cape Town, South Africa: A.A. Balkema), p.112.

2. A. Sparrman, 1785, *A Voyage to the Cape of Good Hope, towards the Antarctic Polar Circle, and Round the World: But Chiefly into the Country of the Hottentots and Caffres, from the Year 1772, to 1776*, 2 Volumes, Vol.1 (London: G.G.J. and J. Robinson), p.367.

3. F. Le Vaillant, 1791, *Travels into the Interior Parts of Africa by the Cape of Good Hope: In the Years 1780, 81, 82, 83, 84, and 85*, 2 Volumes, Vol.1 (Perth, Australia: R. Morrison & Son), p.49.

4. On fire regimes, S.J. Pyne, 2010, *America's Fires: A Historical Context for Policy and Practice* (Durham, NC: Forest History Society), pp.xvi–xvii.

5. On prehistoric fire use, see M.T. Hoffman, 1997, 'Human Impacts on Vegetation,' in R.M. Cowling (ed.), *The Vegetation of Southern Africa* (Cambridge: Cambridge University Press), pp.507–34, 513; M.V. Hall, 1984, 'Man's Traditional and Historical Use of Fire in Southern Africa,' in P. du V. Booysen and N.M. Tainton (eds.), *Ecological Effects of Fire in South African Ecosystems* (Berlin, Germany: Springer Verlag), pp.39–52; H.J. Deacon, 1992, 'Human Settlement,' in R.M. Cowling (ed.), *The Ecology of Fynbos: Nutrients, Fire and Diversity* (Cape Town, South Africa: Oxford University Press), pp.260–70, 263; A.G. Rebelo, C. Boucher, N. Helme, L. Mucina, and M.C. Rutherford, 2006, 'Fynbos Biome,' in L. Mucina and M.C. Rutherford (eds.), *The Vegetation of South Africa, Lesotho and Swaziland*, Strelitzia 19 (Pretoria, South Africa: South African National Biodiversity Institute), pp.52–219, 85.
6. A.B. Smith, 1992, *Pastoralism in Africa: Origins and Development Ecology* (London: Hurst & Co.), pp.193–200.
7. Hoffman, 'Human Impacts on Vegetation,' p.513; A.G. Rebelo et al., 'Fynbos Biome,' p.87.
8. On Van Riebeeck's life, see E.C. Godée Molsbergen, 1968, *Jan van Riebeeck en sy Tyd* (Pretoria, South Africa: J.L. van Schaik); H.B. Thom, 1952, 'Introduction,' in H.B. Thom (ed.), *Journal of Jan van Riebeeck*, Vol.1 (1651–55) (Cape Town, South Africa: A.A. Balkema), pp.xv–xliv, xvi–xxvi.
9. On the Khoikhoi in this period (called by the authors the Khoisan), see R. Elphick and V.C. Malherbe, 1992, 'The Khoisan to 1828,' in R. Elphick and H. Giliomee (eds.), *The Shaping of South African Society, 1652–1840* (Cape Town, South Africa: Maskew Miller Longman), pp.3–65. See also E. Boonzaier, P. Berens, C. Malherbe, and A. Smith, 1996, *The Cape Herders: A History of the Khoikhoi of Southern Africa* (Athens, OH: Ohio University Press).
10. On livestock numbers, see Thom, *Journal of Jan van Riebeeck*, p.372. See also *Journal of Jan van Riebeeck*, 1958, Vol.3, pp.196, 337.
11. L. Guelke, 1992, 'Freehold Farmers and Frontier Settlers, 1657–1780,' in R. Elphick and H. Giliomee (eds.), *Shaping of South African Society*, p.69.
12. On human development, see Hoffman, 'Human Impacts on Vegetation,' pp.507–34; Hall, 'Man's Traditional and Historical Use of Fire,' pp.42–7. On the impacts of animals, see A.G. Rebelo et al., 'Fynbos Biome,' pp.85–7. On the hippo and rhino, see Thom, *Journal of Jan van Riebeeck*, pp.34, 286; for leopard and lion attacks, see *Journal of Jan van Riebeeck*, Vol.1, pp.310, 315 and Vol.3, p.120. On wine and wheat production, see R. Ross, 1989, 'The Cape of Good Hope and the World Economy, 1652–1835,' in R. Elphick and H. Giliomee (eds.), *Shaping of South African Society*, pp.243–80.
13. Thom, *Journal of Jan van Riebeeck*, Vol.1, p.109; Vol.3, pp.51, 53, 59, 112; Jan van Riebeeck et al., 'Resolutions of the Council of Policy of Cape of Good Hope' (hereafter *RCP CGH*), Thursday 24 October 1658, Cape Town Archives Repository, South Africa (hereafter CT AR), Ref.C.1, 385–6; Jan van Riebeeck et al., 'RCP CGH,' Thursday 22 July 1659, CT AR, Ref.C.2, 30–3. On Rossouw's house: A. Cranendonk et al., 'RCP CGH,' Wednesday 20 November 1715, CT AR, Ref.C.34, 102–8.
14. Thom, *Journal of Jan van Riebeeck*, Vol.3, pp.17, 38, 173.
15. Simon van der Stel et al., 'RCP CGH,' Wednesday February 1691, CT AR, Ref.C.21, 2–4; 'RCP CGH,' Thursday 3 April 1691, CT AR, Ref.C.21, 17–23.

16. See: 'RCP CGH,' Cape Town Archives Repository, South Africa, as follows: Ref.C.23 (Monday 2 December 1697); Ref.C.73 (Wednesday 19 August 1711); Ref.C.73 (Thursday 9 October 1725); Ref.C.82 (10 February 1729); Ref.C.126 (19 December 1748); Ref.C.140 (Thursday 16 November 1762); Ref.C.174 (Wednesday 17 January 1787), 12.

17. O.F. Mentzel, 2006 [1787], *A Geographical-Topographical Description of the Famous and (All Things Considered) Remarkable African Cape of Good Hope*, Part I, Translated by H.J. Mandelbrote, First Reprint Series, 4 (Cape Town, South Africa: Van Riebeeck Society), pp.134–5; O.F. Mentzel, 2006, *Life at the Cape in the Mid-Eighteenth Century: Being the Biography of Rudolph Siegfried Allemann, Captain of the Military Forces at the Cape of Good Hope*, translated by M. Greenlees, First Reprint Series, 2 (Cape Town, South Africa: Van Riebeeck Society), pp.101–2. R. Tulbagh et al., 'RCP CGH,' Thursday 20 March 1736, CT AR, Ref.C.100, 29–48.

18. Mentzel, *Geographical-Topographical Description*, pp.133–4.

19. Ibid., pp.93–4. On Mentzel, see F. Hale, 2005, 'Khoi Khoi Culture Refracted through a German Prism: Christian Ludwig Willebrand's Reconstruction in Historical Context,' *South African Journal of Cultural History* 19, 2, 127–34, 135.

20. On fires and shipping: R. Tulbagh et al., 'RCP CGH,' Saturday 1 December 1742, CT AR, Ref.C.120, 354–9. For other relevant RCP CGH, CT AR, see Ref.C.94 (Thursday 24 December 1733), 22–5; Ref.C.33 (26 December 1714), 25–6.

21. T.R. Sim, 1907, *The Forests and Forest Flora of the Colony of the Cape of Good Hope* (Aberdeen: Taylor & Henderson), p.77.

22. In R. Raven-Hart, 1971, *Cape Good Hope 1652–1702: The First Fifty Years of Dutch Colonization as Seen by Callers*, Vol.1 (Cape Town, South Africa: A.A.Balkema), p.213.

23. P. Kolb, 1738, *The Present State of the Cape of Good-hope*, Vol. 1, 2nd ed. (London: Innys and Manby), pp.62–3.

24. Anon, 1984, *Almanach der Africaansche Hoveniers en Landbowers* (Cape Town, South Africa: South African Library), pp.25, 26, 27, 35, 37.

25. Hendrik Swellengrebel et al., 'RCP CGH,' Thursday 19 January 1741, CT AR, Ref.C.116, pp. 35–41. My thanks to Robyn Kampshoff-Pooley for help with translation.

26. See C.L. Wicht and F.J. Kruger, 1983, 'Die Ontwikkeling van Bergveldbestuur in Suid-Afrika,' *South African Forestry Journal* (henceforth *SAFJ*), 86, Special Issue 2 'Our Green Heritage,' 3. This article (in Afrikaans) gives a good overview of the history of conservation in South Africa's humid mountain catchment areas.

27. On livestock and trade, see D. Sleigh, 1993, *Die Buiteposte* (Pretoria, South Africa: Protea Boekhuis).

28. On the botanical explorers, see H.F. Glen and G. Germishuizen (compilers), 2010, *Botanical Exploration of Southern Africa*, 2nd ed., Strelitzia 26 (Pretoria, South Africa: South African National Biodiversity Institute).

29. C.P. Thunberg, 1986 [1793–95], *Travels at the Cape of Good Hope 1772–1775*, ed. V.S. Forbes, translated by J. and I. Rudner (Cape Town, South Africa: Van Riebeeck Society), p.83.

30. Sparrman, *A Voyage to the Cape of Good Hope,* On rhinoceros (renoster) bush: Vol.1, p.254; on locusts and fire, Vol.1, p.367; on the runaway bushfire, Vol.2, pp.320–1.
31. Glen and Germishuizen, *Botanical Exploration of Southern Africa,* pp.263–4; Le Vaillant, *Travels into the Interior Parts of Africa,* pp.60–1.
32. Le Vaillant, *Travels into the Interior Parts of Africa,* p.49.
33. Thom, 1952, 1954, 1958, *Journal of Jan van Riebeeck,* 3 volumes (Cape Town, South Africa: A.A. Balkema), see: Vol.1 (1651–55), 109, 118, 127, 193; Vol.2 (1656–58), p.82; Vol.3 (1659–62), p.112.
34. See notes 14–17 above.

2   Fire at the Cape: British Colonial Rule, 1795–1900

1. J.C. Brown, 1887, *Management of Crown Forests at the Cape of Good Hope under the Old Regime and under the New* (Edinburgh: Oliver & Boyd), p.3.
2. Cited by David Hutchins in Reports of the Conservators of Forests, Cape of Good Hope (C.G.H.) for the year 1893, G.50–94, p.30.
3. Ibid., p.31.
4. On the British arrival, see N. Worden, E. van Heyningen, and V. Bickford-Smith, 1998, *Cape Town: The Making of a City* (Cape Town, South Africa: David Philip), pp.86–7.
5. R. Elphick and H. Giliomee (eds.), 1992, *The Shaping of South African Society, 1652–1840* (Cape Town, South Africa: Maskew Miller Longman), see R. Ross, 'The Cape of Good Hope and the World Economy, 1652–1835,' pp.243–80; R. Elphick and H. Giliomee, 'The Origins and entrenchment of European Dominance at the Cape, 1652–1840,' pp.521–66. On Plettenburg Bay forests, see R.H. Grove, 1987, 'Early Themes in African Conservation: The Cape in the Nineteenth Century,' in D. Anderson and R.H. Grove (eds.), *Conservation in Africa: People, Policies and Practice* (Cambridge: Cambridge University Press), pp.21–39.
6. H. Lichtenstein, 1812, *Travels in Southern Africa in the Years 1803, 1804, 1805, and 1806,* translated by Anne Plumptre (London: Henry Colburn), p.185.
7. W.J. Burchell, 1822, *Travels in the Interior of Southern Africa,* Vol.1, (London: Longman, Hurst, Rees, Orme and Brown), p.38.
8. Ibid.
9. Ibid.
10. Ibid., pp.116–17.
11. Ibid., p.117.
12. R.H. Grove, 1987, 'Early Themes in African Conservation,' pp.21–39.
13. H.F. Glen and G. Germishuizen (compilers), 2010, *Botanical Exploration of Southern Africa,* 2nd ed., Strelitzia 26 (Pretoria, South Africa: South African National Biodiversity Institute), p.327. Regarding the post of Colonial Botanist, see 'Letter from Ludwig Pappe to Sir William Jackson Hooker; from Cape Town, South Africa; 28 Jan 1857,' Royal Botanic Gardens, Kew: Archives: Directors' Correspondence, Vol.59, folio 254.
14. Cited in Brown, *Management of Crown Forests,* p.84.
15. See R.H. Grove, 1987, 'Early Themes in African Conservation,' pp.23–6. Rawson was not, as Grove suggests, ever Governor of the Cape.

16. Ibid., pp.27–8.
17. P. Nobbs, 1956, 'A Pioneer of Botany and Forestry at the Cape of Good Hope,' *Journal of the South African Forestry Association* (henceforth *JSAFA*), 27, 87–9; Grove, 'Early Themes.'
18. On the 'dead letter,' see Brown, *Management of Crown Forests*, pp.84–5.
19. On Brown's conservation campaigns and contribution to debates about environmental degradation at the Cape, see Grove, 'Early Themes,' pp.28–31; W. Beinart, 2003, *The Rise of Conservation in South Africa: Settlers, Livestock, and the Environment 1770–1950* (Oxford: Oxford University Press), pp.99–114. For a summary of Brown's views on fire and its effects, see Brown, *Management of Crown Forests*, pp.1–5; on Pappe, see pp.83–5; for Brown's report on the conservation of forests, see pp.90–113; for a sample of his moralising, see p.258. On the Tswana, see R. Moffat, 1846, *Missionary Labours and Scenes in Southern Africa* (London: John Snow), p.87.
20. J.C. Brown, 1875, *Hydrology of South Africa: Or Details of the Former Hydrographic Condition of the Cape of Good Hope and of Causes of its Present Aridity with Suggestions of Appropriate Remedies for this Aridity* (London: Henry S King & Co.), p.171; Grove, 'Early Themes,' p.29.
21. Grove, 'Early Themes,' p.31.
22. Brown, *Hydrology of South Africa*, pp.177–8.
23. Ibid., 178–93. See also J. Burman, 1971, 'The Great Fire of 1869,' in J. Burman (ed.), *Disaster Struck South Africa* (Cape Town, South Africa: C. Struik), pp.3–18.
24. M.H. Lister, 1957, 'Joseph Storr Lister, the First Chief Conservator of the South African Department of Forestry,' *JSAFA*, 29, 10–18.
25. T.R. Sim, 1907, *Forests and Forest Flora of the Colony of the Cape of Good Hope* (Aberdeen: Taylor & Henderson), pp.84, 88, 91. On the Count, see Lister, 'Joseph Storr Lister,' 15.
26. Report of the Superintendent of Woods and Forests, Cape of Good Hope (henceforth C.G.H. followed by year reported on, then publication number), C.G.H.1881, G1–'82, 30–33, p.45.
27. C.G.H.1883, G34–'84, p.5.
28. C.G.H.1882, G1–'82, pp.5, 15.
29. C.G.H.1883, G34–'84, p.6.
30. C.G.H.1882, G1–'82, pp.5, 15–16, 38.
31. C.G.H.1883, G34–'84, p.21.
32. C.G.H.1884, G32–'85, p.6; C.G.H.1885, G34–'86, pp.6–7, C.G.H.1886, G41–'87, p.6; C.G.H.1887, G45–'88, pp.12–13.
33. On the origins and influences on the 1888 Act, see G.A. Barton, 2002, *Empire Forestry and the Origins of Environmentalism* (Cambridge: Cambridge University Press), pp.102–4. For Hutchins' description of the Midland Conservancy, see Sim, *Forests and Forest Flora*, pp.11–12.
34. Sim, *Forests and Forest Flora*, pp.51, 79; A.M. Avis, 1989, 'A Review of Coastal Dune Stabilization in the Cape Province of South Africa,' *Landscape and Urban Planning*, 18, 55–68. For Harison's report, see C.G.H.1890, p.61.
35. C.G.H.1886, G41–'87, p.12.
36. C.G.H.1892, G27–93, p.16. On buck, C.G.H.1893, G50–94, p.30.
37. C.G.H.1892 27–93, p.17.

38. C.G.H.1892 27–93, pp.17–18. Regarding Lister's recommendation for dense planting, Bennett relates that by the 1930s South African foresters were being criticised for the wide spacing of their plantations. See B.M. Bennett and F.J. Kruger, 2013, 'Ecology, Forestry and the Debate over Exotic Trees in South Africa,' *Journal of Historical Geography*, 42, 100–9, 103.
39. C.G.H.1893 G.50–94, pp.30–31.
40. C.G.H.1893 G.50–94, p.31. At Knysna, James Cooper reported that while the best time to burn was when the 'Berg winds' (off the mountain) were starting up after rains, burning season happened during felling season when the Forest Officer was very busy. Thus protective belts of oaks and well-timed veld burning was the best compromise. In 1890, 22 forest fires caused £507 worth of damage (and £32 to extinguish) in the Knysna forests. C.G.H.1890 G.41–91, p.94.
41. C.G.H.1893 G.50–94, p.31.
42. Ibid., p.32.
43. C.G.H.1894 G51–95, pp.28–29. George Perkins Marsh, 1864, *Man and Nature: Or, Physical Geography as Modified by Human Action* (New York: Charles Scribner), p.273.
44. C.G.H.1896 G17–97.

3   Science, Management and Fire in Fynbos: 1900–45

1. R. Marloth, 1924, 'Notes on the Question of Veld Burning,' *South African Journal of Science* (henceforth *SAJS*), 21, 342–5, 343–4.
2. C.L. Wicht, 1944, 'Controlled Burning in the Jonkershoek Forestal District,' Memorandum to The Conservator of Forests, Cape Town, M.600, 22 March, 1–3, 2.
3. Harry Bolus to Col. Prain, Director's Correspondence, 192, 31–43, Kew Gardens Library and Archive (29 April 1908), p.4. On Bolus, see H.G. Glen and G. Germishuizen (compilers), 2010, *Botanical Exploration of Southern Africa*, 2nd ed., Strelitzia 26 (Pretoria, South Africa: South African National Biodiversity Institute), p.102.
4. Anon, 1914, 'Seventh Award of the South Africa Medal and Grant,' *SAJS*, 11, xxxii–vi; E.P. Phillips, 1930, 'A Brief Historical Sketch of the Development of Botanical Science in South Africa,' *SAJS*, 27, 39–80, 51, 62.
5. Marloth, 'Veld Burning,' 343–4.
6. On his origins, G.W. Gale, 1954, *John William Bews: A Memoir* (Pietermaritzburg, South Africa: University of Natal Press), pp.9–11, and on heather burning, p.15.
7. On Bews's life, see G.W. Gale, 1954, *John William Bews* and M. Gunn and L.E. Codd, 1981, *Botanical Exploration of Southern Africa* (Cape Town, South Africa: A.A. Balkema), pp.93–4.
8. See Geo. Potts, 1923, 'The Plant Succession in the Orange Free State, and the Need for Maintaining a Covering of Vegetation,' *SAJS*, 20, 196–201, 196. On Moss: R.H. Compton, 1930, 'Notes and News,' *Journal of the Botanical Society of South Africa* (henceforth *JBSSA*), 16, 3; Gunn and Codd, *Botanical Exploration*, p.254.

9. Gale, *Bews*, pp.52, 54, 88–9.
10. J.W. Bews, 1912, 'The Vegetation of Natal,' *Annals of the Natal Govern-ment Museum*, 2, 3, 253–331; J.W. Bews, 1913, 'An Ecological Survey of the Midlands of Natal,' *Annals of the Natal Government Museum*, 2, 4, 485–545.
11. J.W. Bews, 1916, 'An Account of the Chief Types of Vegetation in South Africa, with Notes on the Plant Succession,' *The Journal of Ecology*, 4, 129–59. P. Anker, 2001, *Imperial Ecology: Environmental Order in the British Empire, 1895–1945* (Cambridge, MA: Harvard University Press), p.59. On Clements' influence on plant ecology in this period, and his *Research Methods*, see J.W. Bews, 1935, *Human Ecology* (London: Oxford University Press), p.3; D. Worster, 1994, *Nature's Economy: A History of Ecological Ideas*, 2nd ed. (Cambridge: Cambridge University Press), pp.208–36.
12. J.W. Bews, 1918, *The Grasses and Grasslands of South Africa*, (Pietermaritzburg, South Africa: P. Davis & Sons); for Bayer's comment, see Gale, *Bews*, pp.121–2.
13. Gale, *John William Bews*, pp.64–5, for the Smuts quote, 82.
14. H.S.D. du Toit, S.M. Gadd, G.A. Kolbe, A. Stead, and R.J. van Reenen, 1923, 'Final Report of the Drought Investigation Commission,' U.G.49-'23 (Cape Town, South Africa: Government of the Union of South Africa), 3, 5, 62. For Cape botanists criticising veld burning at the 1924 symposium, see *SAJS*, 21, 1924: R. Marloth, 1924, 'Notes on the Question of Veld Burning,' 342–5; M.R. Levyns, 'Some Observations of the Effect of Bush Fires on the Vegetation of the Cape Peninsula,' 346–7; N.S. Pillans, 'Destruction of Indigenous Vegeta-tion by Burning on the Cape Peninsula,' 348–50. See R.S. Adamson, 1927, 'The Plant Communities of Table Mountain: Preliminary Account,' *Journal of Ecology*, 15, 2, 278–309, 304.
15. On Adamson's life, see H.G. Glen and G. Germishuizen (compilers), 2010, *Botanical Exploration*, p.74; P. Star, 2006, 'Ecology: A Science of Nation? The Utilization of Plant Ecology in New Zealand, 1896–1930,' *Historical Records of Australian Science*, 17, 197–207; G. Duncan, 2003, 'Lachenalia salteri *Hyacinthaceae*,' *Curtis's Botanical Magazine*, 20, 4, 208–9.
16. Adamson, 'The Plant Communities of Table Mountain,' 284–5.
17. Ibid., 304.
18. On Levyns, see J.P. Rourke, 1999, 'Plant Systematics in South Africa: A Brief Historical Overview, 1753–1953,' *Transactions of the Royal Society of South Africa* (hereafter *TRSSA*), 54, 1, 179–90; Glen and Germishuizen (compilers), 2010, *Botanical Exploration*, p.266. The dedication is in R.M. Cowling (ed.), 1992, *The Ecology of Fynbos: Nutrients, Fire and Diversity* (Cape Town, South Africa: Oxford University Press).
19. M.R. Michell (later Levyns), 1922, 'Some Observations on the Effects of a Bush Fire on the Vegetation of Signal Hill,' *TRSSA*, 10, 1, 346–7; M.R. Levyns, 1929, 'Veld-burning Experiments at Ida's Valley, Stellenbosch,' *TRSSA*, 17, 2, 61–92.
20. Bews, 'Account of the Chief Types of Vegetation in South Africa,' 147.
21. See A. von Hase, M. Rouget, K. Maze, and N. Helme, 2003, 'A Fine-Scale Conservation Plan for Cape Lowland Renosterveld: Technical Report,' Cape Conservation Unit Report CCU 2/03, p.1.
22. R.M. Cowling and P.M. Holmes, 1992, 'Flora and Vegetation,' in R.M. Cowling (ed.), *The Ecology of Fynbos*, pp.41–2.

23. L. van Sittert, 2003, 'Making the Cape Floral Kingdom: The Discovery and Defence of Indigenous Flora at the Cape ca. 1890–1939,' *Landscape Research*, 28, 113–29; R.H. Compton, 1924, 'News and Notes,' *JBSSA*, 10, 3.

24. R.H. Compton, 1926, 'Veld Burning and Veld Deterioration,' *South African Journal of Natural History*, 6, 5–19, 11. See S. Pooley, 2010, 'Pressed Flowers,' for an extended discussion.

25. Compton, 'Veld Burning and Veld Deterioration,' 6–7, 9–11.

26. Ibid., 13.

27. R.H. Compton, 1927, 'Notes and News,' *JBSSA*, 12, 4–5.

28. H.D. Leppan, J.M. Hector, W.R. Thompson, T.D. Hall, A. Lindsay Robb, and D. Moses, 1932, 'The Grasslands of South Africa: Problems and Possibilities,' University of Pretoria Series 1, 23, (Pretoria, South Africa: University of Pretoria); S. Dubow, 2006, *A Commonwealth of Knowledge: Science, Sensibility, and White South Africa 1820–2000* (Oxford: Oxford University Press), p.180; 'Notes,' 1920, *Journal of the Department of Agriculture (JDA)*, 1, 7, 605–6.

29. W.J. Bond, G.F. Midgley, and F.I. Woodward, 2003, 'What Controls South African Vegetation – Climate or Fire?,' *SAJB*, 69, 1, 1–13, 7–8. On veld burning debates in the agricultural journals: S. Pooley, 2010, 'An Environmental History of Fire in South Africa in the Twentieth Century' (DPhil thesis, University of Oxford), 28–33.

30. Pooley, 'Environmental History of Fire,' 28–33.

31. S. Dubow, 1992, 'Afrikaner Nationalism, Apartheid, and the Question of "Race",' *Journal of African History*, 33, 209–337.

32. Dubow, *Commonwealth of Knowledge*, pp.165, 178. On the 'cult of the veld' and Fitzpatrick, see pp.185–7. See also K. Brown, 2002, 'Progressivism, Agriculture and Conservation in the Cape Colony circa 1902–1908' (DPhil thesis, University of Oxford). For an entertaining account of du Toit's career, see W. Beinart, 2003, *The Rise of Conservation in South Africa: Settlers, Livestock, and the Environment 1770–1950* (Oxford: Oxford University Press), Chapter 7.

33. W. Beinart, 1984, 'Soil Erosion, Conservationism and Ideas about Development: A Southern African Exploration, 1900–1960,' *Journal of Southern African Studies*, 11, 1, 52–83, see 54–61, 67–8. The Smuts quote is on p.68.

34. Anker, *Imperial Ecology*, especially pp.54–65. On Smuts' authorising pasture research stations in Natal, see J.D. Scott, 1948, 'A Contribution to the Study of the Problems of the Drakensberg Conservation Area' (PhD thesis, University of the Witwatersrand), 2. On Smuts' final musings on grass, and T.C. Robertson: S. Bell, 2005, *The Happy Warrior: The Story of T.C. Robertson* (Howick, South Africa: T.C. Robertson Trust), p.265.

35. K. Brown, 2001, 'The Conservation and Utilisation of the Natural World: Silviculture in the Cape Colony, c.1902–1910,' *Environment and History*, 7, 4, 427–47, 430–2.

36. Synthesised from the annual reports of the department, and see H.A. Witt, 1998, 'Trees, Forests and Plantations: An Economic, Social and Environmental Study of Tree-Growing in Natal, 1890–1960' (PhD thesis, University of Natal).

37. Government of the Union of South Africa (U.G.), Forest Department, Report of the Chief Conservator of Forests (henceforth ARFD) 1932/33, U.G.35–1933.
38. W. Schlich, 1906, *Forest Policy in the British Empire, Schlich's Manual of Forestry*, Vol.1, 3rd ed. (Bradbury: Agnew & Co.), p.139.
39. Protection measures discussed in Chapter 2. C.E. Legat, 1913, 'Protection from Fire,' Union of South Africa, Forest Department Memorandum, 25 August 1913.
40. J.S. Henkel, 'Prevention of forest fires,' Memorandum to the District Forest Officer, Cape,' Forest Department Memorandum 7244/146, 13 November 1907; J.S. Storr Lister, 'Memorandum of the Provisions of the Forest Acts and Regulations Framed Thereunder Relative to Forest Fires,' Memorandum 243, 1 February 1908. On evictions and punishments of Africans by foresters in the Eastern Cape, see K. Brown, 2001, 'Conservation and Utilisation of the Natural World'; J.A. Tropp, 2006, *Nature's of Colonial Change: Environmental Relations in the Making of the Transkei* (Athens, OH: Ohio University Press).
41. N.L. King, 1938, 'Historical Sketch of the Development of Forestry in South Africa,' *JSAFA*, 1, 4–16; see 'South Africa' in B.M. Bennett, 2011, 'A Global History of Australian Trees,' *Journal of the History of Biology*, 44, 1, 125–45; Witt, 'Trees, Forests and Plantations'; B.M. Bennett, 2013, 'The Rise and Demise of South Africa's First School of Forestry,' *Environment and History* 19, 63–85.
42. For the United States see S.J. Pyne, 1997, *Fire in America: A Cultural History of Wildland and Rural Fire* (Seattle, WA: University of Washington Press), p.468; J.F.V. Phillips, 1928, 'Plant Indicators in the Knysna Region,' *SAJS*, 25, 202–24; J.F.V. Phillips, 1931, 'Forest Succession and Ecology in the Knysna Region,' *Memoirs of the Botanical Survey of South Africa*, 14.
43. J.F.V. Phillips, 1954, 'A Tribute to Frederic E. Clements and his Concepts in Ecology,' *Ecology*, 35, 2, 114–15.
44. J.F.V. Phillips, 1928, 'The Behaviour of *Acacia Melanoxylon* R. Br. ("Tasmanian Blackwood") in the Knysna Forests: An Ecological Study,' *TRSSA*, 16, 1, 31–43; J.F.V. Phillips, 1928, 'Plant indicators,' *SAJS*, 25, 202–24.
45. On fire in forests, see C.J. Geldenhuys, 1977, 'Bergwind Fires and the Location of Forest Patches in the Southern Cape Landscape, South Africa,' *Journal of Biogeography*, 21, 1, 49–62.
46. J.M. Powell, 2007, ' "Dominion over Palm and Pine": The British Empire Forestry Conferences, 1920–1947,' *Journal of Historical Geography*, 33, 852–77. For West Africa, New Zealand and Canada, see: Anon, 1924, 'Preliminary Reports on the Forest Requirements of the Eastern and Central Provinces, Gold Coast Colony' (Accra, Ghana: Government Press), p.6; R. Heaton Rhodes and L. MacIntosh Ellis, 1923, 'Forests and Forestry in New Zealand: Prepared for the Imperial Forestry Conference, Ottawa, 1923,' p.17; D.R. Cameron, 1928, 'Forest Fire Protection in Canada: Progress since 1923,' in *Proceedings of the Third British Empire Forestry Conference*, pp.721–33.
47. W.R. Fisher, 1907, *Forest Protection, Dr Schlich's Manual of Forestry*, Vol.4, 2nd ed. (Bradbury: Agnew & Co.), pp.638–68.

48. C.E. Lane-Poole, 1928, 'Australia: Commonwealth handbook' (H.J. Green, Government Printer, Canberra, Australia); N.W. Folly, 'Forestry in New South Wales,' in Anon, 1928, 'Forestry Handbook for New South Wales, Forestry Commission of New South Wales' (Sydney, Australia: Alfred James Kent, Government Printer), pp.13–47; S. L. Kessell, 1928, 'Fire Control in Australia,' *Proceedings of the Third British Empire Forestry Conference*, Australia and New Zealand, pp.741–8.

49. E.O. Shebbeare, 1928, 'Fire Protection and Fire Control in India,' *Proceedings of the Third British Empire Forestry Conference*, Australia and New Zealand, pp.715–20; see also C.E. Simmons, 1928, 'Forests Management in British India (exclusive of Burma),' "papers presented" in Third British Empire Forestry Conference: papers presented (Canberra, Australia: H.J. Green, Government Printer), pp.572–81.

50. See, for instance, Cameron, 'Forest Fire Protection in Canada'; J.R. Ainsilie (Nigeria), comments in the section 'Tropical Forestry in Relation to Agriculture,' in Anon, 1935, *Proceedings of the fourth British Empire Forestry Conference* (henceforth *BEFC 4*), pp.217–19; J.B. Clements (Nyasaland), comments in the section 'Tropical Forestry in Relation to Agriculture,' Anon, 1935, *BEFC 4*, pp.214–15; A.V. Galbraith, 1947, 'Statement Prepared by the Forest Commission of Victoria, Australia' on 'Empire Forests and the War' (Melbourne, Australia: J.J. Gourley, Government Printer).

51. ARFD1929/30 U.G.49/1930, p.3; W.E. Watt, 'Forest Protection,'3 and J.D.M. Keet, 1935, 'Soil Erosion and Allied Problems in South Africa,' 6–7, both supplements to the *BEFC 4*, South Africa, 1935 (Pretoria, South Africa: Government Printer, 1936).

52. On the resolution, see ARFD1936/37 U.G.53/1937, p.7; on fire incidence: ARFD1935/36 U.G.53–1936, p.16; on catchments, and surveys, ARFD1935/36 U.G.53–1936, p.13; ARFD1936/37 U.G.53–1937, p.7; J.C. Ross, 1961, 'Report of the Interdepartmental Committee on the Conservation of Mountain Catchments in South Africa' (Pretoria, South Africa: Department of Agricultural Technical Services), 38–9.

53. J.D. Keet, 1935, 'Afforestation in Relation to Natural Flora and the Effect of Afforestation on Water Supplies,' Letter to All Professional Officers, available at: http://digi.nrf.ac.za/dspace/handle/10624/390.

54. Ibid.

55. On the politics of the conference, see B.M. Bennet and F.J. Kruger, 2013, 'Ecology, Forestry and the Debate over Exotic Trees in South Africa,' *Journal of Historical Geography* 42, 100–109. They suggest this conference was the first place that the department's afforestation policies were publically criticised and that this represents a 'rupture, which ... had lasting ramifications on the institutional and intellectual history of the environmental sciences in South Africa.' For J.D. Keet's comments, see 'Forest Influences,' Anon, 1935, *BEFC 4*, pp.114–17.

56. See Phillips' comments, see 'Forest Influences,' in *BEFC 4*, 181; for Pole-Evans' views, see his letter to Smuts of 8 March 1935, asking Smuts to intervene with Reitz, Smuts Collection, University of the Witwatersrand, 23853.

57. On Cajander and Phillips, see Bennet and Kruger, 'Ecology, Forestry and the Debate over Exotic Trees in South Africa,' 105–6.

58. For criticism of Phillips and Smuts, see *BEFC 4* as follows: Mr G. A. Wilmot, 'Union of South Africa,' 135–6; C.E. Lane-Poole (Australia), p.136; P.C. Kotzé, 'Forest Influences,' pp.176–8.
59. Anon, 1935, *BEFC 4*, 14–16; C.L. Wicht, 1948, 'Hydrological Research in South African Forestry,' *JSAFA*, 16, 4–21, 4.
60. On Robertson, see C. Plug, 'Robertson, Mr Colin Charles (forestry),' S2A3 Biographical Database of Southern African Science, available at: http://www.s2a3.org.za/bio/Biograph_final.php?serial=2366, accessed 2 December 2013; on Wicht: Gunn and Codd, *Botanical Exploration*, p.376.
61. C.L. Wicht, 1937, 'Research on Forest Influences: Work being Done at Jonkershoek, Stellenbosch,' extract from *Farming in South Africa*, available at: http://digi.nrf.ac.za/dspace/handle/10624/388.
62. C.L. Wicht, 1939, 'Forest Influences Research Technique at Jonkershoek,' *JSAFA*, 3, 65–80, 65; D. Bands, J.M. Bosch, A.J. Lamb, D.M. Richardson, B.W. van Wilgen, D.B. van Wyk, and D.B. Versfeld, 1987, *Jonkershoek Forestry Research Centre*, Pamphlet 384 (Pretoria, South Africa: Department of Environment Affairs), p.7.
63. C.L. Wicht, 1948, 'A Statistically Designed Experiment to Test the Effects of Burning on a Sclerophyll Scrub Community. 1. Preliminary Account,' *TRSSA*, 31, 479–501, 492–4. On the long view: 1937, 'Research on forest influences,' 3.
64. W.E. Watt, in *BEFC 4*, 186. On average annual expenditure: Republic of South Africa (R.P.): ARFD 1959/60 R.P.20/1961, p.5.
65. C.L. Wicht, 1945, *Preservation of the Vegetation of the South Western Cape*, Report for The Royal Society of South Africa (Cape Town); British Ecological Society's 1944 report is cited on page 9; J.S. Henkel, 1943, 'Preservation of Mountain Vegetation of the South-West Districts of the Cape Province,' Memorandum to the Royal Society of South Africa, pp.1, 3, 5.
66. R.S. Adamson, 1945, 'Characteristics of Vegetation,' in C.L. Wicht, 1945, *Preservation of the Vegetation*, pp.19, 22.
67. Ibid., pp.22–3.
68. Ibid., pp.23–4. On lightning in the annual reports, ARFD1917/18 U.G. 57–1918, p.11. The first explicitly linked to fire in fynbos was in 1933/34: ARFD1933/34 U.G. 45–1934, p.15. Fires were attributed to falling rocks in ARFD1917/18 U.G. 57–1918 and ARFD1919/20 U.G. 7-'21, p.9.
69. Adamson, 'Characteristics of Vegetation,' 24–5.
70. Ibid., pp.24–6; P. Slingsby and A. Jones, 2009, *T.P. Stoekoe: The Man, the Myths, the Flowers* (Cape Town, South Africa: Baardskeerder), pp.127–8, 142.
71. Adamson, 'Characteristics of Vegetation,' 31–2.
72. Ibid., 27.
73. Ibid., 26–9.
74. Ibid., 30–1.
75. Ibid., 38.
76. Ibid., 39–40.
77. Ibid., 40–2.
78. Ibid., 43–5.
79. C.L. Wicht, 1944, 'Controlled Burning in the Jonkershoek Forestal District.' For policy in the early 1940s, see ARFD1939/40 U.G.44–1940, pp.5–6.
80. Ibid., p.2.
81. Ibid.

4 Science, Management and Fire in Fynbos: 1945–99

1. H.H. Bennett, 1945, *Soil Erosion and Land Use in the Union of South Africa* (Pretoria, South Africa: Department of Agriculture and Forestry), p.23.
2. C.L. Wicht, 1948, 'The Role of South African Forestry in the Conservation of Natural Resources,' corrected proof, paper presented at Conference Africaine des Sols, Goma, Congo Belge, November 1948, p.2.
3. F.H. Wroughton, 1948, 'To Burn or Not to Burn,' *JSAFA*, 16, 76–8; C.L. Wicht, 1948, 'Hydrological Research in South African Forestry,' *JSAFA*, 16, 4–21, 12.
4. For the Wicht report's recommendations on prescribed burning, see 40–1. On the Department's decision and failure to implement it, see A.H.W. Seydack and S.J. Bekker, 1993, 'Review of Fynbos Catchment Management Policy,' Unpublished Report, Republic of South Africa, Department of Water Affairs and Forestry, copy from author Seydack, pp.1–2.
5. H.H. Bennett, 1945, 'Soil Erosion': On Veld Burning, pp.9, 22–4, and on technical staff, pp.25–6. See Bennett's farewell message, published in *Libertas* by T.C. Robertson as 'South Africa, I pity you,' cited in S. Bell, 1985, *The Happy Warrior: The Story of T.C. Robertson* (Howick, South Africa: T.C. Robertson Trust), p.85.
6. C.T. te Water, preface to E. Roux, 1946, *The Veld and the Future* (Cape Town, South Africa: The African Bookman), unnumbered page. Roux advocated the use of fire in humid grasslands to manipulate the plant succession, 23–6.
7. See Bell, *The Happy Warrior*, pp.4, 18, 20, 46.
8. Ibid., pp.76–9.
9. Ibid., pp.115–17.
10. C.L. Wicht, 1948, 'Role of South African Forestry,' p.2. The title Director of Forestry changed in 1958 when D.R. de Wet became Secretary for Forestry. A.E. Sonntag, Secretary from 1979, became Deputy Director-General of the Directorate of Forestry and Environmental Conservation in 1982. In the 1980s the Department became a branch of the Department of Environment Affairs.
11. F.J. Kruger, 1979, 'Fire,' in J.A. Day, N.R. Siegfried, and M.L. Jarman (eds.), *Fynbos Ecology: A Preliminary Synthesis*, CSIR Report No.40 (Pretoria, South Africa: CSIR), pp.43–57, 45. On conservancies: ARFD1955/56 U.G.59/1958, p.19; ARFD1962/63 R.P.58/1965, p.25.
12. S. Jones and A. Muller, 1992, *The South African Economy, 1910–90* (Hampshire: Palgrave Macmillan), pp.189, 306. For locomotive fires, see ARFD R.P.20/1961, p.12.
13. For the forested area in 1957/58, see ARFD1957/58 U.G.28/1960, p.17; for staff complement in this period: ARFD1957/58 U.G.28/1960, pp.14–15. On staff shortages: ARFD1954/55 U.G.35/1956, p.1; ARFD1955/56 U.G.59/1958, p.3.
14. ARFD1954/55 U.G.35/1956, pp.1, 3; Seydack and Bekker, 'Review,' p.2.
15. J.C. Ross, 1963, *Soil Conservation in South Africa* (Pretoria, South Africa: Department of Agricultural Technical Services), p.44.
16. Seydack and Bekker, 'Review,' p.1.
17. S.J. Pyne, 1997, *Vestal Fire: An Environmental History, told through Fire, of Europe and Europe's Encounter with the World* (Seattle, WA: University of Washington Press), p.538.

18. H.B. Rycroft, 1947, 'A Note on the Immediate Effects of Veldburning on Stormflow in a Jonkershoek Stream Catchment,' *JSAFA*, 15, 80–8.
19. J.C. Ross, 1961, *Report of the Interdepartmental Committee on the Conservation of Mountain Catchments in South Africa* (Pretoria, South Africa: Department of Agricultural Technical Services), pp.1–2, 12–14, 18, 29.
20. C.L. Wicht, 1958, 'The Management of Water Catchments,' Technical Report for the Department of Forestry, pp.1–2, available at: http://digi.nrf.ac.za/dspace/handle/10624/434.
21. Ibid., pp.2–4.
22. Ibid., pp.6–7.
23. C.L. Wicht, 1960, 'Die invloed van beheerde brand en bebossing op die waterleweringspotensiaal van die bergopvanggebiede in die winter-reënstreek,' confidential technical report, available at: http://digi.nrf.ac.za/dspace/handle/10624/411; C.H. Banks, 1964, 'Further Notes on the Effect of Autumnal Veldburning on Stormflow in the Abdolskloof Catchment, Jonkershoek,' *Forestry in South Africa*, 4, 79–84.
24. Republic of South Africa, Annual Reports of the Department of Forestry (R.P.), R.P.80/1964, pp.1, 10; R.P.10/1967, p.42, and see M. Grut, 1965, *Forestry and Forest Industry in South Africa* (Amsterdam, the Netherlands: A.A. Balkema), p.31.
25. The research stations were: Jonkershoek (Eerste Rivier), Jakkalsrivier in the Caledon district (Bot River), and Zachariashoek in the Paarl district (Wemmer and Berg rivers). Interview with Fred Kruger, Pretoria, South Africa, 6 September 2007; On Keet: M. Gunn and L.E. Codd, 1981, *Botanical Exploration of Southern Africa* (Cape Town, South Africa: A.A. Balkema), p.204; On the Lückhoffs: H.G. Glen and G. Germishuizen (compilers), 2010, *Botanical Exploration of Southern Africa*, 2nd ed., Strelitzia 26 (Pretoria, South Africa: South African National Biodiversity Institute), p.275.
26. H.L. Malherbe, E.R. March, U.W. Nanni, C.E.M. Tidmarsh, F.S. Grevenstein, J.C. Cox, and J.S. Whitmore, 1968, 'Report of the Interdepartmental Committee of Investigation into Afforestation and Water Supplies in South Africa,' Unpublished Report, Plant Sciences Library, University of Oxford, Ref.83313.
27. H.L. Malherbe, E.R. March, U.W. Nanni, C.E.M. Tidmarsh, F.S. Grevenstein, J.C. Cox, and J.S. Whitmore, 1968, 'Report of the Interdepartmental Committee.' On the Forest Act, see ARFD1963/64 R.P.34/1966, p.13.
28. ARFD R.P.18/1972, p.8. On agricultural research and continuing anti-burning prejudice, see J.D. Scott, 1970, 'Pros and Cons of Eliminating Veld Burning,' *Proceedings of the Grassland Society of South Africa*, 5, 23–6; S. Pooley, 2010, 'An Environmental History of Fire in South Africa in the Twentieth Century' (DPhil thesis, University of Oxford), pp.64–75. On the implementation of burning policy, see Seydack and Bekker, 1993, 'Review,' p.4.
29. P.G. Jordaan, 1949, 'Aantekeninge oor die voortplanting en brandperiodes van *Protea mellifera* Thunb.,' *The Journal of South African Botany*, 15, 4, 121–5, 124, 125. In 1935 Watt had considered wildfires every eight to ten years to be acceptable: W.E. Watt, 1935, 'Forest Protection from Fire,' supplement to *BEFC 4*, pp.3–4.
30. ARFD1952/53 U.G.47/1954, p.3; ARFD1953/54 U.G.38/1955, p.3. A.V. Hall, 1959, 'Observations on the Distribution and Ecology of Orchidaceae in the

Muizenberg Mountains, Cape Peninsula,' *Journal of South African Botany*, 25, 265–78.

31. B.W. van Wilgen, 2009, 'The Evolution of Fire and Invasive Plant Management Practice in Fynbos,' *SAJS*, 105, 335–42, 336; S.W. Worth and B.W. van Wilgen, 1988, 'The Blushing Bride: Status of an Endangered Species,' *Veld & Flora*, 74, 122–3, 122.

32. J. Donaldson and A.G. Rebelo, 2009, 'Draft Proposal for Amendment of Appendices I and II: *Orothamnus Zeyheri*,' CITES, Eighteenth Meeting of the Plants Committee, PC18 Inf.11 (Buenos Aires, 17–21 March 2009), 3. Interview with Dr John Rourke, at his home in Constantia, Cape Town, on 26 October 2007.

33. ARFD1971/72 R.P.32/1973, p.7; ARFD1973/74 R.P.42/1975, p.7. On Conservation Forestry research: H.A. Lückhoff and C.L. Wicht, 1975, 'The Influence of Demand and Supply Trends on the Research Programme of the Forest Research Institute, Pretoria, Republic of South Africa,' paper presented at an International Union of Forest Research Organizations (IUFRO) conference, Paris, pp.1–7.

34. On Our Green Heritage, see ARFD1972/73 R.P.38/1974, p.7.

35. C.L. Wicht and F.J. Kruger, 1973, 'Die Ontwikkeling van Bergveldbestuur in Suid-Afrika,' *SAFJ*, 86, Special Issue No.2, 1–17, see 12–16. On landsdiens: http://www.landsdiens.co.za/landsdiens_link.php?id=5.

36. ARFD1970/71 R.P.18/1972, pp.8, 29; ARFD1971/72 R.P.32/1973, p.33; ARFD1972/73 R.P.38/1974, pp.33, 35.

37. C.L. Wicht, 1973, 'Die uitwerking van brand op die fynbosveld,' first of a series of lectures entitled 'Brand in die Fynbosveld,' available at: http://digi.nrf.ac.za/dspace/handle/10624/421.

38. C.L. Wicht, 1973, 'Fire: Evil or Asset,' second in the series 'Brand in die Fynbosveld,' available at: http://digi.nrf.ac.za/dspace/handle/10624/421.

39. C.L. Wicht, 1973, ''n stelsel van veldbrandbeheer,' third in the series 'Brand in die Fynbosveld,' available at: http://digi.nrf.ac.za/dspace/handle/10624/421.

40. J.B. Hagan, 2008, 'Teaching Ecology during the Environmental Age, 1965–80,' *Environmental History*, 13, 704–23, 705–6, 708.

41. D. Worster, 1994, *Nature's Economy: A History of Ecological Ideas*, 2nd ed. (Cambridge: Cambridge University Press), pp.388–411.

42. IBP position statement cited in C. Kwa, 1987, 'Representations of Nature Mediating Between Ecology and Science Policy: The Case of the International Biological Programme,' *Social Studies of Science*, 17, 3, 413–42, 415; On the IBP see also Worster, *Nature's Economy*, pp.372–3; H.A. Mooney, 1998, *The Globalization of Ecological Thought* (Oldendorf/Luhe, Germany: Ecology Institute), pp.32, 48.

43. Mooney, *Globalization of Ecological Thought*, pp.33, 4–5; Jon E. Keeley, W.J. Bond, R.A. Bradstock, J.G. Pausas, and P.W. Rundel, 2012, *Fire in Mediterranean Ecosystems: Ecology, Evolution and Management* (Cambridge: Cambridge University Press), pp.11, 27.

44. H.A. Mooney, 1982, 'Applied and Basic Research in Mediterranean-Climate Ecosystems,' in C.E. Conrad and W.C. Oechel (Technical Coordinators), 'Symposium on Dynamics and Management of Mediterranean-Type Ecosystems,' San Diego State University, General Technical Report PSW-58

(Berkeley, CA: Pacific Southwest Forest and Range Experiment Station), pp.8–12, 8.

45. H.A. Mooney and C.E. Conrad (eds.), 1977, *Proceedings of the Symposium on the Environmental Consequences of Fire and Fuel Management in Mediterranean-Climate Ecosystems*, Tech Report WO-3 (Washington, DC: U.S. Department of Agriculture); F.J. Kruger, 1977, 'Ecology of Cape Fynbos in Relation to Fire,' in H.A. Mooney and C.E. Conrad (eds.), 1977, 'Symposium on the Environmental Consequences of Fire,' pp.230–44.

46. F.J. Kruger, D.T. Mitchell, and J.U.M. Jarvis, (eds.), 1983, 'Preface,' in *Mediterranean-Type Ecosystems: The Role of Nutrients* (Berlin, Germany: Springer-Verlag), pp.v, vi.

47. W.J. Bond, 1983, 'On Alpha Diversity and the Richness of the Cape Flora: A study in Southern Cape Fynbos,' in F.J. Kruger, D.T. Mitchell, and J.U.M. Jarvis (eds.), *Mediterranean-Type Ecosystems*, pp.336–56, 351–2.

48. F.J. Kruger, 1983, 'Plant Community Diversity and Dynamics in Relation to Fire,' in F.J. Kruger, D.T. Mitchell, and J.U.M. Jarvis (eds.), *Mediterranean-Type Ecosystems*, pp.446–72.

49. Ibid., pp.447–48; B.W. van Wilgen, 1980, 'Some Effects of Fire Frequency on Fynbos at Jonkershoek, Stellenbosch, South Africa' (MSc dissertation, University of Cape Town), p.452.

50. Kruger, 'Plant Community Diversity,' pp.448, 450. On the importance of the 'fire regime' framework, and an updated version of its components: Keeley et al., *Fire in Mediterranean Ecosystems*, pp.30–2.

51. P.G. Jordaan, 1965, 'Die Invloed van 'n Winterbrand op die Voortplanting van Vier Soorte van die Proteaceae,' *Tydskrif vir Natuurwetenskappe*, 5, 27–31.

52. On Kruger's influences, personal communication, by email, 31 January 2011. Interview with Brian van Wilgen, Skukuza, 11 September 2007; B.W. van Wilgen, 1980, 'Some Effects of Fire Frequency on Fynbos.'

53. Kruger, 'Plant Community Diversity,' pp.466, 467.

54. F.J. Kruger, 1982, 'Prescribing Fire Frequencies in Cape Fynbos in Relation to Plant Demography,' in C.E. Conrad and W.C. Oechel (Technical Coordinators), 'Symposium on Dynamics and Management of Mediterranean-Type Ecosystems,' San Diego State University, General Technical Report PSW-58 (Berkeley, CA: Pacific Southwest Forest and Range Experiment Station), pp.118–22.

55. Ibid.

56. Interview with W.J. Bond, Lakeside, Cape Town, 27 October 2007.

57. W.J. Bond, 1984, 'Fire Survival of Cape Proteaceae – Influence of Fire Season and Seed Predators,' *Vegetatio*, 56, 65–74, 73. For a summary of the rainfall regions, see G. Forsyth and B.W. van Wilgen, 2007, 'An Analysis of the Fire History Records from Protected Areas in the Western Cape,' CSIR Report CSIR/NRE/ECO/ER/2007/0118/C (Stellenbosch, CSIR), pp.1–2.

58. Worster, *Nature's Economy*, p.375.

59. Interview with W.J. Bond; W.J. Bond and B.W. van Wilgen, 1996, *Fire and Plants* (London: Chapman & Hall).

60. J.L. Harper, 1967, 'A Darwinian Approach to Plant Ecology,' *Journal of Ecology*, 55, 2, 247–70, 248, 253, 255, 259–60, 263.

61. J.L. Harper, 1977, *Population Biology of Plants* (London: Academic Press), pp.xiii–xxii for a summary.

62. Interview with Professor Jeffrey Burley, Green College, Oxford, 5 March 2008; P. de V. Booysen, 1984, 'Preface,' in P. de V. Booysen and N.M. Tainton (eds.), *Ecological Effects of Fire in South African Ecosystems* (Berlin, Germany: Springer Verlag). Trollope encountered American foresters' preoccupation with fire intensity at the 1971 Tall Timbers Fire Ecology conference organised in Tallahassee, Florida, by Edwin Komarek Snr. and dedicated to South African ecologist John Phillips's contribution to fire ecology in Africa. See W.S.W. Trollope, 1984, 'Characteristics of Fire Behaviour,' in P. de V. Booysen and N.M. Tainton (eds.), *Ecological Effects of Fire*, pp.200–17.

63. Interview with Prof. Brian Huntley, South African National Botanical Institute, Kirstenbosch, South Africa, 9 October 2007. For an overview of the Fynbos Biome Project, see Brian J. Huntley, 1992, 'The Fynbos Biome Project,' in R.M. Cowling (ed.), *The Ecology of Fynbos: Nutrients, Fire and Diversity* (Cape Town, South Africa: Oxford University Press), pp.1–5; F.J. Kruger, 1978, 'Description of the Fynbos Biome Project,' National Scientific Programs Unit: CSIR, South African National Scientific Programs Report 28. Regarding MAREP: interview with Christo Marais, Cape Town, 3 October 2008.

64. On the McArthur school see Worster, *Nature's Economy*, p.375. On the influence of Harper and McArthur on thinking about fynbos, interview with W.J. Bond.

65. R.M. Cowling, P.M. Holmes, and A.G. Rebelo, 1992, 'Plant Diversity and Endemism,' in R.M. Cowling (ed.), *Ecology of Fynbos*, 62–112, 72.

66. B.W. van Wilgen, D.M. Richardson, and A.H.W. Seydack, 1994, 'Managing Fynbos for Biodiversity: Constraints and Options in a Fire-prone Environment,' *SAJS*, 90, 322–8, see 322–3. Richard M. Cowling's key paper on this is, 1987, 'Fire and its Role in Coexistence and Speciation in Gondwanan Shrublands,' *SAJS*, 83, 106–12.

67. A.G. Rebelo, 1992 'Preservation of Biotic Diversity,' in R.M. Cowling (ed.), *Ecology of Fynbos*, p.347. R. Noble and R.O. Slatyer, 1980, 'The Use of Vital Attributes to Predict Successional Changes in Plant Communities Subject to Recurrent Disturbances,' *Vegetatio*, 43, 5–21.

68. W.J. Bond, J. Vlok, and M. Viviers, 1984, 'Variation in Seedling Recruitment of Cape Proteaceae after Fire,' *Journal of Ecology*, 72, 209–21; B.W. van Wilgen and M. Viviers, 1985, 'The Effect of Season of Fire on Serotinous Proteaceae in the Western Cape and the Implications for Fynbos Management,' *SAFJ*, 133, 49–53; B.W. van Wilgen, W.J. Bond, and D.M. Richardson, 1992, 'Ecosystem Management,' in R.M. Cowling (ed.), *Ecology of Fynbos*, pp.345–71, 348, 349–50; see also J.J. Midgley, 1989, 'Season of Burn of Serotinous Proteaceae: A Critical Review and Further Data,' *SAJB*, 55, 165–70.

69. van Wilgen et al., 'Ecosystem Management,' p.351; B.W. van Wilgen, 1984, 'Adaptation of the United States Fire Danger Rating System to Fynbos Conditions: Part I. A Fuel Model for Fire Danger Rating in the Fynbos Biome,' *SAFJ*, 61–5; B.W. van Wilgen and R.E. Burgan, 1984, 'Adaptation of the United States Fire Danger Rating System Part II. Historic Fire Danger in the Fynbos Biome,' *SAFJ*, 66–78.

70. ARFD R.P.20/1961, p.12; D.M. Richardson, I.A.W. Macdonald, P.M. Holmes, and R.M. Cowling, 1992, 'Plant and Animal Invasions,' in R.M. Cowling (ed.), *Ecology of Fynbos*, pp.271–308, 270; van Wilgen et al., 'Ecosystem Management,' p.362.

71. Richardson et al., 'Plant and Animal Invasions,' pp.289–90, 299–302.

72. W.S.W. Trollope, 1973, 'Fire as a Method of Controlling Macchia (Fynbos) Vegetation on the Amatole Mountains of the Eastern Cape,' *Proceedings of the Grassland Society of South Africa*, 8, 35–41; Rebelo, 'Preservation of Biotic Diversity,' pp.318, 344. *Cape Times*, 'Hottentots Holland Review,' 11 November 1971, 5.

73. D.R. Woods, 1985, 'An Innovative and Far-sighted Research Support Program,' *BioEssays*, 3, 6, 272–3; Interview with Brian J. Huntley; Interview with W.J. Bond.

74. ARFD1978/79 R.P.39/1980, p.1; ARFD1981/82 R.P.105/1982, pp.186, 191, 249.

75. ARFD1982/83 R.P.85/1984, p.185; ARFD1983/84 R.P.28/1985, p.171; van Wilgen et al., 'Ecosystem Management,' p.364; ARFD1986/87 R.P.119/1987, p.43.

76. ARFD1987/88 R.P.114/1988, pp.21, 31.

77. ARFD1989/90 R.P.29/1991, p.39; author's interviews with Fred J. Kruger and Brian W. van Wilgen.

78. Seydack and Bekker, 'Review,' p.9.

79. Interview with Philip Prins, Newlands Fire Base, Cape Town, 19 October 2007. The first burn on Table Mountain was allowed in 1984.

80. Seydack and Bekker, 'Review,' pp.20, 22; van Wilgen et al., 'Ecosystem Management,' p.364; Kruger et al., 'A Review of the Veld Fires in the Western Cape during 15 to 25 January 2000,' Department of Water Affairs and Forestry Report, p.24.

81. P.J. Brown, P.T. Manders, D.P. Bands, F.J. Kruger, and R.H. Andrag, 1991, 'Prescribed Burning as a Conservation Management Practice: A Case History from the Cederberg Mountains, Cape Province, South Africa,' *Biological Conservation*, 56, 2, 133–50, 147.

82. Seydack and Bekker, 'Review,' pp.22–3.

83. On using fire in plantations, see Pooley, 'Environmental History of Fire,' pp.182, 236–9.

84. Interview with Greg Forsythe, CSIR, Stellenbosch, 15 October 2007; F.J. Kruger, G.G. Forsyth, L.M. Kruger, K. Slater, D.C. Le Maitre, and J. Matshate, 2006, 'Classification of Veldfire Risk in South Africa for the Administration of the Legislation regarding Fire Management,' in D.X. Viegas (ed.), *Proceedings of 5th International Conference on Forest Fire Research*, 27–30 November 2006, Portugal (Amsterdam, the Netherlands: Elsevier B.V.), p.1; Interview with Zane Erasmus, by telephone, 25 October 2007; C.N. de Ronde, 1999, '1998: A Year of Destructive Wildfires in South Africa,' *International Forest Fire News*, 20, 73–8, 74.

85. 'National Veld and Forest Fire Act No.101 of 1998,' *Republic of South Africa: Government Gazette*, 401, 19515 (Cape Town), pp.4–6, 8. Interview with Zane Erasmus, Fire Manager, CapeNature, 25 October 2007. On the role of forestry c.2013, see http://www.daff.gov.za/.

86. 'National Veld and Forest Fire Act,' pp.7–8; C. Willis et al., 'The Development of a National Fire Danger Rating System'; Kruger et al., 'Classification of Veldfire Risk.'

87. Kruger et al., 'A Review of the Veld Fires,' p.23.

5   Fire Geography and Urbanisation on the Cape Peninsula

1. D.G.D. Riley, 1989, 'City Engineer's Statement,' *Annual Report of the City Engineer (ARCE) for 1988/89*, City of Cape Town, 1–2.

2. On the Peninsula's geology, see D. MacPhee and M. de Wit, 2003, 'How the Cape got its Shape,' Map, 1st edition (Cape Town, South Africa: Map Studio/CIGCES/Department of Land Affairs). See also http://web.uct.ac.za/depts/geolsci/cape.htm, accessed on 28 October 2013.

3. For a detailed account of the Peninsula's topography and geology, see R.M. Cowling, I.A.W. MacDonald, and M.T. Simmons, 1996, 'The Cape Peninsula, South Africa: Physiographical, Biological and Historical Background to an Extraordinary Hot-Spot of Biodiversity,' *Biodiversity and Conservation*, 5, 527–50.

4. Ibid., 536–44. For (contrasting) plant species numbers and endemism, see B.W. van Wilgen, 2012, 'Evidence, Perceptions, and Trade-offs Associated with Invasive Alien Plant Control in the Table Mountain National Park, South Africa,' *Ecology and Society* 17, 3; A.G. Rebelo, P.M. Holmes, C. Dorse, and J. Wood, 2011, 'Impacts of Urbanization in a Biodiversity Hotspot: Conservation Challenges in Metropolitan Cape Town,' *SAJB*, 77, 20–35, 25.

5. Ibid., 545.

6. Ibid., 538, 540–3.

7. L. Mucina, M.C. Rutherford, and L.W. Powrie, 2005, 'Vegetation Map of South Africa, Lesotho and Swaziland,' 1:1,000,000 (Pretoria, South Africa: South African National Biodiversity Institute); A.G. Rebelo, C. Boucher, N. Helme, L. Mucina, and M.C. Rutherford, 2006, 'Fynbos Biome,' in L. Mucina and M.C. Rutherford (eds.), *The Vegetation of South Africa, Lesotho and Swaziland* (Pretoria, South Africa: South African National Biodiversity Institute), p.107.

8. Rebelo et al. 'Fynbos Biome,' p.82; G.G. Forsyth and B.W. van Wilgen, 2008, 'The Recent Fire History of the Table Mountain National Park and its Implications for Fire Management,' *Koedoe*, 50, 1, 3–9, 7; A.G. Rebelo, P.M. Holmes, C. Dorse, and J. Wood, 2011, 'Impacts of Urbanization in a Biodiversity Hotspot: Conservation Challenges in Metropolitan Cape Town,' *SAJB*, 77, p.25.

9. Rebelo et al., 'Fynbos Biome,' pp.183, 83; Forsyth and van Wilgen, 2008, 'Recent Fire History of the Table Mountain National Park,' 7; Rebelo et al., 'Impacts of urbanization,' p.25.

10. A.H.W. Seydack, S.J. Bekker, and A.H. Marshall, 2007, 'Shrubland Fire Regime Scenarios in the Swartberg Mountain Range, South Africa: Implications for Fire Management,' *International Journal of Wildland Fire*, 16, 81–95.

11. On the fire history of the Cape of Good Hope section of the Peninsula, see R.F. Powell, 2013, 'Long-term Vegetation Change in the Cape of Good

Hope Section of Table Mountain National Park, in Response to Climate, Fire and Land Use' (MSc thesis, University of Cape Town). For Table Mountain National Park, see G.G. Forsyth and B.W. van Wilgen, 2007, 'An Analysis of the Fire History Records from Protected Areas in the Western Cape,' *CSIR Report CSIR/NRE/ECO/ER/2007/0118/C*, 4–9.

12. D. Southey, 2009, 'Wildfire in the Cape Floristic Region: Exploring Vegetation and Weather as Drivers of Fire Frequency' (MSc thesis, University of Cape Town).

13. On impounding livestock, see Cape of Good Hope, Reports of the Conservators of Forests (C.G.H.): C.G.H. 1902 G.55–1903, p.6; Minutes of His Worship the Mayor (henceforth MM), MM1901, MM1903, MM1923, p.x; MM1924, p.iii; MM1927, p.vi; MM1929, p.6; MM1932, p.3. On transport and electrification, see V. Bickford-Smith, E. Van Heyningen, and N. Worden, 2009, *Cape Town in the Twentieth Century: An Illustrated Social History* (Cape Town, South Africa: David Philip), p.63.

14. Regarding the Upper Gardens dairy, grass and Ericas, see C.A. Lückhoff, 1951, *Table Mountain: Our National Heritage after Three Hundred Years* (Cape Town, South Africa: A.A. Balkema), fig.50 for the dairy herd, and pp.56, 47. The map is 'Cape Peninsula,' Chief Directorate: Surveys and Mapping, Commemorative Reprints of First Map Series, 1:25,000 (first printed 1934, reprinted October 2000), Sheets I and II. Regarding impounded livestock: MM1940, Annual Report of the Director of Parks and Gardens (ARDPG) 32, 33; MM1943, ARDPG, 35; *Cape Times*, 'Great Bush Fire on Mountain Slopes,' 17 January 1944, front page.

15. W.T. Stead (ed.), 1902, *The Last Will and Testament of Cecil John Rhodes, with Elucidatory Notes, to which are Added Some Chapters Describing the political and Religious Ideas of the Testator* (London: Review of Reviews Office), pp.13, 16.

16. Population is discussed in detail in Chapter 9.

17. C.G.H.1903 G.26–1904, pp.4–5.

18. K. Wall, 2008, 'Water Supply: Reshaper of Cape Town's Local Government a Century Ago,' in J.E. van Zyl, A.A. Ilemobade, and H.H. Jacobs (eds.), *Proceedings of the 10th Annual Water Distribution Systems Analysis Conference*, WDSA2008, Kruger National Park, South Africa.

19. Bickford-Smith et al., *Cape Town in the Twentieth Century*, pp.62, 67, 70; MM1948, Annual Report of the City Electrical Engineer, Map OH/PT153, unnumbered insert after page 19.

20. N. Worden, E. van Heyningen, and V. Bickford-Smith, 1998, *Cape Town, the Making of a City: An Illustrated Social History* (Cape Town, South Africa: David Philip), pp.162, 214; Bickford-Smith et al., 2009, *Cape Town in the Twentieth Century*, pp.62, 67; C.G.H.1905 G.50–1906, pp.19, 26; Union of South Africa, ARFD 1931/32 U.G.37–1932, p.10.

21. V. Bickford-Smith, E. Van Heyningen, and N. Worden, 1999, *Cape Town in the Twentieth Century: An Illustrated Social History* (Cape Town, South Africa: David Philip), p.63; *Cape Times*, 'Scenic Road on Mountain: Immediate Construction Contemplated to Provide Relief Work Jobs for Additional 350 Men,' 9 February 1938, 13; *Cape Times*, 'Ambitious Roads Scheme,' 10 November 1943, 5.

22. *Cape Times*, 'Cigarettes thrown from Cars,' Letters to the Editor, 8 February 1938, 17.

23. *Cape Times*, 'Clifton Bush Fire,' 20 January 1934, 11; *Cape Times*, 'Signal Hill Blaze,' 24 January 1938, 13.
24. S.S. Morris, 1958, 'Keep Pace with Progress – Suggested Improvements to the University Education of Civil Engineers,' *Transactions of the South African Institution of Civil Engineers*, 15, 374–77. For his biography: Ibid., 'Our Authors,' 244.
25. *Cape Times*, '£1/2 Million New Scenic Drives Round Cape,' 29 January 1960, front page; Bickford-Smith et al., *Cape Town in the Twentieth Century*, pp.152–4; *Cape Times*, 'New Concrete Link,' 11 November 1971, front page; *Cape Times*, 'Row Brews over Freeway Plan,' 19 November 1971, 9.
26. *Cape Times*, 'Peninsula's Ugly Face,' Letters to the Editor, 27 October 1971, 7.
27. On the history of Hout Bay's fishing industry, see L. van Sittert, 1988, ' "Slawe van die Fabriek" – The State, Monopoly Capital and the Subjugation of Labour in the Hout Bay Valley Crayfish Fishing Industry, 1946–1956,' *Studies in the History of Cape Town*, 6 (Cape Town, South Africa: Centre for African Studies, University of Cape Town), pp.112–49.
28. There were notable fires in Devil's Peak plantation in 1899, 1902, 1920, 1922, 1932, 1935 and 1952. See Chapter 9.
29. The largest occurred in January 2000, see F.J. Kruger P. Reid, M. Mayet, W. Alberts, J.G. Goldammer, K. Tolhurt, and S. Parker, 2000, 'A Review of the Veld Fires in the Western Cape during 15 to 25 January 2000,' *Department of Water Affairs and Forestry Report*, 62–8.
30. For example, in February 1938: ARFD1937/38 U.G.46–1938, pp.17, 26.
31. Notably in 1942 and January 1992: ARCE1942, p.35; *Cape Times*, 'Battle to Put Out Huge Blaze,' 18 January 1992, front page.
32. Including a big fire in 1999, data courtesy of Lance van Sittert.
33. Kruger et al., 'Review of the Veld Fires,' 62–8.
34. Rebelo et al.'s survey of the wider region (referred to as 'City of Cape Town' but embracing a 2460km² administrative region) identifies 450 threatened and near-threatened IUCN Red List plant species, with the most threatened being Cape Flats Sand Fynbos. On this and for the figures cited, see Rebelo et al., 'Impacts of Urbanization,' 26.

6   Conserving Table Mountain

1. Cited in W.T. Stead (ed.), 1902, *The Last Will and Testament of Cecil John Rhodes* (London: Review of Reviews Office), p.16.
2. J.C. Smuts (ed.), 1942, 'The Spirit of the Mountain,' *Plans for a Better World: Speeches of Field-Marshal The Right Honourable J.C. Smuts* (London: Hodder & Stoughton), pp.45–51, 49, 51.
3. Ibid., p.51.
4. Cape of Good Hope, Reports of the Conservators of Forests (C.G.H.), 1901, G.38–1902, p.12.
5. MM, including various branch reports, Chancellor Oppenheimer Library, University of Cape Town, Government Publications, G682 VC2 s.2247: MM1929, Report of the Director of Parks and Gardens, p.3; MM1915, Report of the Superintendent of Tree Planting, p.ii.

6. *Cape Times*, 'Letters to the Editor,' 1 February 1950, 8; *Cape Times*, 'Letters to the Editor,' 2 February 1950, 8.
7. L. van Sittert, 2003, 'The Bourgeois Eye Aloft: Table Mountain in the Anglo Urban Middle Class Imagination, c.1891–1952,' *Kronos*, 29, 161–90; W. Beinart, 1994, *Twentieth-Century South Africa* (Oxford: Oxford University Press), p.7.
8. Ibid., p.163.
9. Ibid., p.165.
10. Ibid., p.173.
11. Ibid., p.174.
12. Ibid., p.177.
13. Ibid. On Rhodes: Stead, *Last Will and Testament of Cecil John Rhodes*, pp.16, 86, 128.
14. See S. Dubow, 2006, *A Commonwealth of Knowledge: Science, Sensibility, and White South Africa 1820–2000* (Oxford: Oxford University Press), p.5; P. Anker, 2002, *Imperial Ecology: Environmental Order in the British Empire, 1895–1945* (Cambridge, MA: Harvard University Press), p.65.
15. van Sittert, 'Bourgeois Eye,' p.186.
16. H. Pearson, 1910, text of speech appended to a letter to Colonel Pierie, Director of Kew, Miscellaneous Reports: South Africa: National Botanic Garden, 1910–28, Kew Gardens Library and Archive, 302, R6 (4 November 1910), pp.1–13, see pp.1–4, 13.
17. T.P. Stoekoe, 1921, 'Letters to the Director of Kew, Directors' Correspondence, Vol.182, 96–108, No.102, Kew Gardens Library and Archive (12 October 1921), p.1.
18. For the Smuts quote, see C.A. Lückhoff, 1951, *Table Mountain: Our National Heritage after Three Hundred Years* (Cape Town, South Africa: A.A. Balkema), p.74, and see van Sittert, 'Bourgeois Eye,' p.190.
19. R. Marloth, 1924, 'Notes on the Question of Veld Burning,' *SAJS*, 21, 343–4.
20. T.P. Stoekoe, 1921, Letters to the Director of Kew, Nos. 100 and 102 (21 July 1921 and 12 October 1921), pp.1 and 2, respectively.
21. R.H. Compton, 1924, 'Notes and News,' *JBSSA*, 10, 3.
22. R.H. Compton, 1927, 'Notes and News,' *JBSSA*, 13, 4, 5. On Kruger National Park: J. Carruthers, 1995, *The Kruger National Park: A Social and Political History* (Pietermaritzburg, South Africa: University of Natal Press).
23. R.H. Compton, 1928, 'Notes and News,' *JBSSA*, 14, 5. On the Rural Amenities Bill: Hansard HC Debate 23 January 1931, 247, p.545.
24. A.W. Hill, 1929, Sir Arthur Hill's Diary of his South Africa Trip, Kew Gardens Library and Archive, AWH/1/8, see 11–18 November.
25. Ibid., 18 November 1929.
26. R.H. Compton, 1929, 'Notes and News,' *JBSSA*, 15, 4; A.R. Hill, 1931, 'Botany in South Africa: Sir Arthur Hill's Report,' *JBSSA*, 17, 6–9, 8.
27. A. Notten and L. van der Walt, 2008, '*Leucadendron argenteum* (L.) R.Br.,' Kirstenbosch Botanical Garden, available at: http://www.plantzafrica.com/plantklm/leucadadendronargent.htm, accessed on 17 May 2010.
28. *Cape Times*, 'Fire Patrols on Mountain,' 24 January 1934, 10.
29. *Cape Argus*, '£250,000 Havoc in Great Fire,' 26 December 1935, 11.
30. For example, *Cape Times*, as follows: 'Forest Fires,' 2 March 1920, 6; 'Letters to the Editor,' 4 March 1920, 8; 'Letters,' 11 March 1982, 10; *Cape Argus*,

'The Tragedy of Mountain Fires'; MM1949, City Lands and Forests Branch (CLFB), 7.

31. *Cape Argus*, 'To Safeguard Mountain,' Letters to the Editor, 27 December 1935, 12; *Cape Argus*, 'The Mountain Fire,' 11.

32. *Cape Argus*, 'The Mountain Fire,' 14.

33. D.J. Strohmaier, 2005, *Drift Smoke: Loss and Renewal in a Land of Fire* (Reno, NV: University of Nevada Press), pp.103–4, 114–16.

34. *Cape Argus*, '£250,000 Havoc in Great Fire,' 26 December 1935, 11; *Cape Argus*, 'Was it Started Deliberately?,' 11.

35. *Cape Argus*, 'The Fire Controversy,' Editorial, 4 January 1936, 8.

36. Ibid.

37. Union of South Africa, ARFD, U.G.53–1936, p.6.

38. C.G.H.1899 G.24–1900, p.14.

39. C.G.H.1900 G.39–1901, p.13; C.G.H.1901 G. 38–1902, p.12; C.G.H.1905 G.50–1906, p.26.

40. On Sir Lionel Phillips and his influence, see Dubow, *Commonwealth of Knowledge*, pp.189–93; R.H. Compton, 1936, 'News and Notes,' *JBSSA*, 22, 2. For the description of Lady Phillips and her quote, see *Cape Argus*, 'To Safeguard Mountain,' 11–12.

41. *Cape Times*, '£2,000,000 a Year from Visitors,' 9 February 1938, 6; *Cape Argus*, 'To Safeguard Mountain,' 12; *Cape Argus*, 'The Mountain Fire,' Letters to the Editor, 27 December, 1935, 14.

42. *Cape Argus*, 'To Safeguard the Mountain,' 12.

43. *Cape Argus*, 'Mountain under State Protection,' 28 December, 1935, 11.

44. Discussed in detail in Chapter 3.

45. *Cape Times*, 'Grass More Precious than Gold,' 5 February 1938, 25.

46. Cited in *Cape Argus*, 'To Safeguard the Mountain,' 27 December 1935.

47. *Cape Times*, 'Up in Smoke,' Editorial, 9 February 1938, 12.

48. *Cape Times*, as follows: 'Terrific Fire Sweeps Peninsula Mountains, Constantiaberg Ablaze,' 7 February 1938, 13; 'The Great Fire Sweeps Down into Hout Bay: Residents Flee in Terror,' 8 February 1938, 13–14; 'Up in Smoke,' 12. On the Wildlife Protection Society, and Smith's Farm, see J. Pringle, 1982, *The Conservationists and the Killers* (Cape Town, South Africa: T.V. Bulpin and Books of Africa), pp.160–70.

49. *Cape Times*, as follows: 'Scenic Road on Mountain,' 9 February 1938, 13; 'New Road on the Mountainside,' Letters to the Editor, 10 February 1938, 17; 'New Roads on Mountainside,' Letters to the Editor, 11 February 1938, 17; 'Another Blaze on Devil's Peak,' 24 January 1944, front page.

50. *Cape Times*, 'Journey's End for Quarries,' Editorial, 29 November 1940, 10.

51. *Cape Times*, 'Evacuation Plan for Cape Town,' 13 January 1942, 11.

52. *Cape Times*, 'Start of War on Soil Erosion,' 20 November 1943, 8; *Cape Times*, 'The Enemy within,' Editorial, 4 December 1943, 6; W.J. Bond, P. Slingsby, 1984, 'Collapse of an Ant-Plant Mutualism: The Argentine Ant (*Iridomyrmex humilis*) and Myrmecochorous Proteaceae,' *Ecology*, 65, 1031–7.

53. *Cape Times*, 'National Veld Trust,' Editorial, 7 January 1943, 4.

54. *Cape Times*, 'Stop those Mountain Fires,' 27 November 1942, 6.

55. *Cape Times*, 'Franschhoek Mountain Blaze,' 23 December 1942, 9.

56. *Cape Times*, 'By their Fruits,' Editorial, 14 September 1943, 4.

57. R.H. Compton, 1942, 'Notes and News,' *JBSSA*, 28, 4–5; 29 (1943), pp.3–4.
58. On Forestry Department support for fire suppression, see *Cape Times*, 'Mountain Fire Menace: Suppression Plans,' 23 October 1943, 8. J.S. Henkel, 1943, 'Preservation of Mountain Vegetation of the South-West Districts of the Cape Province,' Memorandum to the Royal Society. On the commission: *Cape Times*, 'South-Western Cape Mountains: Flora and Fauna Preservation,' 12 November 1943, 7.
59. C.L. Wicht, 1945, *Preservation of the Vegetation of the South Western Cape* (Cape Town, South Africa: Royal Society of South Africa), pp.7–9, 53. This is discussed in Chapter 3.
60. See J. Foster, 2003, '"Land of Contrasts"' or "Home we have always known?": The SAR&H and the Imaginary Geography of White South African Nationhood, 1910–1930,' *Journal of Southern African Studies*, 29, 3, 657–80.
61. R.H. Compton, 1946, 'Notes and News,' *JBSSA*, 32, 8.
62. Lückhoff, *Table Mountain*, pp.143–4.
63. Cited in Ibid., pp.136–7.
64. Annual Report of the City Engineer (ARCE) 1979, 'Nature Conservation,' p.19.
65. D. Gordon Bagnall and H. Louw, 1962, *Bokkie the Grysbuck*, ed. Douglas Hey (Cape Town, South Africa: Department of Nature Conservation & Cape Peninsula Fire Protection Committee).
66. Ibid., pp.21, 28, 33, 37.
67. D. Hey, 1963, 'The Conservation of Flora in the Province of the Cape of Good Hope,' *Lantern*, 13, 68–9; On nature reserves, see ARCE1979, 'Nature Conservation,' p.19.
68. *Cape Times*, as follows: 'Row Brews over Freeway Plan,' 19 November 1971, 9; 'Kirstenbosch Protest Mounts/"Devastating Plan", says Rycroft,' 20 November 1971, front page; 'Exco Backs Down on Freeway,' 24 November 1971, front page.
69. *Cape Times*, Editorial, 22 November 1971, 8.
70. *Cape Times*, 'Row Brews over Freeway Plan,' 9; *Cape Times*, 'Kirstenbosch Protest Mounts,' front page.
71. *Cape Times*, 'Stop New Freeway Scheme,' Letters to the Editor, 29 November 1971, 11.
72. *Cape Times*, 'Morris Defends Freeway Plan,' 26 November 1971, 6–7.
73. *Cape Times*, '"Bystander" by Bob Molloy,' 9 December 1971, 14; *Cape Times*, '"Bystander" with Bob Molloy,' 10 December 1971, 16.
74. *Cape Times*, 'Indignation Mounting over Sandy Cove,' 11 December 1971, 11.
75. *Cape Times*, 'Fire Sweeps Cape Point: A Mile of Rare Flora Destroyed,' 8 November 1971, front page.
76. *Cape Times*, 'Fires could Ruin Flora,' 18 April 1973, 2.
77. *Cape Times*, 'Police Investigate Week-end Fires,' 24 April 1973, front page.
78. *Cape Times Weekend Magazine*, 'Ban the Outdoor Braaivleis,' 28 April 1973, 2.
79. ARCE1979, Parks and Forests Branch (ARPFB), 18–20; *African Wildlife* citations in Pringle, *The Conservationists*, pp.240, 242; On golden gladiolus in

2014, pers. comm. Patricia Holmes, 9 February 2014; E.J. Moll, B. McKenzie, D. McLachlan, and B.M. Campbell, 1978, 'A Mountain in a City – The Need to Plan the Human Usage of the Table Mountain National Monument, South Africa,' *Biological Conservation*, 13, 2, 117–31, 117–18; M. Shroyer, D. Kilian, and J. Jakelman, 1998, 'A Wilderness in an Urban Setting: Planning and Management of the Cape Peninsula National Park, Cape Town, South Africa,' Paper presented at 6th World Wilderness Congress (Bangalore, India 24–29 October 1998).

80. Annual Reports of the Department of Forestry, Republic of South Africa, 1980/81 R.P.96/1981, 185–6; ARCE1982, 'Cape Town Mountain Fires,' 13–14; ARCE1982, ARPFB, p.34; ARCE1983/84, ARPFB, p.5.

81. Moll et al., 'A Mountain in a City,' 123; ARCE1984/85, p.63.

82. *Cape Times*, 'Help Save our Mountain,' 20 December 1986, front page; *Cape Times*, 'You Can't Lock the Mountain,' Editorial, 20 December 1986, 8.

83. *Cape Times*, 'Help Save our Mountain'; *Cape Times*, 'Save Mountain Fund Started,' *Cape Times*, 22 December 1986, front page; Shroyer et al., 'A Wilderness in an Urban Setting.'

84. *Cape Times*, 'Council to Explain Fire Action Today,' 11 February, 1991, 3; ARCE1992/93, ARPFB, p.34.

85. My account of these negotiations is based on SAPA, 'Environment – Table Mountain,' by Charl de Villiers, 20 January 1995; Shroyer et al., 'A Wilderness in an Urban Setting'; Anon, 1999, 'Cape Peninsula National Park Integrated Environmental Management System: Management Plan,' final draft, 20, available at: http://www.capepeninsula.co.za, accessed on 19 January 2010.

86. N.R. Mandela, 1998, 'Message from President Nelson Mandela on the Occasion of World Environment Day,' 5 June, available at: http://www.info.gov.za/speeches/1998/98609_0w0709810057.htm, accessed on 15 November 2013.

7 Afforestation, Plant Invasions and Fire

1. Union of South Africa, ARFD U.G.: 1935/36 U.G.53–1936, p.7.

2. T. Pakenham, 2007, *In Search of Remarkable Trees* (London: Weidenfeld & Nicolson), p.164.

3. T.R. Sim, 1907, *The Forests and Forest Flora of the Colony of the Cape of Good Hope* (Aberdeen: Taylor & Henderson), pp.51, 79; A.M. Avis, 1989, 'A Review of Coastal Dune Stabilization in the Cape Province of South Africa,' *Landscape and Urban Planning*, 18, 55–68; C. Joubert, 1996, 'Commercial Forestry on Signal Hill and Lion's Head, Cape Town,' *SAFJ*, 175, 43–54, 45; Cape of Good Hope, Reports of the Conservators of Forests (C.G.H.), C.G.H.1890, p.61.

4. On the building boom and afforestation, see C.G.H.1902 G.55–1903, pp.2, 4.

5. C.G.H. as follows: 1905 G.50–1906, p.27; 1908 G.35–1909, p.12; 1901 G.38–1902, p.13; 1910 U.G.30–1911, p.12; Annual Report of the City Engineer (ARCE) 1986/87, p.52.

6. On World War I shortages, see ARFD1916/17 U.G.3–1918, p.3; ARFD1917/18 U.G.57–1918, pp.2–3.

7. ARFD1919/20 U.G.7–'21, p.4; Republic of South Africa (R.P.), ARFD 1962/63 R.P.58–1965, p.26; 1966/67 R.P.37–1968, p.92; On Langa, see V. Bickford-Smith, E. Van Heyningen, and N. Worden, 1999, *Cape Town in the Twentieth Century: An Illustrated Social History* (Cape Town, South Africa: David Philip), pp.87–8.

8. *Cape Times*, 'Newlands Forests keep down rates,' 4 February 1944, p.7; MM1946, p.16.

9. 'Tokai and Cecilia Plantation Fact Sheet,' SANParks, available at: http://www.sanparks.org/parks/table_mountain/environment/tokai.php, accessed on 19 January 2010.

10. Derived from annual reports of the Department of Forestry and G.L. Shaughnessy, 1980, 'Historical Ecology of Alien Woody Plants in the Vicinity of Cape Town, South Africa' (PhD thesis, University of Cape Town), pp.260, 267, 310.

11. On the building boom and afforestation, see C.G.H.1902 G.55–1903, pp.2, 4; C. Joubert, 1996, 'Commercial Forestry on Signal Hill and Lion's Head, Cape Town,' *SAFJ*, 175, 43–54, 46.

12. *Cape Times*, 'Newlands Forests Keep Down Rates,' 4 February 1944, 7; MM1915, Annual Report of the Superintendent of Tree Planting (ARSTP), p.ii.

13. MM, including Annual Reports of the City Forest Officer (ARCFO): MM1941, ARCFO, p.42; MM1952, p.10; MM1973, ARCFO, p.59; MM1974, ARCFO, p.62; MM1975, ARCFO, p.62. See also C. Joubert, 1996, 'Commercial Forestry,' 50–1. On the use of the Newlands Forests, and timber for ammunition boxes, see *Cape Times*, 'Newlands Forests Keep Down Rate,' 4 February 1944, 7.

14. For fires stopped at Tokai and Cecilia plantations, see G.55–1902, pp.9, 16. For Devil's Peak, see C.G.H.1902 G55–1903, pp.14–15; ARFD1935/36 U.G.53–1936, pp.6, 16.

15. ARFD1919/20 U.G. 7–'21, pp.8–9; *Cape Times*, 'Caused by Flower Pickers?' 13 January 1934, 13.

16. *Cape Times*, as follows: 'Mountain Blaze Threatens Homes,' 8 February 1974, front page; 'Smokeout,' 3; 'Council to explain fire action today,' 11 February, 1991, 3.

17. A.W. Crosby, 1986, *Ecological Imperialism: The Biological Expansion of Europe 900–1900* (Cambridge: Cambridge University Press); R.H. Grove, 1995, *Green Imperialism: Colonial Expansion, Tropical Island Edens and the Origins of Environmentalism, 1600–1860* (Cambridge: Cambridge University Press). On South African silviculture and Australian trees, see B.M. Bennet, 2011, 'A Global History of Australian Trees,' *Journal of the History of Biology*, 44, 125–45. For thoughtful discussions of plant transfers, embracing biological and social dimensions, see W. Beinart and K. Middleton, 2004, 'Plant Transfers in Historical Perspective: A Review Article,' *Environment and History* 10, 3–29; C.A. Kull and H. Rangan, 2008 'Acacia Exchanges: Wattles, Thorn Trees, and the Study of Plant Movements,' *Geoforum*, 39, 1258–72.

18. On terminology for invasive introduced plants, see P. Coates, 2003, 'Editorial Postscript: The Naming of Strangers in the Landscape,' *Landscape Research*, 28, 131–7. For the definition of invasive plants and perceptions on the

Peninsula, see B.W. van Wilgen, 2012, 'Evidence, Perceptions, and Trade-offs Associated with Invasive Alien Plant Control in the Table Mountain National Park, South Africa,' *Ecology and Society* 17, 2 and D.M. Richardson, P. Pyšek, and J.T. Carlton, 2011, 'A Compendium of Essential Concepts and Terminology in Invasion Ecology,' in D.M. Richardson (ed.), *Fifty Years of Invasion Ecology: The Legacy of Charles Elton* (Oxford: Wiley-Blackwell), pp.409–20. For historical and cultural dimensions, see B.M. Bennett, 2014, 'Model Invasions and the Development of National Concerns over Invasive Introduced Trees: Insights from South African History,' *Biological Invasions*, 16, 3, 499–512; B.M. Bennett and F.J. Kruger, 'Ecology, Forestry and the Debate over Exotic Trees in South Africa,' *Journal of Historical Geography*, 42, 100–9; S. Pooley, 2010, 'Pressed Flowers: Notions of Indigenous and Alien Vegetation in South Africa's Western Cape, c.1902–45,' *Journal of Southern African Studies*, 36, 599–618; L. van Sittert, 2000, ' "The Seed Blows about in Every Breeze": Noxious Weed Eradication in the Cape Colony, 1860–1909,' *Journal of Southern African Studies*, 26, 655–74; J. Carruthers, L. Robin, J. Hattingh, C. Kull, H. Rangan, and B. van Wilgen, 2011, 'A Native at Home and Abroad: The History, Politics, Ethics and Aesthetics of *Acacia*,' *Diversity and Distributions*, 17, 810–21.

19. B.W. van Wilgen, 2012, 'Evidence, Perceptions, and Trade-offs,' 3.

20. C.G.H.1900 G.39–1901, p.12; C.G.H.1902, p.17; C.G.H.1905, pp.17, 19, 25.

21. Notably R.H. Compton, 1924, 'Notes and News,' *JBSSA*, 10, 3.

22. On MacOwan and Marloth, see C.H. Stirton, 1983 [1978], *Plant Invaders: Beautiful, but Dangerous: A Guide to the Identification and Control of Twenty-six Plant Invaders of the Province of the Cape of Good Hope* (Cape Town, South Africa: Department of Nature and Environment Conservation), p.149. Harold Compton wrote on the need to control invasive introduced plants in 1924 in 'Notes and News,' *JBSSA*, 10, 3. R.S. Adamson, 1927, 'The Plant Communities of Table Mountain: Preliminary Account,' *The Journal of Ecology*, 15, 278–309. On Fairbridge, see L. van Sittert, 2003, 'Making the Cape Floral Kingdom: The Discovery and Defence of Indigenous Flora at the Cape ca. 1890–1939,' *Landscape Research*, 28, 1,113–29, 115. Sir Arthur Hill's diary of his South Africa Trip is in Kew Gardens Library and Archive, AWH/1/8 (1929), see 11–18 November.

23. ARFD1935/36 U.G.53–1936, p.7.

24. For example, *Cape Argus*, 'Mountain under State Protection,' 28 December, 1935, 11; *Cape Argus*, 'To Safeguard Mountain,' 27 December, 1935, 12.

25. *Cape Argus*, 'To Safeguard Mountain,' 27 December, 1935, 11.

26. *Cape Argus*, 'Was it started deliberately?' 26 December 1935, 11.

27. ARFD1936/37 U.G.53–1937, p.25; C.A. Lückhoff, 1951, *Table Mountain: Our National Heritage after Three Hundred Years* (Cape Town, South Africa: A.A. Balkema), p.128.

28. Phillips quote from *Cape Argus*, 'To Safeguard Mountain,' 27 December, 1935, 11; see also *Cape Argus*, 'Letters on the Fire: Lady Phillips's Suggestion,' 28 December, 1934, 11. On the Phillipses, see S. Dubow, 2006, *A Commonwealth of Knowledge: Science, Sensibility, and White South Africa 1820–2000* (Oxford: Oxford University Press), pp.189–93. For Reitz, see *Cape Argus*, 'Mountain under State Protection,' 28 December, 1935, 11.

29. ARFD1935/36 U.G.53–1936, p.7.

30. Lückhoff, *Table Mountain*, p.116; MM1940, ARCFO, p.42; *Cape Times*, 'Large Area of Bush Burnt at Glencairn,' 21 January, 1942, 9; *Cape Times*, 'Second Veld Fire at Glencairn,' 28 December, 1943, 5; R.H. Compton, 1944, 'Notes and News,' *JBSSA*, 30, 5.
31. Wicht cited in B.W. Van Wilgen, 2009, 'The Evolution of Fire and Invasive Plant Management Practice in Fynbos,' *SAJS*, 105, 335–42, 338.
32. Lückhoff, *Table Mountain*, p.116.
33. *Cape Times*, 'Letters to the Editor,' 1 February 1950, 8.
34. Lückhoff, *Table Mountain*, 115–17.
35. MM1951, pp.9–10.
36. R.S. Adamson, 1953, 'Can We Preserve the Cape Flora?' *JBSSA*, 39, 11–12.
37. E.J. Moll and T. Trinder-Smith, 1992, 'Invasion and Control of Alien Woody Plants on the Cape Peninsula Mountains, South Africa 30 Years On,' *Biological Conservation*, 60, 135–43,135–7.
38. Control of Alien Vegetation Committee, 1959, *The Green Cancers in South Africa: The Menace of Alien Vegetation* (Cape Town, South Africa: Citadel Press); see M.E. Boehi, 'Being/becoming the "Cape Town Flower Sellers": The Botanical Complex, Flower Selling and Floriculture in Cape Town' (MA thesis, University of the Western Cape), 23; ARCE1979, Annual Report of the Parks and Forests Branch (ARPFB), p.19.
39. ARFDs R.P.20/1961, p.12; R.P.80/1964, pp.8–9; R.P.34/1966, p.10; R.P.10/1967, p.22.
40. ARFD1972/73 R.P.38/1974, p.9.
41. See ARCE 1963–76.
42. E.J. Moll and T. Trinder-Smith, 1992, 'Invasion and Control of Alien Woody Plants on the Cape Peninsula Mountains, South Africa 30 Years On,' *Biological Conservation*, 60, 135–43, see 137–9.
43. ARCE1979, ARPFB, p.19.
44. J.G. Brand, 'Foreword,' ARCE1976, p.1; ARCE1976, ARPFB, p.34.
45. ARCE1982, 'Cape Town Mountain Fires: Enemy or Friend?' p.14.
46. ARCE1982, 'Cape Town Mountain Fires,' p.14.
47. Moll and Trinder-Smith, 1992, 'Invasion and Control,' 137–42.
48. Van Wilgen, 'The Evolution of Fire and Invasive Plant Management,' 339–40.
49. Interview with Prof. David Richardson, Centre for Invasion Biology, Stellenbosch, 15 October, 2007; I.A.W. Macdonald, F.J. Kruger, and A.A. Ferrar, 1986, *The Ecology and Management of Biological Invasions in Southern Africa* (Cape Town, South Africa: Oxford University Press); D.M. Richardson, I.A.W. Macdonald, P.M. Holmes, and R.M. Cowling, 1992, 'Plant and Animal Invasions,' in R.M. Cowling (ed.), *The Ecology of Fynbos: Nutrients, Fire and Diversity* (Cape Town, South Africa: Oxford University Press), pp.271–308.
50. *Cape Times*, 'Fires could Ruin Flora,' 18 April 1973, 2.
51. van Wilgen, 2009, 'The Evolution of Fire and Invasive Plant Management,' 339–40.
52. ARCE1991/92, pp.7–8. On fuel levels, see B.W. van Wilgen and D.M. Richardson, 1985, 'The Effect of Alien Shrub Invasions on Vegetation Structure and Fire Behaviour in South African Fynbos Shrublands: A Simulation Study,' *Journal of Applied Ecology* 22, 955–66. On costs, see F.J. Kruger, P. Reid, M. Mayet, W. Alberts, J.G. Goldammer, K. Tolhurt, and S. Parker, 2000, 'A

Review of the Veld Fires in the Western Cape During 15 to 25 January 2000,' *Department of Water Affairs and Forestry Report*, 24.

53. ARCE1995/96, p.24.
54. Interview with Guy Preston, Pearson Hall, Kirstenbosch, 19 October 2007.
55. Interview with Guy Preston; B.W. van Wilgen, R.M. Cowling, and C.J. Burgers, 1996, 'Valuation of Ecosystem Services,' *BioScience*, 46, 184–9, 184, 189.
56. D.C. Le Maitre, B.W. Van Wilgen, R.A. Chapman, and D.H. Kelly, 1996, 'Invasive Plants and Water Resources in the Western Cape Province, South Africa; Modelling the Consequences of a Lack of Management,' *Journal of Applied Ecology*, 33, 161–72; Interview with David Le Maitre, CSIR Natural Resources, Stellenbosch, South Africa, 15 October 2007; Interview with Guy Preston.
57. Kruger et al., 'A Review of the Veld Fires,' 29, 30.
58. T. Pakenham, 2007, *In Search of Remarkable Trees: On Safari in Southern Africa* (London: Weidenfeld & Nicolson), pp.10, 138, 150, 164–7; *Cape Times*, ' "Tree Taliban" Ignores Pleas to Save Green Heritage,' 6 February 2007, 3; *Sunday Argus*, 'Author Blames "Eco-Fascists" for Tree-Culling,' 7 October 2007, 11.
59. Alan Nash, 'The World Goes By,' *Cape Times*, 22 November 1943, 4; Poem 'Gold to Burn' by R. Baur, in *Cape Times*, 27 October 1943, 4.
60. See Stephen's poems 'The Balcony,' 'Above Camps Bay,' 'October Song' and 'A Farewell' in S. Watson, 1995, *Presence of the Earth: New Poems* (Cape Town, South Africa: David Philip); Stephen Watson, 1989, *Cape Town days, and Other Poems* (Cape Town, South Africa: Cecil Skotnes and Clarke's Bookshop).
61. Orion Planning, 2002, 'Cape Peninsula National Park: Signal Hill, Kloof Nek and Tafelberg Road Initial Development Framework Report,' prepared for South African National Parks/Cape Peninsula National Park, 30.
62. R.M. Cowling, I.A.W. MacDonald, and M.T. Simmons, 1996, 'The Cape Peninsula, South Africa: Physiographical, Biological and Historical Background to an Extraordinary Hot-Spot of Biodiversity,' *Biodiversity and Conservation*, 5, 527–50, 537–8; L. Mucina, M.C. Rutherford, and L.W. Powrie, 2005, 'Vegetation Map of South Africa, Lesotho and Swaziland,' 1:1,000,000 (Pretoria, South Africa: South African National Biodiversity Institute).
63. MM1911, pp.47–8; MM1924, p.ii; MM1929, p.3.
64. Alan Nash, 'The World Goes By,' *Cape Times*, 30 December 1943, 4.
65. MM1939, p.31; MM1952, p.10; MM1953, p.10; ARCE1967, p.30; ARCE1977, p.40; ARCE1979, p.19; ARCE1982, p.131, ARCE1986/87, p.52.
66. For a discussion, see S. Pooley, 2010, 'Pressed Flowers: Notions of Indigenous and Alien Vegetation in South Africa's Western Cape.' For Pearson, Smuts, and Mandela, see Chapter 6.
67. For exchanges on the management of introduced trees and conflicts between ecological and cultural heritage priorities, see Anon, 2001, 'Draft Conservation Development Framework and Associated Maps: Comment and Responses Report, Synthesis of Written Submissions by Settlement Planning Services and South African National Parks,' prepared for South African National Parks by de Villiers Brownlie Associates, 21–2, 24.

8 Socio-Economic Causes of Fire: Population, Utilisation and Recreation

1. M.R. Levyns, 1924, 'Some Observations of the Effect of Bush Fires on the Vegetation of the Cape Peninsula,' *SAJS*, 21, 346–7, 346.
2. G.G. Forsyth and B.W. van Wilgen, 2008, 'The Recent Fire History of the Table Mountain National Park and its Implications for Fire Management,' *Koedoe*, 50, 1, 3–9, 6, 7.
3. *Cape Times*, 'Native Influx,' 17 February 1943, 8; *Cape Times*, 'Native Influx,' 25 February 1943, 5; *Cape Times*, 'Holiday Ban on Muizenberg,' 25 November 1943, front page; *Cape Times*, 'The Crowded Peninsula,' 2 December 1943, 4.
4. V. Bickford-Smith, E. van Heyningen, and N. Worden, 2009, *Cape Town in the Twentieth Century: An Illustrated Social History* (Cape Town, South Africa: David Philip), pp.182–3, 238.
5. V. Bickford-Smith, E. Van Heyningen, and N. Worden, 1999, *Cape Town in the Twentieth Century: An Illustrated Social History* (Cape Town, South Africa: David Philip) pp.119, 182–3, 238.
6. Annual Report of the City Electrical Engineer for 1947 and 1948, 'Cape Peninsula Area of Supply,' Map O.H./P.T. 153 (1947), unnumbered insert after p.19.
7. On linking population and conservation, see A.G. Rebelo, 1992, 'Preservation of Biotic Diversity,' in R.M. Cowling (ed.), *The Ecology of Fynbos: Nutrients, Fire and Diversity* (Cape Town, South Africa: Oxford University Press), pp.308–44, 339; B.W. van Wilgen, 1996, 'Management of the Natural Ecosystems of the Cape Peninsula: Current Status and Future Prospects,' *Biodiversity and Conservation*, 5, 671–84, 671. On Cape Coloured views, see L. Green, 2007, 'Changing Nature: Working lives on Table Mountain, 1980–2000,' in S. Field, F. Swanson, and R. Meyer (eds.), *Imagining the City: Memories and Culture in Cape Town* (Cape Town, South Africa: HSRC Press), pp.173–90, 176–7.
8. 'City of Cape Town: Statistics and Population Census,' 1996 and 2001 census data available at: http://www.capetown.gov.za/en/stats/1996census/Pages/SuburbProfiles.aspx, accessed on 2 March 2014. Imizamo Yethu is not listed separately in 1996.
9. 'Electricity and Lighting for Red Hill Residents,' City of Cape Town Media Release No.89/2005 (6 May 2005), available at: http://www.capetown.gov.za/en/MediaReleases/Pages/ElectricityAndLightingForRedHillResidents.aspx, accessed on 2 March 2014.
10. On flower sellers and wildflower legislation, see L. van Sittert, 2003, 'Making the Cape Floral Kingdom: The Discovery and Defence of Indigenous Flora at the Cape ca. 1890–1939,' *Landscape Research*, 28, 1, 113–29, see 119–20; M.E. Boehi, 2010, 'Being/becoming the "Cape Town Flower Sellers": The Botanical Complex, Flower Selling and Floriculture in Cape Town,' (MA thesis, University of the Western Cape).
11. 'Rural Amenities Bill,' House of Commons Debate, Hansard, 247, 23 January 1931, 537–71, 545.
12. R.H. Compton, 1924, 'News and Notes,' *JBSSA*, 10, 4–5.

13. See Ibid., and MM 1939, p.40, MM1940, p.45; MM1943, p.35; MM1944, p.31. R.H. Compton, 1930, 'Notes and News,' *JBSSA*, 16, 4.
14. *Cape Argus*, 'Mountain under State Protection,' 28 December, 1935, 11.
15. *Cape Times*, 'Sites for Sales of Wild Flowers Sellers Saved from Prosecution,' 1 January 1938, 24.
16. *Cape Times*, 'Sally Starke studies the New Ordinances and Sheds a Tear for the Flower Sellers of Adderley Street,' 6 January 1938, 7.
17. Ibid.
18. *Cape Times*, 'The Tokai Fire,' Letters to the Editor, 10 February 1938, 17.
19. Flower pickers quote cited in Bickford-Smith et al., 2009, *Cape Town*, p.116. For persisting concern over flower-picking, see D.J. McDonald and R.M. Cowling, 1995, 'Towards a Profile of an Endemic Mountain Fynbos Flora: Implications for Conservation,' *Biological Conservation*, 72, 1–12, 8.
20. Levyns, 1924, 'Some Observations of the Effect of Bush Fires,' 346; N.S. Pillans, 'Destruction of Indigenous Vegetation by Burning on the Cape Peninsula,' *SAJS*, 21, 348–50, 348.
21. *Cape Argus*, 'Mountain under State Protection,' 11.
22. C.A. Lückhoff, 1951, *Table Mountain: Our National Heritage after Three Hundred Years* (Cape Town, South Africa: A.A. Balkema), p.117.
23. *Cape Times*, 'Up in Smoke,' 9 February 1938, 12.
24. *Cape Times*, 'Fire Lighter Caught,' 7 January 1944, 7; *Cape Times*, 'Incendiaries,' 11 January 1944, 4.
25. *Cape Argus*, 'Mountain under State Protection,' 11; *Cape Times*, 'Up in Smoke.'
26. *Cape Times*, 'Lion's Head Fire,' 22 January 1909, 7.
27. *Cape Times*, 'The World Goes By: Week-end Fires,' 3 January 1934, 10.
28. *Cape Times*, 'Huge Fire Sears Apostles,' 21 April 1973, front page.
29. *Cape Times*, 'Too Little Action on Fire, says Councillor,' 9 March 1982, front page.
30. V. Bickford-Smith, 2009, 'Creating a City of the Tourist Imagination: The Case of Cape Town, "The Fairest Cape of Them All,"' *Urban Studies*, 46, 9, 1763–85, 1768–70.
31. MM1912, ARSTP, 2.
32. Bickford-Smith, 'Creating a City,' 1770–71.
33. R.H. Compton, 1924, 'Notes and News,' *JBSSA*, 10, 3.
34. See R.H. Compton, 1928, 'Notes and News,' *JBSSA*, 14, 5; Bickford-Smith et al., *Cape Town*, p.63; Bickford-Smith, 'Creating a City,' 1770; *Cape Times*, 'Clifton Bush Fire,' 20 January 1934, 11.
35. Compton, 1924, 'Notes and News,' 10, 3.
36. *Cape Argus*, 'To Safeguard the Mountain,' 27 December, 1935, 11, 12.
37. Compton, 1924, 'Notes and News,' 10, 3; *Cape Argus*, as follows: 'To Safeguard the Mountain,' 27 December, 1935, 11, 12; 'The Tragedy of Mountain Fires,' Letters to the Editor, 26 December, 1935, 14; 'The Mountain Fire,' and 'Mountain Fires – Where Does the Blame Lie?,' Letters to the Editor, 27 December 1935, 14; '£250,000 Havoc in Great Fire,' 26 December, 1934, 12; 'The Fire Controversy,' Editorial, 4 January 1935, 8.
38. *Cape Argus*, 'To Safeguard Mountain,' 12; Union of South Africa, ARFD U.G., 1937/38 U.G.46 – 1938, 17, 26; MM1940, Annual Report of the City Lands and Forests Branch (CLFB), 40.

39. MM1948, CLFB, pp.9–10; MM1949, pp.7–8.
40. *Cape Times*, '£1/2 Million New Scenic Drives Round Cape: Added Attraction for Tourists,' 29 January 1960, front page; *Cape Times*, 'Record Year for Nature Reserve,' 12 January 1961, 6.
41. Republic of South Africa, ARFD R.P. 1972/73 R.P.38/1974, 7.
42. *Cape Times*, 'Exodus from Reef to Cape,' 24 December 1971, 3.
43. Annual Report of the City Engineer (ARCE), 1979, Parks and Forests Branch (ARPFB), 18–20; E.J. Moll, B. McKenzie, D. McLachlan, and B.M. Campbell, 1978, 'A Mountain in a City – The Need to Plan the Human Usage of the Table Mountain National Monument, South Africa,' *Biological Conservation*, 13, 2, 117–31, 127.
44. *Cape Times*, 'Make Your Picnic Fire Proof!' 27 April 1973, 13.
45. ARFD1975/76 R.P.39/1977, p.18; ARCE1979, ARPFB, pp.18–20.
46. ARCE1982, p.36.
47. ARCE1982, 'Cape Town Mountain Fires: Enemy or Friend?' 14; ARCE1982, ARPFB, p.34; ARCE1984/85, p.63.
48. *Cape Times*, 'Lessons from the Fire,' Editorial, 19 December 1986, 12; Anon, 2000, 'Draft Conservation Development Framework for the Cape Peninsula National Park,' prepared for South African National Parks by Settlement Planning Services, A4.
49. ARFD1978/79 R.P.39/1980, p.91; ARFD1979/80 R.P.51/1981, p.243.
50. ARCE1990/91, p.43; ARCE1991/92, p.34; ARCE1992/93, p.34; ARCE1993/94, p.34; ARCE1994/95, p.34, ARCE1995/96, p.24.
51. Data courtesy of Lance van Sittert, 3 October 2008.
52. Bickford-Smith, 'Creating a City,' 1773; Anon, 2000, 'Draft Conservation Development Framework,' 20.
53. J.E. Keeley, C.J. Fotheringham, and M. Morais, 1999, 'Re-examining Fire Suppression Impacts on Brushland Fire Regimes,' *Science*, 284, 1829–32, 1829; J.E. Keeley and C.J. Fotheringham, 2001, 'Historic Fire Regime in Southern California Shrublands,' *Conservation Biology*, 15, 6, 1536–48, 1545.

9   Fire on the Cape Peninsula, 1900–2000

1. F.J. Kruger, M. Mayet, W. Alberts, J.G. Goldammer, K. Tolhurt, and S. Parker, 2000, 'A Review of the Veld Fires in the Western Cape during 15 to 25 January 2000,' *Department of Water Affairs and Forestry Report*, 75, 101.
2. MM1902, p.50; MM1945, p.2; Annual Report of the Fire Brigade (ARFB) 1949–53; MM1949, pp.9–10; MM1986/87, p.16.
3. For names and dates of various departments, see Selected References.
4. MM1903, p.72; MM1914, p.107.
5. MM1929, p.21; MM1941, ARFB, p.6.
6. MM1933, ARFB, p.16.
7. *Cape Times*, 'Crossroads Fire Fiasco,' Editorial, 23 December 1986, 8.
8. MM1935, p.28.
9. MM1900, p.xcvii; MM1947, Annual Report of the City Lands and Forests Branch (CLFB), p.10.
10. *Cape Times*, 'Great Bush Fires,' 19 February 1909.

11. MM1909, pp.27, 29; *Cape Times*, 'Big Bush Blaze on Lion's Head,' 21 January 1909; *Cape Times*, 'Lion's Head Fire,' 22 January 1909.

12. MM1915, Report of the Superintendent of Tree Planting, p.2.

13. MM1921, ARFB, p.i; *Cape Times*, 'Forest Fires,' 2 March 1920, 6; *Cape Times*, 'Need for Legislation,' Letters to the Editor, 4 March 1920.

14. Annual Report of the Forest Department (ARFD), U.G. (No reference printed), 1921/22, p.5.

15. MM1923, p.108.

16. R.H. Compton, 1929, 'Notes and News,' *JBSSA*, 15, 4.

17. MM1930, p.7 'Bush Fires'; *Cape Times*, 'Fire Patrols on Mountain,' 24 January 1934, 10; ARFD1931/32 U.G.37–1932, p.44.

18. *Cape Times* as follows: 'Five Fierce Bush Fires,' 3 January 1934; 'Clifton Bush Fire,' 20 January 1934; 'Fire Patrols on Mountain,' 24 January 1934.

19. ARFD1935/36 U.G.53–1936, pp.6, 16; *Cape Argus*, '£250,000 Havoc in Great Fire,' 26 December 1935; *Cape Argus*, 'To Safeguard Mountain,' 27 December 1935.

20. *Cape Times*, 'Terrific Fire Sweeps Peninsula Mountains, Constantiaberg Ablaze,' 7 February 1938, 13; *Cape Times*, 'The Great Fire Sweeps Down into Hout Bay: Residents Flee in Terror,' 8 February 1938, 13.

21. *Cape Times*, as follows: 'Terrific Fire Sweeps Peninsula Mountains'; 'Great Fire Sweeps down into Hout Bay'; 'Tragedy of Bush Fires,' Letters to the Editor, 8 February 1938, 17; 'Causes of Mountain Fires,' Letters to the Editor, 11 February 1938, 17; 'Fires in Hout Bay Valley,' Letters to the Editor, 16 February 1938, 17.

22. *Cape Times*, 'Irreparable Damage Done by Great Bush Fire,' 9 February 1938, 15; *Cape Times*, 'Bush and Forest Fires Menace,' 11 February 1938, 14. On Working on Fire, see http://www.workingonfire.org/.

23. *Cape Times*, 'Up in Smoke,' Editorial, 9 February 1938.

24. MM1940, CLFB, p.41; C.A. Lückhoff, 1951, *Table Mountain: Our National Heritage after Three Hundred Years* (Cape Town, South Africa: A.A. Balkema), p.138. For forestry contributions: ARFD1939/40 U.G.44–1940, p.32; ARFD1948/49 U.G.24–1950, p.19.

25. *Cape Times*, 'Silvermine Valley Ablaze,' 31 January 1942, 11; *Cape Times*, 'Extensive Bush Fires above Clifton,' 14 February 1942, 13.

26. *Cape Times*, as follows: 'Fire Devastation at Gordon's Bay,' 18 November 1942, 9; 'Stop those Mountain Fires,' Editorial, 27 November 1942, 6; 'Franschhoek Mountain Blaze,' 23 December 1942, 9.

27. MM1940, CLFB, p.41. See also Lückhoff, *Table Mountain*, p.138; MM1947, CLFB, p.10; *Cape Times*, 'Beauties of the Peninsula,' 2 December 1942, 9; Brunt quote from MM1941, CLFB, pp.45, 46.

28. On the CPS, see V. Bickford-Smith, E. Van Heyningen, and N. Worden, 2009, *Cape Town in the Twentieth Century: An Illustrated Social History* (Cape Town, South Africa: David Philip), p.93.

29. MM1940, CLFB, p.6; MM1947, CLFB 1946, p.10.

30. *Cape Times*, as follows: 'Native Influx,' 17 February 1943, 5; 'Native Influx,' 25 February 1943, 5; 'The Crowded Peninsula,' Editorial, 2 December 1943, 4.

31. *Cape Times*, as follows: 'Two Large Bush Fires at Oudekraal,' 20 December 1943; 'Reef Fire at Lakeside,' 22 December 1943; 'Fish Hoek Bush Fire,' 23 December 1943; 'Glencairn Valley Swept by Fire,' 24 December 1943; 'Second

Veld Fire at Glencairn,' 28 December 1943; 'Fire Lighter Caught,' 7 January 1944; '200 School Cadets Fight Bush Fire,' 8 January 1944; 'Great Bush Fire on Mountain Slopes,' 17 January 1944; 'Bush Fire at Clifton,' 18 January 1944; 'Big Mountain Fire at Simon's Town,' 21 January 1944, front page. On Brunt, see MM1947, p.10.

32. Lückhoff, *Table Mountain*, pp.136–7.
33. *Cape Times*, 'Royal Family Win Cape Town' and 'National Road Cut by Fire,' front page, 18 February 1947.
34. MM1948, CLFB, p.10; ARFD1946/47 U.G.26–1948, p.3; *Cape Times*, as follows: 'National Road Cut by Fire; Hout Bay Homes Threatened,' 18 February 1947, front page; 'Extinguished after Three Days; Hout Bay Fire Battle,' 20 February 1947, 5; 'Too Many Fires,' Editorial, 19 February 1947, 6.
35. MM1947, CLFB 1946, pp.9–10.
36. MM1949, CLFB, p.7; MM1949, ARFB, p.1; *Cape Times*, 'Slogan Against Bush Fires,' 11 January 1949, 7; *Cape Times*, 'Winners of Fire Slogan Contest,' 5 February 1949, front page.
37. Lückhoff, *Table Mountain*, p.139.
38. Ibid., pp.140–2.
39. MM1950, CLFB, p.11; ARFD1948/49 U.G.24/1950, p.4.
40. ARFD1949/50 U.G.30/1951, p.5; MM1951, CLFB, p.9.
41. MM1952, Annual Report of the City Engineer (ARCE) 1951, pp.9–11.
42. ARCE1955, p.23.
43. On Forests, Parks and Gardens department, see MM1954, ARCE1953, p.23.
44. ARFD1952/53 U.G.47/1954, p.3.
45. Ibid., p.6.
46. ARFD1958/59 U.G.69/1960, p.9.
47. ARFD1957/58 U.G.28/1960, pp.1, 4.
48. *Cape Times*, 'Weekend of Fire Fighting for Western Cape,' 7 November 1960, 3; *Cape Times*, 'Two Houses Burnt Down in Kommetjie Fire,' 22 November 1960, front page.
49. ARFD1960/61 R.P.45/1964, p.8.
50. *Cape Times*, 'Red Hill Fire Sweeps across Battery,' 3 January 1961, front page; *Cape Times*, '100 Square Miles of "Black Cape Carpets",' 4 January 1961, 2.
51. *Cape Times*, 'These Men are Ready to Beat Mountain Fires,' 7 December 1960, 13.
52. Ibid., and *Cape Times*, 'Record Year for Nature Reserve,' 12 January 1961, 6.
53. See page 153.
54. ARFD1961/62 R.P.80/1964, pp.1, 7; ARFD1965/66 R.P.44/1967, p.12.
55. ARFD1963/64 R.P.34/1966, p.3; ARFD1965/66 R.P.44/1967, p.7.
56. ARCE1970, ARFB, p.31; ARCE1971, ARFB, p.35; ARCE1972, ARFB, p.38; ARCE1973, ARFB, pp.59–60; ARCE1974, ARFB, pp.64–5; ARCE1975, ARFB, pp.62–4; *Cape Times*, 'Exodus from Reef to Cape,' 24 December 1971; G.L. Shaughnessy, 1980, 'Historical Ecology of Alien Woody Plants in the Vicinity of Cape Town, South Africa,' (PhD thesis, University of Cape Town), 307.
57. *Cape Times*, 'Mountain Fire Hazard the Worst Ever,' 10 April 1973, 13.
58. *Cape Times*, 'Huge Fire Sears Apostles,' 21 April 1973, front page; *Cape Times*, 'Police Investigate Weekend Fires,' 24 April 1973, front page.
59. Ibid., and *Cape Times*, 'Drastic Fire Action Needed,' 26 April 1973, 13.
60. *Cape Times*, ' "No" to Plane Fire-Fighters,' 21 February 1974.

61. ARFD1975/76 R.P.39/1977, p.18; ARFD1977/78 R.P.43/1979, p.26; ARCE1978, Annual Report Parks and Forests Branch (ARPFB), p.51.
62. *Cape Times*, 'Mountain Blaze Threatens Homes,' 8 February 1974.
63. ARFD1975/76 R.P.39/1977, p.6.
64. C.H. Feinstein, 2005, *An Economic History of South Africa* (Cambridge: Cambridge University Press), pp.226–7.
65. J.G. Brand, 'Foreword,' ARCE1976, p.1.
66. ARCE1977, p.41; ARCE1979, pp.18, 59.
67. ARCE1979, CLFB, pp.18–20.
68. *Cape Times*, as follows: 'Fire Rages on Devil's Peak,' 8 March 1982, front page; 'Houses Ablaze in Huge Fire,' 8 March 1982, front page; 'Too Little Action on Fire, says Councilor,' 9 March 1982, front page; ARFD1981/82 R.P.105/1982, p.176.
69. *Cape Times*, 'Too Little Action on Fire,' 3.
70. *Cape Times*, 'City Fire-Fighting Approach Criticized,' 9 March 1982, 3.
71. *Cape Times*, 'Letters,' 11 March 1982, 10.
72. ARCE1982, ARPFB, p.14; *Cape Times*, 'Forty Winks for a Few Weary Firemen,' 9 March 1982, 4; *Cape Times*, 'Fire and the Public,' Editorial, 9 March 1982, 12.
73. *Cape Times*, 'Criticism of Fire-Fighting "Unjustified",' 13 March 1982, front page, 2.
74. *Cape Times*, 'Sinking Rand and Gold Hit Economy,' 10 March 1982, front page.
75. *Cape Times*, as follows: 'Raging Veld Fires, Adrift Yachts as Gale Lashes City,' 17 December 1986, 3; ARCE1986/87, 'The Table Mountain Fire of December 1986,' 14–18; 'Ring of Fire,' 18 December 1986, front page; 'Task Force Stands Watch,' 19 December 1986, 3.
76. ARCE1986/1987, 'The Table Mountain Fire,' 16.
77. *Cape Times*, 'Ring of Fire'; *Cape Times*, 'Task Force Stands Watch,' 3; ARCE1986/1987, 'The Table Mountain Fire,' 16.
78. *Cape Times*, 'Fire Report Promised,' 19 December 1986, 3.
79. *Cape Times*, 'Lessons from the Fire,' Editorial, 19 December 1986, 12.
80. Ibid.
81. ARCE1986/87, 'The Table Mountain Fire,' p.18; ARCE1989/90, ARPFB, p.45.
82. ARCE1986/87, ARPFB, p.51.
83. G.G. Forsyth and B.W. van Wilgen, 2008, 'The Recent Fire History of the Table Mountain National Park and its Implications for Fire Management,' *Koedoe*, 50, 1, 3–9, 6.
84. *Cape Times*, 'Smokeout: Mountain Fire Chokes City,' (9 February 1991), front page, 3; ARCE1990, p.20.
85. Ibid.; *Cape Times*, 'Blazes Caused by Arsonist?,' 12 February 1991, 5; *Cape Times*, '60 Fire-Fighters Stay on Alert at Sandy Bay,' 13 February 1991, 9.
86. *Cape Times*, as follows: 'Fire Aftermath,' Editorial, 12 February 1991, 8; 'Fire Probe: Council "Quite Satisfied",' 13 February 1991, 9; ARCE1994/95, ARPFB, p.32; 'Lion's Head Fire: German Tourists Airlifted to Safety,' 24 March 1995, front page. The *kramat* is the burial place of Sheikh Mohamed Hassen Ghaibie Shah.
87. *Cape Times*, 'Council to Explain Fire Action Today,' 11 February 1991, 3.

88. *Cape Times*, 'Mountain Bushfire Rages out of Control,' 16 January 1992, 2; *Cape Times*, 'Battle to Put out Huge Blaze,' 18 January 1992, front page.
89. Data supplied courtesy of Lance van Sittert, 3 October 2008.
90. Vivian Bickford-Smith, 2009, 'Creating a City of the Tourist Imagination: The Case of Cape Town, "The Fairest Cape of Them All",' *Urban Studies*, 46, 9, 1763–85; Anon, 2000, 'Draft Conservation Development Framework for the Cape Peninsula National Park,' prepared for South African National Parks by Settlement Planning Services, 20.
91. Forsyth and van Wilgen, 'The Recent Fire History of the Table Mountain'; Kruger et al., 'Review of the Veld Fires,' 29, 30.
92. Interview with Philip Prins, Newlands Fire Base, Cape Town, 19 October 2007.
93. My account of these fires is based primarily on Kruger et al., 'Review of the Veld Fires'; L. Aupiais and I. Glenn (eds.), 2000, *The Cape of Flames: The Great Fire of January 2000* (Cape Town, South Africa: Inyati Publishing).
94. Kruger et al., 'Review of the Veld Fires,' 62–8.
95. Ibid., 72–3.

Conclusion

1. See W. Beinart, 2003, *The Rise of Conservation in South Africa: Settlers, Livestock, and the Environment 1770–1950* (Oxford: Oxford University Press), Chapter 11, on 'debating degradation over the long term.'
2. F.J. Kruger, P. Reid, M. Mayet, W. Alberts, J.G. Goldammer, K. Tolhurt, and S. Parker, 2000, 'A Review of the Veld Fires in the Western Cape during 15 to 25 January 2000,' *Department of Water Affairs and Forestry Report*, 108, 117.

Appendix 1: Cape Peninsula Vegetation, Climate, Weather and Fire

1. A.G. Rebelo, C. Boucher, N. Helme, L. Mucina, and M.C. Rutherford, 2006, 'Fynbos Biome,' in L. Mucina and M.C. Rutherford (eds.), *The Vegetation of South Africa, Lesotho and Swaziland* (Pretoria, South Africa: South African National Biodiversity Institute), pp.53–219, 139–40, 183.

Appendix 2: Fire Causes

1. MM1943, ARFB, 11.

Selected References

Books

Anderson, D., and R.H. Grove, 1987, *Conservation in Africa: People, Policies and Practice* (Cambridge: Cambridge University Press).

Anker, P., 2002, *Imperial Ecology: Environmental Order in the British Empire, 1895–1945* (Cambridge, MA: Harvard University Press).

Aupiais, L., and I. Glenn (eds.), 2000, *The Cape of Flames: The Great Fire of January 2000* (Cape Town, South Africa: Inyati Publishing).

Barton, G.A., 2002, *Empire Forestry and the Origins of Environmentalism* (Cambridge: Cambridge University Press).

Beinart, W., 1994, *Twentieth-Century South Africa* (Oxford: Oxford University Press).

——, 2003, *The Rise of Conservation in South Africa: Settlers, Livestock, and the Environment 1770–1950* (Oxford: Oxford University Press).

Bell, S., 2005, *The Happy Warrior: The Story of T.C. Robertson* (Howick, South Africa: T.C. Robertson Trust).

Bews, J.W., 1918, *The Grasses and Grasslands of South Africa* (Pietermaritzburg, South Africa: P. Davis & Sons).

Bickford-Smith, V., E. Van Heyningen, and N. Worden, 1999, *Cape Town in the Twentieth Century: An Illustrated Social History* (Cape Town, South Africa: David Philip).

Bond, W.J., and B.W. van Wilgen, 1996, *Fire and Plants* (London: Chapman & Hall).

Boonzaier, E., P. Berens, C. Malherbe, and A. Smith, 1996, *The Cape Herders: A History of the Khoikhoi of Southern Africa* (Athens, OH: Ohio University Press).

Booysen, P. du V., and N.M. Tainton (eds.), 1984, *Ecological Effects of Fire in South African Ecosystems* (Berlin, Germany: Springer Verlag).

Brown, J.C., 1875, *Hydrology of South Africa* (London: Henry S King & Co.).

——, 1887, *Management of Crown Forests at the Cape of Good Hope under the Old Regime and under the New* (Edinburgh: Oliver & Boyd).

Burchell, W.J., 1822, *Travels in the Interior of Southern Africa*, Vol.1, (London: Longman, Hurst, Rees, Orme and Brown).

Burman, J., 1971, *Disaster Struck South Africa* (Cape Town, South Africa: C. Struik).

Carruthers, J., 1995, *The Kruger National Park: A Social and Political History* (Pietermaritzburg, South Africa: University of Natal Press).

Control of Alien Vegetation Committee, Kirstenbosch, 1959, *The Green Cancers in South Africa: The Menace of Alien Vegetation* (Cape Town, South Africa: Citadel Press).

Cowling, R.M. (ed.), 1992, *The Ecology of Fynbos: Nutrients, Fire and Diversity* (Cape Town, South Africa: Oxford University Press).

——, 1997, *The Vegetation of Southern Africa* (Cambridge: Cambridge University Press).

Crosby, A.W., 1986, *Ecological Imperialism: The Biological Expansion of Europe 900–1900* (Cambridge: Cambridge University Press).

Dubow, S., 2006, *A Commonwealth of Knowledge: Science, Sensibility, and White South Africa 1820–2000* (Oxford: Oxford University Press).

Elphick, R., and H. Giliomee (eds.), 1992, *The Shaping of South African Society, 1652–1840* (Cape Town, South Africa: Maskew Miller Longman).

Fairhead, J., and M. Leach, 1996, *Misreading the African Landscape: Society and Ecology in a Forest-Savanna Mosaic* (Cambridge: Cambridge University).

Feinstein, C.H., 2005, *An Economic History of South Africa* (Cambridge: Cambridge University Press).

Field, S., F. Swanson, and R. Meyer (eds.), 2007, *Imagining the City: Memories and Cultures in Cape Town* (Cape Town, South Africa: HSRC Press).

Fisher, W.R., 1907, *Forest Protection, Dr Schlich's Manual of Forestry*, Vol.4, 2nd ed. (Bradbury: Agnew & Co.).

Gale, G.W., 1954, *John William Bews: A Memoir* (Pietermaritzburg, South Africa: University of Natal Press).

Glen, H.F., and G. Germishuizen (compilers), 2010, *Botanical Exploration of Southern Africa*, 2nd ed., Strelitzia 26 (Pretoria, South Africa: South African National Biodiversity Institute).

Godée Molsbergen, E.C., 1968, *Jan van Riebeeck en sy Tyd* (Pretoria, South Africa: J.L. van Schaik).

Gordon Bagnall, D., and H. Louw, 1962, *Bokkie the Grysbuck*, ed. Douglas Hey (Cape Town, South Africa: Department of Nature Conservation & Cape Peninsula Fire Protection Committee).

Grove, R.H., 1995, *Green Imperialism: Colonial Expansion, Tropical Island Edens and the Origins of Environmentalism, 1600–1860* (Cambridge: Cambridge University Press).

Grut, M., 1965, *Forestry and Forest Industry in South Africa* (Amsterdam, the Netherlands: A.A. Balkema).

Gunn, M., and L.E. Codd, 1981, *Botanical Exploration of Southern Africa* (Cape Town, South Africa: A.A. Balkema).

Harper, J.L., 1977, *Population Biology of Plants* (London: Academic Press).

Hodge, J., 2007 *Triumph of the Expert: Agrarian Doctrines of Development and the Legacies of British Colonialism* (Athens, OH: Ohio University Press).

Hodge, J., and B. Bennett (eds.), 2012, *Science and Empire: Knowledge and Networks of Science in the British Empire 1800–1970* (London: Palgrave Macmillan).

Jones, S., and A. Muller, 1992, *The South African Economy, 1910–90* (Hampshire: Palgrave Macmillan).

Keeley, J.E., W.J. Bond, R.A. Bradstock, J.G. Pausas, and P.W. Rundel, 2012, *Fire in Mediterranean Ecosystems: Ecology, Evolution and Management* (Cambridge: Cambridge University Press).

Kolb, P., 1738, *The Present State of the Cape of Good-Hope*, Vol.1, 2nd ed. (London: Innys and Manby).

Kruger, F.J., D.T. Mitchell, and J.U.M. Jarvis (eds.), 1983, *Mediterranean-Type Ecosystems: The Role of Nutrients* (Berlin, Germany: Springer-Verlag).

Kull, C.A., 2004, *Isle of Fire: The Political Ecology of Landscape Burning in Madagascar* (Chicago, IL: Chicago University Press).

Le Vaillant, F., 1791, *Travels into the Interior Parts of Africa by the Cape of Good Hope, in the years 1780, 81, 82, 83, 84 and 85*, Vol.1, 2 Volumes, (Perth, Australia: R. Morrison & Son).

Lichtenstein, H., 1812, *Travels in Southern Africa in the years 1803, 1804, 1805, and 1806*, translated by Anne Plumptre (London: Henry Colburn).

Lückhoff, C.A., 1951, *Table Mountain: Our National Heritage after Three Hundred Years* (Cape Town, South Africa: A.A. Balkema).

Macdonald, I.A.W., F.J. Kruger, and A.A. Ferrar, 1986, *The Ecology and Management of Biological Invasions in Southern Africa* (Cape Town, South Africa: Oxford University Press).

Mentzel, O.F., 2006 [1787], *A Geographical-Topographical Description of the Famous and (All Things Considered) Remarkable African Cape of Good Hope*, Part I, translated by H. J. Mandelbrote, First Reprint Series, 4 (Cape Town, South Africa: Van Riebeeck Society).

——, *Life at the Cape in the Mid-Eighteenth Century, being the Biography of Rudolph Siegfried Allemann, Captain of the Military Forces at the Cape of Good Hope*, translated by M. Greenlees, First Reprint Series, 2 (Cape Town, South Africa: Van Riebeeck Society).

Moffat, R., 1846, *Missionary Labours and Scenes in Southern Africa* (London: John Snow).

Mooney, H.A., 1998, *The Globalization of Ecological Thought* (Oldendorf/Luhe, Germany: Ecology Institute).

Mucina, L., and M.C. Rutherford (eds.), 2006, *The Vegetation of South Africa, Lesotho and Swaziland*, Strelitzia 19 (Pretoria, South Africa: South African National Biodiversity Institute).

Pakenham, T., 1992, *Scramble for Africa* (London: Abacus).

——, 2007, *In Search of Remarkable Trees: On Safari in Southern Africa* (London: Weidenfeld & Nicolson).

Perkins Marsh, G.P., 1864, *Man and Nature: Or, Physical Geography as Modified by Human Action* (New York: Charles Scribner).

Pringle, J., 1982, *The Conservationists and the Killers* (Cape Town, South Africa: T.V. Bulpin and Books of Africa).

Pyne, S.J., 1997, *Fire in America: A Cultural History of Wildland and Rural Fire* (Seattle, WA: University of Washington Press).

——, 1997, *Vestal Fire: An Environmental History, told through Fire, of Europe and Europe's Encounter with the World* (Seattle, WA: University of Washington Press).

——, 2010, *America's Fires a Historical Context for Policy and Practice* (Durham, NC: Forestry History Society).

Raven-Hart, R., 1971, *Cape Good Hope 1652–1702: The First Fifty Years of Dutch Colonization as Seen by Callers*, Vol.1, (Cape Town, South Africa: A.A. Balkema).

Reitz, D., 1975 [1929], *Commando: A Boer Journal of the Boer War* (London: Faber and Faber).

Richardson, D.M. (ed.), 2011, *Fifty Years of Invasion Ecology: The Legacy of Charles Elton* (Oxford: Wiley-Blackwell).

Ross, J.C., 1963, *Soil Conservation in South Africa* (Pretoria, South Africa: Department of Agricultural Technical Services).

Roux, E., 1946, *The Veld and the Future* (Cape Town, South Africa: The African Bookman).

Schlich, W., 1906, *Forest Policy in the British Empire, Schlich's Manual of Forestry*, Vol.1, 3rd ed. (Bradbury: Agnew & Co.).

Scott, J.C., 1999, *Seeing like a State: How Certain Schemes to Improve the Human Condition Have Failed* (New Haven, CT: Yale University Press).

Sim, T.R., 1907, *The Forests and Forest Flora of the Colony of the Cape of Good Hope* (Aberdeen: Taylor & Henderson).

Sleigh, D., 1993, *Die Buiteposte* (Pretoria, South Africa: Protea Boekhuis).

Slingsby, P., and A. Jones, 2009, *T.P. Stoekoe: The Man, the Myths, the Flowers* (Cape Town, South Africa: Baardskeerder).

Smith, A.B., 1992, *Pastoralism in Africa: Origins and Development Ecology* (London: Hurst & Co.).

Smuts, J.C., 1942, 'The Spirit of the Mountain,' in J.C. Smuts, *Plans for a Better World: Speeches of Field-Marshal The Right Honourable J.C. Smuts* (London: Hodder & Stoughton).

Sparrman, A., 1785, *A Voyage to the Cape of Good Hope, towards the Antarctic Polar Circle, and Round the World: But Chiefly into the Country of the Hottentots and Caffres, from the Year 1772, to 1776*, 2 Volumes, Vol.1, (London: G.G.J. and J. Robinson).

Stead, W.T. (ed.), 1902, *The Last Will and Testament of Cecil John Rhodes, with Elucidatory Notes, to which are Added Some Chapters Describing the Political and Religious Ideas of the Testator* (London: Review of Reviews Office).

Stirton, C.H., 1983 [1978], *Plant Invaders: Beautiful, But Dangerous: A Guide to the Identification and Control of Twenty-six Plant Invaders of the Province of the Cape of Good Hope* (Cape Town, South Africa: Department of Nature and Environment Conservation).

Strohmaier, D.J., 2005, *Drift Smoke: Loss and Renewal in a Land of Fire* (Reno, NV: University of Nevada Press).

Thom, H.B. (ed.), 1952 (1651–55), 1954 (1656–58), 1958 (1656–58), *Journal of Jan van Riebeeck*, 3 volumes (Cape Town, South Africa: A.A. Balkema).

Thunberg, C.P., 1986 [1793–95], *Travels at the Cape of Good Hope 1772–1775*, ed. V.S. Forbes, translated by J. and I. Rudner (Cape Town, South Africa: Van Riebeeck Society).

Tropp, J.A., 2006, *Nature's of Colonial Change: Environmental Relations in the Making of the Transkei* (Athens, OH: Ohio University Press).

Van Sittert, L., 1988, 'Slawe van die Fabriek' – The State, Monopoly Capital and the Subjugation of Labour in the Hout Bay Valley Crayfish Fishing Industry, 1946–1956,' pages 112–49 in *Studies in the History of Cape Town*, 6, (Cape Town, South Africa: Centre for African Studies, University of Cape Town).

Watson, S., 1989, *Cape Town Days, and Other Poems* (Cape Town, South Africa: Cecil Skotnes and Clarke's Bookshop).

——, 1995, *Presence of the Earth: New Poems* (Cape Town, South Africa: David Philip).

Worden, N., E. van Heyningen, and V. Bickford-Smith, 1998, *Cape Town, the Making of a City: An Illustrated Social History* (Cape Town, South Africa: David Philip).

Worster, D., 1994, *Nature's Economy: A History of Ecological Ideas*, 2nd ed. (Cambridge: Cambridge University Press).

Journal articles

Abbreviations

Biological Conservation (BiolCon)
Environment and History (E&H)
(The) *Journal of Ecology* (JoE)
Journal of South African Botany (JSAB)
Journal of Southern African Studies (JSAS)
Journal of the Botanical Society of South Africa (JBSSA)
Journal of the South African Forestry Association (JSAFA)
South African Forestry Journal (SAFJ)
South African Journal of Botany (SAJB)
South African Journal of Science (SAJS)
Transactions of the Royal Society of South Africa (TRSSA)

Adamson, R.S., 1927, 'The Plant Communities of Table Mountain: Preliminary Account,' *JoE*, 15, 2, 278–309.
——, 1953, 'Can We Preserve the Cape Flora?' *JBSSA*, 39, 11–12.
Anon, 1914, 'Seventh Award of the South Africa Medal and Grant,' *SAJS*, 11, xxxii–xxxvi.
——, 1920, 'Notes,' *Journal of the Department of Agriculture*, 1, 7, 605–6.
Avis, A.M., 1989, 'A Review of Coastal Dune Stabilization in the Cape Province of South Africa,' *Landscape and Urban Planning*, 18, 55–68.
Banks, C.H., 1964, 'Further Notes on the Effect of Autumnal Veldburning on Stormflow in the Abdolskloof Catchment, Jonkershoek,' *Forestry in South Africa*, 4, 79–84.
Beinart, W., 1984, 'Soil Erosion, Conservationism and Ideas about Development: A Southern African Exploration, 1900–1960,' *JSAS*, 11, 1, 52–83.
——, 2000, 'African History and Environmental History,' *African Affairs*, 99, 269–302.
Beinart, W., K. Brown, and D. Gilfoyle, 2009, 'Experts and Expertise in Africa Revisited,' *African Affairs*, 108, 413–33.
Beinart, W., and K. Middleton, 2004, 'Plant Transfers in Historical Perspective: A Review Article,' *E&H*, 10, 3–29.
Bennett, B.M., 2011, 'A Global History of Australian Trees,' *Journal of the History of Biology*, 44, 1, 125–145.
——, 2011, 'Naturalising Australian Trees in South Africa: Climate, Exotics and Experimentation,' *JSAS*, 27, 265–80.
——, 2013, 'The Rise and Demise of South Africa's First School of Forestry,' *E&H*, 19, 63–85.
——, 2014, 'Model Invasions and the Development of National Concerns over Invasive Introduced Trees: Insights from South African History,' *Biological Invasions*, 16, 3, 499–512.
Bennett, B.M., and F.J. Kruger, 2013, 'Ecology, Forestry and the Debate over Exotic Trees in South Africa,' *Journal of Historical Geography*, 42, 100–9.
Bews, J.W., 1912, 'The Vegetation of Natal,' *Annals of the Natal Government Museum*, 2, 3, 253–331.

——, 1913, 'An Œcological Survey of the Midlands of Natal,' *Annals of the Natal Government Museum*, 2, 4, 485–545.

——, 1916, 'An Account of the Chief Types of Vegetation in South Africa, with Notes on the Plant Succession,' *JoE*, 4, 129–59.

Bickford-Smith, V., 2009, 'Creating a City of the Tourist Imagination: The Case of Cape Town, "The Fairest Cape of Them All",' *Urban Studies*, 46, 9, 1763–85.

Bond, W.J., 1984, 'Fire Survival of Cape Proteaceae – Influence of Fire Season and Seed Predators,' *Vegetatio*, 56, 65–74.

Bond, W.J., G.F. Midgley, and F.I. Woodward, 2003, 'What Controls South African Vegetation – Climate or Fire?' *SAJB*, 69, 1, 1–13.

Bond, W.J., J. Vlok, and M. Viviers, 1984, 'Variation in Seedling Recruitment of Cape Proteaceae after Fire,' *JoE*, 72, 209–21.

Bond, W.J., and P. Slingsby, 1984, 'Collapse of an Ant-Plant Mutualism: The Argentine Ant (*Iridomyrmex humilis*) and Myrmecochorous Proteaceae,' *Ecology*, 65, 1031–7.

Bowman, D.M.J.S., J. Balch, P. Artaxo, W.J. Bond, M.A. Cochrane, C.M. D'Antonio, R. DeFries, F.H. Johnston, J.E. Keeley, M.A. Krawchuk, C.A. Kull, M. Mack, M.A. Moritz, S. Pyne, C.I. Roos, A.C. Scott, N.S. Sodhi, T.W. Swetnam, 2011, 'The Human Dimension of Fire Regimes on Earth,' *Journal of Biogeography*, 38, 2223–36.

Brown, K., 2001, 'The Conservation and Utilisation of the Natural World: Silviculture in the Cape Colony, c.1902–1910,' *E&H*, 7, 4, 427–47.

Brown, P.J., P.T. Manders, D.P. Bands, F.J. Kruger, and R.H. Andrag, 1991, 'Prescribed Burning as a Conservation Management Practice: A Case History from the Cederberg Mountains, Cape Province, South Africa,' *BiolCon*, 56, 2, 133–50.

Carruthers, J., 2011, 'Trouble in the Garden: South African Botanical Politics ca.1870–1950,' *SAJB*, 77, 258–67.

Carruthers, J., L. Robin, J. Hattingh, C. Kull, H. Rangan, and B. van Wilgen, 2011, 'A Native at Home and Abroad: The History, Politics, Ethics and Aesthetics of *Acacia*,' *Diversity and Distributions*, 17, 810–21.

Coates, P., 2003, 'Editorial Postscript: The Naming of Strangers in the Landscape,' *Landscape Research*, 28, 131–7.

Compton, R.H., 1926, 'Veld Burning and Veld Deterioration,' *South African Journal of Natural History*, 6, 5–19.

——. 'News and Notes,' *JBSSA*, as follows:
1924, 10, 3–5.
1927, 12, 4–5.
1927, 13, 4–5.
1928, 14, 5.
1929, 15, 4.
1930, 16, 3.
1930, 16, 4.
1936, 12, 2.
1944, 30, 5.

Cowling, R.M., 1987, 'Fire and its Role in Coexistence and Speciation in Gondwanan Shrublands,' *SAJS*, 83, 106–12.

Cowling, R.M., I.A.W. MacDonald, and M.T. Simmons, 1996, 'The Cape Peninsula, South Africa: Physiographical, Biological and Historical Background to an Extraordinary Hot-spot of Biodiversity,' *Biodiversity and Conservation*, 5, 527–50.

de Ronde, C.N., 1999, '1998: A Year of Destructive Wildfires in South Africa,' *International Forest Fire News*, 20, 73–8.

Dubow, S., 1992, 'Afrikaner Nationalism, Apartheid, and the Question of "Race",' *Journal of African History*, 33, 209–337.

Duncan, G., 2003, '*Lachenalia Salteri* Hyacinthaceae,' *Curtis's Botanical Magazine*, 20, 4, 208–9.

Forsyth, G.G., and B.W. Van Wilgen, 2008, 'The Recent Fire History of the Table Mountain National Park and its Implications for Fire Management,' *Koedoe*, 50, 1, 3–9.

Geldenhuys, C.J., 1977, 'Bergwind Fires and the Location of Forest Patches in the Southern Cape Landscape, South Africa,' *Journal of Biogeography*, 21, 1, 49–62.

Hagan, J.B., 2008, 'Teaching Ecology during the Environmental Age, 1965–80,' *Environmental History*, 13, 704–23.

Hale, F., 2005, 'Khoi Khoi Culture Refracted through a German Prism: Christian Ludwig Willebrand's Reconstruction in Historical Context,' *South African Journal of Cultural History*, 19, 2, 127–34.

Hall, A.V., 1959, 'Observations on the Distribution and Ecology of Orchidaceae in the Muizenberg Mountains, Cape Peninsula,' *JSAB*, 25, 265–78.

Harper, J.L., 1967, 'A Darwinian Approach to Plant Ecology,' *JoE*, 55, 2, 247–70.

Hey, D., 1963, 'The Conservation of Flora in the Province of the Cape of Good Hope,' *Lantern*, 13, 68–9.

Hill, A.R., 1931, 'Botany in South Africa: Sir Arthur Hill's Report,' *JBSSA*, 17, 6–9.

Jordaan, P.G., 1949, 'Aantekeninge oor die voortplanting en brandperiodes van *Protea mellifera* Thunb.,' *JSAB*, 15, 4, 121–5.

——, 1965, 'Die Invloed van 'n Winterbrand op die Voortplanting van Vier Soorte van die Proteaceae,' *Tydskrif vir Natuurwetenskappe*, 5, 27–31.

Joubert, C., 1996, 'Commercial Forestry on Signal Hill and Lion's Head, Cape Town,' *SAFJ*, 175, 43–54.

Keeley, J.E., and C J. Fotheringham, 2001, 'Historic Fire Regime in Southern California Shrublands,' *Conservation Biology*, 15, 6, 1536–48.

Keeley, J.E., C.J. Fotheringham, and M. Morais, 1999, 'Re-examining Fire Suppression Impacts on Brushland Fire Regimes,' *Science*, 284, 1829–32.

Kepe, T., 2005, 'Grasslands Ablaze: Vegetation Burning by Rural People in Pondoland, South Africa,' *South African Geographical Journal*, 87, 10–17.

King, N.L., 1938, 'Historical Sketch of the Development of Forestry in South Africa,' *JSAFA*, 1, 4–16.

Kull, C.A., and H. Rangan, 2008, 'Acacia Exchanges: Wattles, Thorn Trees, and the Study of Plant Movements,' *Geoforum*, 39, 1258–72.

Kwa, C., 1987, 'Representations of Nature Mediating Between Ecology and Science Policy: The Case of the International Biological Programme,' *Social Studies of Science*, 17, 3, 413–42.

Le Maitre, D.C., B.W. Van Wilgen, R.A. Chapman, and D.H. Kelly, 1996, 'Invasive Plants and Water Resources in the Western Cape Province, South Africa; Modelling the Consequences of a Lack of Management,' *Journal of Applied Ecology*, 33, 161–72.

Levyns, M.R., 1924, 'Some Observations of the Effect of Bush Fires on the Vegetation of the Cape Peninsula,' *SAJS*, 21, 346–7.

——, 1929, 'Veld-burning Experiments at Ida's Valley, Stellenbosch,' *TRSSA*, 17, 2, 61–92.

Lister, M.H., 1957, 'Joseph Storr Lister, the First Chief Conservator of the South African Department of Forestry,' *JSAFA*, 29, 10–18.

Marloth, R., 1924, 'Notes on the Question of Veld Burning,' *SAJS*, 21, 342–5.

Marris, E., 2009, 'Ragamuffin Earth,' *Nature*, 460, 450–3.

McDonald, D.J., and R.M. Cowling, 1995, 'Towards a Profile of an Endemic Mountain Fynbos Flora: Implications for Conservation,' *BiolCon*, 72, 1–12.

McLeod, R., 2000, ' "Introduction" to "Nature and Empire: Science and the Colonial Enterprise",' *Osiris*, 15, 1–13.

Michell, M.R., 1922, 'Some Observations on the Effects of a Bush Fire on the Vegetation of Signal Hill,' *TRSSA*, 10, 1, 346–47.

Midgley, J.J., 1989, 'Season of Burn of Serotinous Proteaceae: A Critical Review and Further Data,' *SAJB*, 55, 165–70.

Moll, E.J., B. McKenzie, D. McLachlan, and B.M. Campbell, 1978, 'A Mountain in a City – The Need to Plan the Human Usage of the Table Mountain National Monument, South Africa,' *BiolCon*, 13, 2, 117–31.

Moll, E.J., and T. Trinder-Smith, 1992, 'Invasion and Control of Alien Woody Plants on the Cape Peninsula Mountains, South Africa 30 Years On,' *BiolCon*, 60, 135–43.

Morris, S.S., 1958, 'Keep Pace with Progress – Suggested Improvements to the University Education of Civil Engineers,' *Transactions of the South African Institution of Civil Engineers*, 15, 374–7.

Neely, A.H., 2010, ' "Blame it on the Weeds": Politics, Poverty, and Ecology in the New South Africa,' *JSAS*, 36, 869–87.

Nobbs, P., 1956, 'A Pioneer of Botany and Forestry at the Cape of Good Hope,' *JSAFA*, 27, 87–9.

Noble, R., and R.O. Slatyer, 1980, 'The Use of Vital Attributes to Predict Successional Changes in plant Communities Subject to Recurrent Disturbances,' *Vegetatio*, 43, 5–21.

Phillips, E.P., 1930, 'A Brief Historical Sketch of the Development of Botanical Science in South Africa,' *SAJS*, 27, 39–80.

Phillips, J.F.V., 1928, 'Plant Indicators in the Knysna Region,' *SAJS*, 25, 202–24.

——, 1928, 'The Behaviour of *Acacia melanoxylon* R. Br. ('Tasmanian Blackwood') in the Knysna Forests: An Ecological Study,' *TRSSA*, 16, 1, 31–43.

——, 1931, 'Forest Succession and Ecology in the Knysna Region,' *Memoirs of the Botanical Survey of South Africa*, 14, 1–327.

——, 1936, 'Fire in Vegetation: A Bad Master, a Good Servant, and a National Problem,' *JSAB*, 2, 35–45.

——, 1954, 'A Tribute to Frederic E. Clements and his Concepts in Ecology,' *Ecology*, 35, 2, 114–15.

Pillans, N.S., 1924, 'Destruction of Indigenous Vegetation by Burning on the Cape Peninsula,' *SAJS*, 21, 348–50.

Pooley, S., 2009, 'Jan van Riebeeck as Pioneering Explorer and Conservator of Natural Resources at the Cape of Good Hope (1652–62),' *E&H*, 15, 3–33.

——, 2010, 'Pressed Flowers: Notions of Indigenous and Alien Vegetation in South Africa's Western Cape, c.1902–1945,' *JSAS*, 36, 599–618.

——, 2010, 'Recovering the Lost History of Fire in South Africa's Fynbos,' *Environmental History*, 17, 1, 55–83.

Potts, G., 1923, 'The Plant Succession in the Orange Free State, and the Need for Maintaining a Covering of Vegetation,' *SAJS*, 20, 1, 196–201.

Powell, J.M., 2007, ' "Dominion over Palm and Pine": The British Empire Forestry Conferences, 1920–1947,' *Journal of Historical Geography*, 33, 852–77.

Pyne, S.J., 2008, 'Spark and Sprawl,' *Forest History Today*, (Fall), 4–10.

Rebelo, A.G., P.M. Holmes, C. Dorse, and J. Wood, 2011, 'Impacts of Urbanization in a Biodiversity Hotspot: Conservation Challenges in Metropolitan Cape Town,' *SAJB*, 77, 20–35.

Rourke, J.P., 1999, 'Plant Systematics in South Africa: A Brief Historical Overview, 1753–1953,' *TRSSA*, 54, 1, 179–90.

Rycroft, H.B., 1947, 'A Note on the Immediate Effects of Veldburning on Stormflow in a Jonkershoek Stream Catchment,' *JSAFA*, 15, 80–8.

Scott, J.D., 1970, 'Pros and Cons of Eliminating Veld Burning,' *Proceedings of the Grassland Society of South Africa*, 5, 23–6.

Seydack, A.H.W., S.J. Bekker, and A.H. Marshall, 2007, 'Shrubland Fire Regime Scenarios in the Swartberg Mountain Range, South Africa: Implications for Fire Management,' *International Journal of Wildland Fire*, 16, 81–95.

Shlisky, A., R. Meyer, J. Waugh, and K. Blankenship, 2008, 'Fire, Nature, and Humans: Global Challenges for Conservation,' *Fire Management Today*, 68, 36–42.

Star, P., 2006, 'Ecology: A Science of Nation? The Utilization of Plant Ecology in New Zealand, 1896–1930,' *Historical Records of Australian Science*, 17, 197–207.

Trollope, W.S.W., 1973, 'Fire as a Method of Controlling Macchia (Fynbos) Vegetation on the Amatole Mountains of the Eastern Cape,' *Proceedings of the Grassland Society of South Africa*, 8, 35–41.

Van Sittert, L., 2000, ' "The Seed Blows about in Every Breeze": Noxious Weed Eradication in the Cape Colony, 1860–1909,' *JSAS*, 26, 655–74.

——, 2003, 'Making the Cape Floral Kingdom: The Discovery and Defence of Indigenous Flora at the Cape ca. 1890–1939,' *Landscape Research*, 28, 1, 113–29.

——, 2003, 'The Bourgeois Eye Aloft: Table Mountain in the Anglo Urban Middle Class Imagination, c.1891–1952,' *Kronos*, 29, 161–90.

Van Wilgen, B.W., 1984, 'Adaptation of the United States Fire Danger Rating System to Fynbos Conditions: Part I. A Fuel Model for Fire Danger Rating in the Fynbos Biome,' *SAFJ*, 129, 1, 61–5.

Van Wilgen, B.W., 1996, 'Management of the Natural Ecosystems of the Cape Peninsula: Current Status and Future Prospects,' *Biodiversity and Conservation*, 5, 671–84.

——, 2009, 'The Evolution of Fire and Invasive Plant Management Practice in Fynbos,' *SAJS*, 105, 335–42.

——, 2012, 'Evidence, Perceptions, and Trade-offs Associated with Invasive Alien Plant Control in the Table Mountain National Park, South Africa,' *Ecology and Society*, 17, 3.

Van Wilgen, B.W., and D.M. Richardson, 1985, 'The Effect of Alien Shrub Invasions on Vegetation Structure and Fire Behaviour in South African Fynbos Shrublands: A Simulation Study,' *Journal of Applied Ecology*, 22, 955–66.

Van Wilgen, B.W., D.M. Richardson, and A.H.W. Seydack, 1994, 'Managing Fynbos for Biodiversity: Constraints and Options in a Fire-prone Environment,' *SAJS*, 90, 322–8.

Van Wilgen, B.W., and M. Viviers, 1985, 'The Effect of Season of Fire on Serotinous Proteaceae in the Western Cape and the Implications for Fynbos Management,' *SAFJ*, 133, 49–53.

Van Wilgen, B.W., and R.E. Burgan, 1984, 'Adaptation of the United States Fire Danger Rating System: Part II. Historic Fire Danger in the Fynbos Biome,' *SAFJ*, 129, 1, 66–78.

Van Wilgen, B.W., R.M. Cowling, and C.J. Burgers, 1996, 'Valuation of Ecosystem Services,' *BioScience*, 46, 184–9.

Wade-Chambers, D., and R. Gillespie, 2000, 'Locality in the History of Colonial Science,' *Osiris*, 15, 221–40.

Wicht, C.L., 1939, 'Forest Influences Research Technique at Jonkershoek,' *JSAFA*, 3, 65–80.

——, 1948, 'A Statistically Designed Experiment to Test the Effects of Burning on a Sclerophyll Scrub Community. 1. Preliminary Account,' *TRSSA*, 31, 479–501.

——, 1948, 'Hydrological Research in South African Forestry,' *JSAFA*, 16, 4–21.

Wicht, C.L., and F.J. Kruger, 1973, 'Die Ontwikkeling van Bergveldbestuur in Suid-Afrika,' *SAFJ*, 86, Special Issue 2 'Our Green Heritage,' 1–17.

Woods, D.R., 1985, 'An Innovative and Far-sighted Research Support Program,' *BioEssays*, 3, 6, 272–73.

Worth, S.W., and B.W. van Wilgen, 1988, 'The Blushing Bride: Status of an Endangered Species,' *Veld & Flora*, 74, 122–3.

Wroughton, F.H., 1948, 'To Burn or Not to Burn,' *JSAFA*, 16, 76–8.

Conference Proceedings

British Empire Forestry Conferences

Third Conference, Australia and New Zealand, 1928, (Canberra, Australia: H.J. Green, Government Printer, 1928).

Cameron, D.R., 'Forest fire protection in Canada: Progress since 1923,' pp.721–33.

Kessell, S.L., 'Fire control in Australia,' pp.741–48.

Shebbeare, E.O., 'Fire protection and fire control in India,' pp.715–20.

Simmons, C.E., 'Forests Management in British India (exclusive of Burma),' pp.572–81.

Fourth Conference, South Africa, 1935 (Pretoria, South Africa: Government Printer, 1936).

Ainslie, J.R., (Nigeria), comments in the section 'Tropical forestry in relation to Agriculture,' pp.217–19.

Clements, J.B., (Nyasaland), comments in the section 'Tropical forestry in relation to Agriculture,' pp.214–15.

Keet, J.D.M., 'Soil erosion and allied problems in South Africa,' supplement to the Proceedings.

Watt, W.E., 'Forest Protection,' supplement to the Proceedings.

Other Conference Proceedings and talks

Kruger, F.J., 1977, 'Ecology of Cape Fynbos in Relation to Fire,' in H.A. Mooney and C.E. Conrad (eds.), 'Symposium on the Environmental Consequences of Fire,' 230–244.

——, 1982, 'Prescribing Fire Frequencies in Cape Fynbos in Relation to Plant Demography,' in C.E. Conrad and W.C. Oechel (Technical Coordinators), 'Symposium on Dynamics and Management of Mediterranean-Type Ecosystems,' 118–122.

Kruger, F.J., G.G. Forsyth, L.M. Kruger, K. Slater, D.C. Le Maitre, and J. Matshate, 2006, 'Classification of Veldfire Risk in South Africa for the Administration of the Legislation regarding Fire Management,' in D.X. Viegas (ed.), *5th International Conference on Forest Fire Research*, 27–30 November 2006, Portugal (Amsterdam, the Netherlands: Elsevier B.V.).

Lückhoff, H.A., and C.L. Wicht, 1975, 'The Influence of Demand and Supply Trends on the Research Programme of the Forest Research Institute, Pretoria, Republic of South Africa,' paper presented at an International Union of Forest Research Organizations (IUFRO) conference, Paris, 1–7.

Wall, K., 2008, 'Water Supply: Reshaper of Cape Town's Local Government a Century Ago,' in J.E. van Zyl, A.A. Ilemobade, and H.H. Jacobs (eds.), *Proceedings of the 10th Annual Water Distribution Systems Analysis Conference*, WDSA2008, Kruger National Park, South Africa.

Wicht, C.L., 1948, 'The Role of South African Forestry in the Conservation of Natural Resources,' corrected proof, paper presented at Conference Africaine des Sols, Goma, Congo Belge.

Wicht, C.L., 1973, 'Brand in die Fynbosveld,' three lectures available from National Research Foundation, South Africa, at: http://digi.nrf.ac.za/dspace/.

Archives

Cape Town Archives Repository, South Africa (CT AR)

Resolutions of the Council of Policy of Cape of Good Hope (*RCP CGH*), as follows:
24 October 1658, Ref.C.1
22 July 1659, CT AR, Ref.C.2
Wednesday February 1691, Ref.C.21
3 April 1691, CT AR, Ref.C.21
2 December 1697, Ref.C.23
19 August 1711, Ref.C.73
20 November 1715, Ref.C.34
26 December 1714, Ref.C.33
9 October 1725, Ref.C.73
10 February 1729, Ref.C.82
24 December 1733, Ref.C.94
20 March 1736, Ref.C.100
19 January 1741, Ref.C.116
1 December 1742, Ref.C.120
19 December 1748, Ref.C.126

16 November 1762, Ref.C.140
17 January 1787, Ref.C.174

State Forestry Department Annual Reports

Series reconstructed from personal copies and libraries as follows:

Rhodes House Library, University of Oxford
Republic of South Africa: Department of Forestry: Annual Report 1951–67, Shelf No.610.44 s. 5.

Plant Sciences Library, University of Oxford
BN/South Africa/Department of Environment Affairs: Report, Annual–Forestry Branch; and on Microfilm, 'Bulletins,' P3, BN/Rumania, 'Forestry and Environmental Conservation Branch Annual Report.
(This library now rehoused in the Radcliffe Science Library)
Reports series (with abbreviations used in text) as follows:
Cape of Good Hope (C.G.H.): Department of Agriculture, Reports of the Conservators of Forests, Cape Colony (to 1909).
Government of the Union of South Africa (U.G.): Forest Department: Report of the Chief Conservator of Forests (ARFD), 1910–58/59.
Republic of South Africa (R.P.): Annual Reports of the Department of Forestry (ARFD), 1959/60–78/79.
R.P.: Annual Report of the Department of Water Affairs and the Department of Forestry and Environmental Conservation (ARFD), 1979/80–83/84.
R.P.: Annual Report of the Department of Environment Affairs (ARFD), 1984/85–94.

Municipal Sources

Chancellor Oppenheimer Library, University of Cape Town, Government Publications Department.
Minute of His Worship the Mayor, Cape Town (MM): 1900–53
Shelf Ref.: G682 VC2 s.2247

Incorporating Annual Reports of the:
Chief Officer of the Metropolitan Fire Brigade (ARFB).
Superintendent of Public Gardens and Tree Planting (ARPGTP), 1900–09
Superintendents of Public Gardens, and Tree Planting (separately), 1910–28
Director of Parks and Gardens (ARDPG), c.1928–53.
City Lands and Forests Branch (CLFB), c.1939–53.

The Corporation of the City of Cape Town: Annual Report of the City Engineer (ARCE), Shelf Ref. G 682 VC ED(ENGI)

Incorporating the Annual Reports of:
Forests, Parks and Gardens (ARFPG) c.1954–60.
Parks and Forests Branch (ARPFB) 1961–94/95.
Annual Report of the Municipal Fire Brigade (ARFB)
Issued separately, 1954–1992/93, not archived, available from Cape Town Fire & Rescue Services.

Newspapers

Cape Times and *Cape Argus* newspaper reports accessed on microfilm at the African Studies Library, University of Cape Town, Ref. BZA 76/72, MP 1025, and Rhodes House Library, Oxford, call number 620.18 t. 2.

Royal Botanic Gardens, Kew, Library and Archive

Directors' Correspondence
Bolus, H., Harry Bolus to Colonel Prain, 192, 31–43, 29 April 1908.
Hill, A.W., 1929, Sir Arthur Hill's diary of his South Africa Trip, AWH/1/8.
Pappe, L., 'Letter from Ludwig Pappe to Sir William Jackson Hooker; from Cape Town, South Africa; 28 Jan 1857,' Vol.59, folio 254.
Stoekoe, T.P., 1921, letters to the Director of Kew, Vol.182, 96–108, No.s 100, 102 (21 July and 12 October).
Miscellaneous Reports: South Africa: National Botanic Garden, 1910–28.
Pearson, R.H., 1910, text of speech appended to a letter to Colonel Pierie, Director of Kew, 302, R6 (4 November 1910), 1–13.

Reports, Briefings, Pamphlets and Memorandums

Anon, 1924, 'Preliminary Reports on the Forest Requirements of the Eastern and Central Provinces, Gold Coast Colony' (Accra, Ghana: Government Press).
——, 1928, 'Forestry Handbook for New South Wales, Forestry Commission of New South Wales' (Sydney, Australia: Alfred James Kent, Government Printer).
——, 2000, 'Draft Conservation Development Framework for the Cape Peninsula National Park,' prepared for South African National Parks by Settlement Planning Services, A4.
——, 2001, 'Draft Conservation Development Framework and Associated Maps: Comment and Responses Report, Synthesis of Written Submissions by Settlement Planning Services and South African National Parks,' prepared for SANParks by de Villiers Brownlie Associates.
Bands, D.P., J.M. Bosch, A.J. Lamb, D.M. Richardson, B.W. van Wilgen, D.B. van Wyk, and D.B. Versfeld, 1987, *Jonkershoek Forestry Research Centre* Pamphlet 384 (Pretoria, South Africa: Department of Environment Affairs).
Conrad, C.E., and W.C. Oechel (Technical Coordinators), 1982, 'Symposium on Dynamics and Management of Mediterranean-Type Ecosystems,' San Diego State University, General Technical Report PSW-58 (Berkeley, CA: Pacific Southwest Forest and Range Experiment Station).
Day, J.A., N.R. Siegfried, and M.L. Jarman, 1979, *Fynbos Ecology: A Preliminary Synthesis*, CSIR Report No.40 (Pretoria, South Africa: CSIR).
Donaldson, J., and A.G. Rebelo, 2009, 'Draft Proposal for Amendment of Appendices I and II: *Orothamnus zeyheri*,' CITES, Eighteenth Meeting of the Plants Committee, PC18 Inf.11 (Buenos Aires, 17–21 March 2009).
du Toit, H.S.D., S.M. Gadd, G.A. Kolbe, A. Stead, and R.J. van Reenen, 1923, 'Final Report of the Drought Investigation Commission,' U.G.49-'23 (Cape Town, South Africa: Government of the Union of South Africa).

Forsyth, G.G., and B.W. van Wilgen, 2007, 'An Analysis of the Fire History Records from Protected Areas in the Western Cape,' CSIR Report CSIR/NRE/ECO/ER/2007/0118/C (Stellenbosch, South Africa: CSIR).

Galbraith, A.V., 1947, 'Statement Prepared by the Forest Commission of Victoria, Australia' on 'Empire Forests and the War' (Melbourne: J.J. Gourley, Government Printer).

Henkel, J.S., 1907, 'Prevention of Forest Fires,' Memorandum to the District Forest Officer, Cape, Forest Department Memorandum 7244/146, 13 November 1907.

——, 1943, 'Preservation of Mountain Vegetation of the South-West Districts of the Cape Province,' Memorandum to the Royal Society of South Africa.

Keet, J.D.M., 1935, 'Afforestation in Relation to Natural Flora and the Effect of Afforestation on Water Supplies,' Letter to All Professional Officers, available at: http://digi.nrf.ac.za/dspace/handle/10624/390.

Kruger, F.J., 1978, 'Description of the Fynbos Biome Project,' National Scientific Programs Unit: CSIR, South African National Scientific Programs Report 28.

Kruger, F.J., P. Reid, M. Mayet, W. Alberts, J.G. Goldammer, K. Tolhurt, and S. Parker, 2000, 'A Review of the Veld Fires in the Western Cape during 15 to 25 January 2000,' Department of Water Affairs and Forestry Report.

Lane-Poole, C.E., 1928, 'Australia: Commonwealth handbook' (Canberra, Australia: H.J. Green, Government Printer).

Legat, C.E., 1913, 'Protection from Fire,' Union of South Africa, Forest Department Memorandum, 25 August 1913.

Leppan, H.D., J.M. Hector, W.R. Thompson, T.D. Hall, A. Lindsay Robb, and D. Moses, 1932, 'The Grasslands of South Africa: Problems and Possibilities,' University of Pretoria Series 1, 23 (Pretoria, South Africa: University of Pretoria).

Malherbe, H.L., E.R. March, U.W. Nanni, C.E.M. Tidmarsh, F.S. Grevenstein, J.C. Cox, and J.S. Whitmore, 1968, 'Report of the Interdepartmental Committee of Investigation into Afforestation and Water Supplies in South Africa,' Unpublished Report, accessed at Plant Sciences Library, University of Oxford, Ref.83313.

Orion Planning, 2002, 'Cape Peninsula National Park: Signal Hill, Kloof Nek and Tafelberg Road Initial Development Framework Report,' prepared for South African National Parks/Cape Peninsula National Park.

Rhodes, R., L. Heaton, and MacIntosh Ellis, 1923, 'Forests and Forestry in New Zealand,' prepared for the Imperial Forestry Conference, Ottawa, Canada.

Ross, J.C., 1961, 'Report of the Interdepartmental Committee on the Conservation of Mountain Catchments in South Africa' (Pretoria, South Africa: Department of Agricultural Technical Services).

Seydack, A.H.W., and S.J. Bekker, 1993, 'Review of Fynbos Catchment Management Policy,' Unpublished Report, Republic of South Africa, Department of Water Affairs and Forestry.

Storr Lister, J.S., 1908, 'Memorandum of the Provisions of the Forest Acts and Regulations framed thereunder relative to Forest Fires,' Memorandum 243, 1 February 1908.

von Hase, A., M. Rouget, K. Maze, and N. Helme, 2003, 'A Fine-Scale Conservation Plan for Cape Lowland Renosterveld: Technical Report,' Cape Conservation Unit Report CCU 2/03.

Wicht, C.L., 1944, 'Controlled Burning in the Jonkershoek Forestal District,' Memorandum to The Conservator of Forests, Cape Town, M.600, 22 March, 1–3.

——, 1945, *Preservation of the Vegetation of the South Western Cape*, Report for the Royal Society of South Africa (Cape Town, South Africa: Royal Society of South Africa).

——, 1958, 'The Management of Water Catchments,' Technical Report for the Department of Forestry, 1–2, available at: http://digi.nrf.ac.za/dspace/handle/10624/434.

——, 1960, 'Die invloed van beheerde brand en bebossing op die waterlew-eringspotensiaal van die bergopvanggebiede in die winterreënstreek,' confidential technical report, available at: http://digi.nrf.ac.za/dspace/handle/106 24/411.

Theses

Boehi, M.E., 2010, 'Being/Becoming "the Cape Town Flower Sellers" ' (MA thesis, University of the Western Cape).

Brown, K. 2002, 'Progressivism, Agriculture and Conservation in the Cape Colony circa 1902–1908' (DPhil thesis, University of Oxford).

Pooley, S., 2010, 'An Environmental History of Fire in South Africa in the Twentieth Century' (DPhil thesis, University of Oxford).

Powell, R.F., 2013, 'Long-term Vegetation Change in the Cape of Good Hope Section of Table Mountain National Park, in Response to Climate, Fire and Land Use' (MSc Thesis, University of Cape Town).

Scott, J.D., 1948, 'A Contribution to the Study of the Problems of the Drakensberg Conservation Area' (PhD thesis, University of the Witwatersrand).

Shaughnessy, G.L., 1980, 'Historical Ecology of Alien Woody Plants in the Vicinity of Cape Town, South Africa' (PhD thesis, University of Cape Town).

Southey, D., 2009, 'Wildfire in the Cape Floristic Region: Exploring Vegetation and Weather as Drivers of Fire Frequency' (MSc thesis, University of Cape Town).

Van Wilgen, B.W., 1980, 'Some Effects of Fire Frequency on Fynbos at Jonkershoek, Stellenbosch, South Africa,' (MSc dissertation, University of Cape Town).

Witt, H.A., 1998, 'Trees, Forests and Plantations: An Economic, Social and Environmental Study of Tree-Growing in Natal, 1890–1960' (PhD thesis, University of Natal).

Index of People

Locators **bold** refers major entry on the subject; Locators in *italics* refers illustration; locators followed by fn refers footnote.

Acocks, John, **54**, 83
Adamson, Robert Stephen, 49, **52–3**, 55, **73–6**, 130, 152, 170, 173
Allemann, Lt. Siegfried, 21
Asmal, Kader, 176–8

Baker, Herbert, 125–6, 135
Banks, C.H., 89
Bayer, Adolf, 51
Beadon-Bryant, Chief Conservator (Burma), 66
Beinart, William, x, 3, 4, 136
Bekker, Stefanus, 86, 111
Bennett, Brett, x, 246fn. 38
Bennett, Hugh Hammond, 80, 81
Bews, John William, **49**, *50*, 51, 52, 54, 56, 57, 64
Blake, G.P., 117, 129
Bloomberg, David, 218, 223
Bolus, Harriet, 90
Bolus, Harry, **47–8**
Bond, John, ix
Bond, William, ix, x, 2, 8, 58, 100, **103–4**, 106
Boocock, J.J., *62*, 205
Botha, Louis, 59
Botha, Pieter Willem (P.W.), 219
Brandis, Dietrich, 66
Bräsler, Nicky, 221
Brown, Karen, x, 60–1
Brown, Rev. John Croumbie, 29, **33–4**, 36, 43, 56, 230, 245fn. 19
Brunt, Hugo, 151, 207, **209**, 210
Bruwer, P.J., 215
Burchell, William, 30–2
Burley, Jeffrey, 105
 see also Oxford Forestry Institute

Cajander, Aimo, 69
Campbell, Bruce, 157
Chaplin, Drummond (Sir), 189
Clements, Frederic, 49, **51**, 58, 64, 87, 95, 101, 104, 231
Cody, Martin, 104
Compton, R. Harold, 49, 53, **55–7**, 73, 79, 141, 142, 151, 152, 155, 188, 192–3, 205
Cowling, Richard, x, 55, 106, 107, 120, 180, 256fn. 66
Crosby, Alfred, 168

Daitz, David, 225
Davidoff, H., 83
De Klerk, W.A., 155–6
De Regné, Count Médéric de Vasselot, **36–7**, 163
De Smidt, Reg, 128, 207
De Villiers, Dawie, 159
De Wet, Jacobus, 109
Diederichs, Nicolaas, 157
Drury, William, 98
Du Toit, Heinrich, 59–60
Dubow, Saul, x, 5, 58

Edwards, Gwen, 172
Erasmus, Zane, x, 113–14
Evans, Stanley, 221

Fairbridge, Dorothea, 170
Fenn, John, 106
Fitzpatrick, Percy, 59, 248fn. 32
Forsyth, Greg, x, 120, 121, 184, 195, 196, 224, 226
Foster, W.C., 193
Fouché, Jacobus, 95
Fourcade, Henri Georges, 36, 39, 75
Frost, William, 201

Fuggle, Richard, 159
Fuller, Buckminster, 99

Galbraith, A.V., 67
Gale, George, 51
Gamble, J.G., 138
Geldenhuys, Coert J., 64
Geldenhuys, F.E., 61, *62*, 71
Geldenhuys, Peter, 221
Gill, Malcolm, 101
Godbold, Ken, 156
Gohl, Colin, 210, 213–14
Grisebach, August, 100
Grohl, Colin, 210, 213–14
Grove, Richard H., 32, 33
Groves, Richard, 101

Hall, Anthony, 93, 173–5
Hardenberg, Lionel, 220
Harison, Captain Christopher, 35, 37–9
Harper, John, 104–5, 106
Henkel, John Spurgeon, 63, **73–5**, 76, 77, 151
Hertzog, Barry, 149
Hey, Douglas, 154, 157, 158
Heywood, A.W., 36
Holling, Crawford Stanley 'Buzz,' 98
Hooker, William (Sir), 32, 33
Huntley, Brian, 106, 159
Hutchins, David Ernest, 29, 36, 39, 40–3, 61, 63, 64, 73, 126, 127, 145

Jameson, Leander Starr, 61
Jarman, Frank, 146, *147* (memorial to)
Jeppe, Julius, 142, 143, 206
Jolly, N.W., 65
Jordaan, Pieter, 91–2, 101

Kahn, Frank, 159
Keeley, Jon E., 8, 196
Keet, Johan, 61, *62*, **68–9**, 90, 95
Kensit, William, 48
Kessell, S.L., 65–6
Kolb, Peter, 22, **23**, 27
Komarek, Edwin (Snr), 256fn. 62
Kotzé, P.C., *62*, **69–70**

Kruger, Frederick J., x, **90**, *92*, 93, 95, 100, 101, 102, 103, 106, 108–9, 111, 120, 175

Lane-Poole, Charles, 65, 69
Le Maitre, David, x, 106, 177
Le Page, H.C., 82
Le Roux, P.J., 86
Le Vaillant, Francois, 15, 26, 31
Legat, Charles, 61, 63, 71
Levyns, Margaret, 52, **53–4**, 55, 184, 190
Lichtenstein, (Martin) Heinrich, **30**, 35
Lindeman, Raymond, 97–8
Linnaeus, Carl, 24
Lister, Joseph Storr, **35–6**, 38, 39, 43, 61, 64, 246fn. 38
Louw, Gene, 221
Lückhoff, Carl, 90, 94, 125, **153**, 171–3, 214
Lückhoff, Hilmar, 90, 94, 95, 103
Lückhoff, James, 90

MacArthur, Robert ('MacArthur school'), 104, 106
MacOwan, Peter, 93, 170
Malan, D.F., 83
Malherbe, H.L. de W., 90
Mandela, Nelson, 159, 161, 176, 182
Marloth, Rudolf, 47, **48**, *49*, 52, 54, 55, 75, 90, 93, 139, 170
Marsh, George Perkins, 42–3
May, Robert, 98
McArthur, Alan, 102
McLachlan, D., 174
Mentzel, Otto, 21–2, 27
Michell, Margaret, 53, *see* Levyns, Margaret
Milner, Alfred (Sir), 59
Möbius, Karl, 87
Moll, Eugene, 157, 159, 174–5, **225**
Molloy, Bob, 156
Mooney, Harold A., 99, 175
Morris, Solomon, **129**, 154, 156, 217–18
Moss, Charles E., 49
Mucina, Ladislav, 120

Nagel, Heinrich (Lt), 223
Nisbet, Ian, 98
Noble, Ian, 107

Odum, Eugene, 97–8
Odum, Howard Washington, 98

Pakenham, Thomas, 162–3, 179
Palmer, Eve, 82
Pappe, Carl Wilhelm Ludwig,
    32–4, 230
Pearson, Henry Harold, 49, 55, 138,
    172, 182
Phillips, E.P., 48
Phillips, Florence (Lady), 146, 171
Phillips, John F.V., **64, 69**, 256fn. 62
Phillips, Lionel (Sir), 146
Pillans, Neville, 190
Pinchot, Gifford, 61
Pirow, Oswald, 150
Pole-Evans, Illtyd Buller, 49, 57, 68, **69**
Porter, Harold, 93
Pothier, R., 162, 172
Preston, Guy, x, 176–8
Prins, Philip, x, 227
Pyne, Stephen J., 10, 16

Raimondo, John, 159
Reitz, Deneys, 59, 68, 171, 188
Rhodes, Cecil, **125–6**, 132, 135, **137**,
    138, 146–8, 164, 200
Richardson, David, x, 106, 108–9,
    110, 175
Riley, D.G.D., 117
Rist, Peter, 224–5
Robertson, Colin C., **71**
Robertson, Thomas Chalmers (T.C.),
    60, **82–3**
Rose-Innes, James, 142, 143
Ross, J.C., 86, 87
Rothermel, Richard, 102
Rourke, John, x, 94
Roux, Edward, 81–2, 252fn. 6
Royal family (British), 210–11
Rutherford, Michael, 120
Rycroft, Brian (H.B.), 86, 155

Salter, T.M., 130
Sampson, Arthur, 76

Schimper, Andreas, 100
Schlich, Wilhelm, 61, 65
Schmidt, Jolyon, 227–8
Seydack, Armin, x, 86, 106, 111, 121
Shebbeare, EO (forester, India), 66
Shelford, Victor E., 87
Siegfried, Roy, 159
Sim, J.T.R., 151
Simmons, Benjamin R., *62*, 171, 193
Skaife, Sydney, 73, 149, 157, 189, 207,
    211, 218
Skotnes, Pippa, 179
Slatyer, Ralph, 107
Smith, A.B., 18
Smith, E.J., 218
Smuts, Jan, **51**, 57, 59, 60, 68–70, 82,
    87, 135–9, 150, 152, 182, **232**
    and grasses and pasture, 51, 60, 150,
        248fn. 34
    and science, 60, 87, 138
    holism, 87–8
    spiritual attitude to nature, 60,
        135–6, 138
    versus afforestation, 68–70, 171
Southey, Diane, 124
Sparrman, Anders, 15, 18, 24–**5**,
    26, 27
Specht, Raymond, 100
Steenkmap, W.P. (M.P.), 149
Stoekoe, Thomas, 93, 141
Strauss, J.G.N, 83
Strohmaier, David J., 144
Strydom, Johannes, 191
Swellengrebel, Hendrik, 23

Tansley, Alfred, 52, 87, 88, 97
Taylor, Hugh, 175
Te Water, Charles, 81
Theron, Cas, 227
Theunissen, Carl, 227–8
Thorpe, Stanley W., 201
Thunberg, Carl, **24–5**, 93
Tredgold, Thomas, 117
Trollope, Anthony, 192
Trollope, Winston, x, 106, 109,
    256fn. 62
Truter, Jan Andries, 82
Twain, Mark, 192

Van der Merwe, Philip, 156
Van der Stel, Simon, 21, 71, 96
Van Oordt, G.A., 153
Van Rensburg, C.J.J., 82, 148–9
Van Riebeeck, Jan, **18–19**, 20, 77, 139, 153
Van Sittert, Lance, x, 136–7, 139
Van Wilgen, Brian W., x, 8, 93, 101, *102*, 104, 106, 108, 110, 120, 121, 177, 184, 195–6, 224, 226
Visser, Chris (MPC), 219

Wagner, Gerhard, 106
Walsh family (Kogelberg), 93
Watson, Stephen, 2, 179
Watt, William E., *62*, 73, 84, 253fn. 29
Wicht, Christiaan Lodewyk, 47, 61, 70–4, 76–9, 80–1, 84, **87–9**, 91, *92*, 93, 95, **96–7**, 103, 152–3, 172, 231, 232
Wiley, John, 158, 218
Worster, Donald, 104
Wroughton, F.H., 80

Index of Places

See General index for countries, and 'mountain ranges 'and 'rivers'.

Adelaide (Eastern Cape), 35
Assegaaiboskloof, 93

Bakkeman's Kloof, 211
Banhoek region, 79
Baviaanskloof, 31, 149, 207
Betty's Bay, 93
Bishopsford, 211
Brakkloofrant, 226

Cahora Bassa, Mozambique, 103
California, 10, 99, 101, 169, 184, 196
    of or related to, 72, 76, 184
Campagna Swamps (Italy), 36
Camp's Bay, 117, 130–1, *160*, 173,
    179, 191, 194, 204, 207, 218,
    222–3
Cape Flats, 24, 32, 35–6, **39–40**, 76,
    82, 119, 125, **127–31**, 142, 152,
    160, 163, 164, 170, 180, 185–7,
    189, 200, 201, 211, 226, 260fn. 34
Cape Peninsula, *see General index*
Cape Point, 2, 29, 41, 119, 130, 142,
    149, 156, *160*, 162, 170, 192, 193
Cape, regions of
    Eastern, 4, 35, 63, 73, 109, 249fn. 40
    Southern, 30, 35, 89, 91, 104
    Southwestern, 4, 15, 16, 19, 47, 73,
        79, 88, 90–1, 96, 110, 148
    Western, *17*–19, 27, 96, 101,
        110–13, 124, 151, 174, 176–7,
        182, 196, 210, 230, 231
Cape Town, suburbs and areas of
    Bakoven, 130, *160*, 201
    Bantry Bay, 130
    Bishopscourt, 129, 131, 151
    Blouberg, 30, 117, 130
    Camps Bay, *see* Camp's Bay
    City Bowl, *see* City Bowl
    Claremont, 131, 179, 186, 188, 201

Clifton, 128, 209, 210
Deer Park (Old), 130, 173, 224
Diep River, 18, 131
District Six, 127, 129, 224
Green Point, 130, 191
Heathfield, 131
Highlands Estate, 173, 200, 219
Kenilworth, 131
Kloof Nek, *see* Kloof Nek
Lakeside, 129, 186, 201, 210
Maitland, 125
Milnerton, 117, 130, 206
Mowbray, 131, 212
Muizenberg, *see* Muizenberg
Newlands, *see* Newlands
Observatory, 131
Oranjezicht, 130, 163, **221–3**
Paarden Eiland, 130
Pinelands, 131, *160*, 164, 201
Plumstead, 131
Rondebosch, *see* Rondebosch
Rosebank, 131, 207
Salt River, *see* Salt River
Sea Point, 29, 41, 127, *160*, 201
Southern suburbs, *see* Southern
    Suburbs
Tamboerskloof, 130
Upper Gardens, 125
Vredehoek, *see* Vredehoek
Woodhead Glen, 222
Woodstock, 41, 128, 130, 164
Wynberg, *see* Wynberg
see also General index: informal
    settlements; townships
Capri, 227
Chapman's Peak, *160*, 167, 173, 206,
    209, 226–8
City Bowl, 42, 119, 125, 127–31, *160*,
    172–3, 178, 180, 198, 200, 201,
    204, 214, 220, *222*

Constantia, 2, 118, 131–2, 156, *160*,
    165, 167, 173, 179, *183*, 189, 190,
    203, 216, *226*, 228
Constantia Nek, 127, 129, 131, 132,
    141, *160*, 167, 173, 178, 228

Da Gama Park, 133
Devil's Peak, 22, 31, 32, 38, 118, 121,
    125, 127, 130–1, **143**, **145–7**, 149,
    150, 157, 159, *160*, 164–71, 173,
    174, 176, 191, 201, 205, *206*, 210,
    215, 216, 217, 219, **220**, 222, 223,
    224–5, 227, 260fn. 28
Dido Valley, 190, 210
Du Toitskloof, 110

Eastern Cape, *see* Cape, regions of
Eerste River, 71, 79, 253fn. 25
Elgin, 68, 146

False Bay, 29, 118, 129, 133, 145, 218
Fernwood Estate, 148–9
Fish Hoek, 119, 133, *160*, 172, 173,
    208, 210, 212, 219
Franschhoek, 93, 151, 209

George, *17*, 33, 37, 103–4
Glencairn, 133, *160*, 172, 190, 226
Gordon's Bay, 209
Groote Schuur, 126, 167, 200, 227

Hangberg (or Sentinel), 224
Helderberg, 79, 109, 164
Hottentotsholland (and Hottentots
    Holland), *17*, 18, 79, 96, 141, 211
Hout Bay, 118, 125, 131–2, 149,
    156–7, *160*, 164–5, 175, 187, 189,
    192, 203, 207, 211, 213, 215, 216,
    218, 224, 226, 228, 260fn. 27
Humansdorp (region of), 37

Jonkershoek, *see* General index

Kalk Bay, 133, *160*, 185, 201, 226
Karbonkelberg, 118, *160*
Kasteels Poort, *see* Table Mountain,
    parts of
Klaas Jagersberg, 226

Kloof Nek, 127, 130, 166, 169, 179,
    192, 215, 220, **221-2**, 224
Knysna, *17*, 33, 35, 37, 38, 41, 64,
    69, 246fn. 40
Kogelberg, 75, 93, 111
Kommetjie, 125, 133, *160*, 173, 187,
    215, 224, 226, 228

Lion's Head, 22, 54, **118**, 130, 150,
    154, *160*, 162, 165–7, 170, 173,
    179, 191, 194, *203*, 204, 205, 220,
    225, 226, 236
Llandudno, 118, 156, 224

Malindidzimu Hill, 137
Milnerton, 117, 130, 206
Misty Cliffs, 228
Muizenberg, 93, 118, 125, 128, 129,
    131, 132, 133, 154, 156, *160*, 167,
    172, 173, 175, 185, 192, 194, 201,
    202–4, 226, *see General index:*
    Cape Peninsula Mountains

Newlands, 26, 119, 127, 129, 131, 132,
    151, 154, *160*, 166, 172, 173, 178,
    181, 198, 203, 218, 220,
    223, 225, *237*
Nieuwberg, 68
Noordhoek, 119, 125, 129, **132–3**,
    160, 187, 192, 215,
    226, 227

Ocean View, 133, *160*, 224, 226
Orange Kloof, 119, 166, 167, 181,
    202, 203, 211, 213,
    220, 228
Orkney Islands (Scotland), 49
Oudekraal, 210
Outeniqua Mountains,*see General
    index:* mountain ranges

Paarl, 23,
    211, 253fn. 25
Papenboom Estate, 143
Pietermaritzburg, 49, 73
Plantations, *see General index*
Pringle Bay, 93

Red Hill, 133, 142, *160*, **187**, 193, **215**, **227–8**
Roeland Street (Cape Town), 41, 166, 191, 201
Rondebosch, 26, 40, 125, 131, 145, 203

Saasveld, 103–4
Salt River, 41, 125, 128, 188, 189, 201
St James, 185, 207
Scarborough, 133, *160*, 228
Sea Point, *see* Cape Town, suburbs and areas of
Signal Hill, 125, 130–1, 150, 154, *160*, 166–7, 180, 194, 203–4, 220, 224, 225, 236
Silvermine, 125, 129, 133, 154, *160*, 179, 181, 187, 193–5, 198–201, 203, 209, 213, 216, 220, 224, 226–9, *237*
Simon's Town, 24, 25, 127, 131, 133, *160*, 187, 190, 200, 208, 210, 212, 213, 215, 218, 219, 227–8
Slangkop, 133, 226
Smith's Farm, 149
Somerset West, 129, 189
Southern suburbs (Cape Town), 119, 127, 129, **131–2**, 145, 167, 220, 221
Steenberg, 1, 11, 132, *160*, 205, 207, 227
Steenbras valley, 127, 174
Stellenbosch, *17*, 19, 70, 71, 95, 100, 142, 151, 174, 175, 189, *see also General index*: universities
Swellendam, *17*, 25, 35, 93

Table Bay, 19, 20, 22, 30, 118, 129, 130, 131, 145, 161, 173, 189, 210
Table Mountain (parts of/places on)
   Constantia Nek, 127, 129, 131, 132, 141, 167, 173, 178, 228
   Kasteels Poort, 125, 191, 218
   Knife Edge, 223
   Kloof Nek, *see* Kloof Nek
   Tafelberg Road, 127, 194, 219–22, 224
   The Saddle, 118, 131, 223
   Window Buttress, 141
   see also General index: Cape Peninsula Mountains; Table Mountain; Twelve Apostles
Tharandt (Germany), 70
Tokai, 36, 41, 63, 131, 132, 149, *160*, 163–7, 175, 178–9, *183*, 189, 203, 205, 207, 211, 213–16, 226, 227, *see General index*:'Tokai Arboretum
Transvaal Province, 4, 82, 90, 126, 146
Tsitsikamma, 35
Twelve Apostles, 117, 118, 130, 173, 218

Van Staden's River valley, 34
Vlakkenberg, 132, 165, 211, 226
Vredehoek, 125, 130, **219–21**, 223–4
Vredehoek Estate, 125

Woodhead Pipe Track, 205
Woodstock, 41, 128, 130, 164
Wynberg, 34, 127, 131, 156, *160*, 164, 170, 201

Index of Plants and Animals Species, Genera

Plants

Acacia (wattle), ix, 36, 39, 62, 64, 69, 70, 108, 133, 142, 157, 163–4, 168, **169**, 170, 172–5, 179–80, 181, 215
   Acacia cyclops, 169, 173–5
   Acacia decurrens, 169
   Acacia longifolia, 169, 172
   Acacia melanoxylon, 64
   Acacia pycnantha, 169
   Acacia saligna, 36, 157, 169, 172, 174, 175, 215
aloes, 38
Andropogoneae, *see* grasses
Arctotis, 130
Ash trees, *see* Kabul Ash
Asparagus capensis, 70
Australian myrtle, *see Leptospermum laevigatum*

blue gum, *see Eucalyptus, Eucalyptus globulus*
Blushing Bride, *see Serruria florida*

camphor trees, 2, 179
casuarina trees, 39, 163, 169
cluster pine, *see* pines, *Pinus pinaster*
common sugarbush, *see* Proteas, *Protea repens*

dew flowers, *see Drosanthemum* species
Digitaria smutsii, *see* grasses
Disa uniflora, 145, 180
Drosanthemum species, 130

Elytropappus rhinocerotis, 25, 28, 34, 53–4, 76, 121
   see also General index: renosterveld
ericas (heaths, *Ericaceae*), 31, 75, 120, 125

Eucalyptus (eucalypts gums), 36, 38, 62, 69, 70, 126, 145, 162, 164, 166–71, 172, 174, 178–82, 216, 231
   Eucalyptus corynocalyx, 169
   Eucalyptus diversicolor, 169
   Eucalyptus globulus, 36, 38
   Eucalyptus pilularis, 169
   Eucalyptus resinifera, 169

fig marigold, *see Mesembryanthemum* species
Fraxinus, *see* Kabul Ash

Gladiolus aureus, 157
golden gladiolus, *see Gladiolus aureus*
grasses (species, genus)
   Andropogoneae, 58
   Digitaria smutsii, 60
   Themeda triandra, 52, 58
gum trees, *see Eucalyptus*

hakeas, 39, 53, 108, 142, 163, 169–70, 172–5, 180, 181
   Hakea gibbosa, 173, 174
   Hakea pectinata, 170
   Hakea sericea, 174
heaths, *see* ericas

Kabul Ash, 145
keur bushes, *see Virgilia* species
Keurboom, *see Virgilia* species

Lachenalia pendula, 130
Leptospermum laevigatum, 168
Leucospermum conocarpodendron (yellow pincushion), 34
Leucadendron argenteum (silver tree), 34, 40, 42, 120, 138, 142, 180, 202

Marsh Rose, *see Orothamnus zeyheri*
Mesembryanthemum species, 36
Mimetes, 75, 107, 119
    Mimetes splendidus, 107
    Mimetes stokoei, 75

oak trees, *see Quercus*
orchids, 93, 136
Orothamnus zeyheri, **93–4**, 141, 232

palm trees, 179
Penaeaceae, 75
pines (trees, species, genus), 2, 3, 36,
    40, 43, 62, 65, 68, 108, 126, 132,
    142–6, *148–9*, 162–7, 169–171,
    173–5, 178–82, *183*, 202, 204–7,
    216, 217, 222, 223, 231, 236
    Pinus halapensis, 166
    Pinus insignis, 145
    Pinus pinaster (cluster), 40, 53, 70,
        143, 164, 166, **169–71**, 173–5
    Pinus pinea (stone pine), 2, *148–9*,
        179, 205
    Pinus radiata, 169
    Pinus rigida Pitch pines, 126
    Pinus sylvestris (scotch), 40, 145
poplar trees, 38, 172
Port Jackson wattle, *see Acacia* (wattle),
    Acacia saligna
proteas, 31, 32, 54, 70, 75, 91–2, 104,
    107–8, 119, 120, 141, 157,
    180–1, 236
    Protea grandiceps, 32
    Protea mellifera, 91
    Protea repens, **91–2**, 180
    Protea nitida, 127

Quercus, 38, 41, 62, 145, 146, 163,
    172, 173, 179, *183*,
    246fn.40

red sugarbush, *see Proteas, Protea
    grandiceps*
renoster bush, or bos (variously
    Rhinoster, rhenoster, rhinoceros),
    see Elytroppapus rhinocerotis
restios (Restionaceae), 56, 120
rhinoceros bush, *see Elytroppapus
    rhinocerotis*
rooigras, *see* grasses, *Themeda triandra*
rooikrans, *see* Acacia (wattle), *Acacia
    cyclops*

scotch pine, *see* pines, *Pinus sylvestris*
Serruria florida, **93**, 141, 232
Shorea robusta, 66
silver tree, *see Leucadendron argenteum*
stone pines, *see* pines, *Pinus pinea*

Tasmanian blackwood, *see* Acacia
    (wattle), *Acacia melanoxylon*
teak (sal) *see Shorea robusta*
Themeda grasses, *see* grasses

venothera (*Oenothera* species),
    142, 170
Virgilia species, 37
Vygies, *see Drosanthemum* species

wag 'n bietjie' thorn, *see Asparagus
    capensis*
'wait a bit' thorn, *see Asparagus
    capensis*
wagon tree, *see* Proteas, *Protea nitida*
watsonias, 16, 47, 48, 117, 130
wattle, *see Acacia*

Animals and Insects

Alcelaphus buselaphus caama, 16
antelope, **16**, **109**, 126, 153, 166, 215
ants, *see* insects

Cape sugarbird, *see Promerops
    cafer*, 119

eland, *see Taurotragus oryx*
Equus quagga quagga, 16
Equus zebra, 16, 149

Grysbuck, *see Raphicerus
    melanotis*

hippopotamus, 20

insects, 26, 119–20, 150, 153
    ants, 119–20, 150
    Argentine ants, *see Linepithema*
        humile
    mosquitoes, 150

Lion, 20, 235
leopard, 20
Linepithema humile, 150

quagga, *see Equus quagga quagga*

Raphicerus melanotis, 109, 153,
        215, 216
red hartebeest, *see Alcelaphus*
        buselaphus caama
rhinoceros, 16, 20

Taurotragus oryx, 16, 109

zebra, *see Equus zebra*

General Index

Aesthetics, of the natural world, ix, 27, 60, 68, 69, 95, 128–31, 135–7, 138, 142–6, 149–52, 154, 156–7, 161, 162–3, 171–2, 178–81, 192, **202**, 232, 234

afforestation, 35, 37, **42–3**, 67–70, 72, **84–5**, 89–90, 124, 142–6, 162–8, 166, 170–1, 178, 181, 203, 206, 217, 230–1, 232, **233**, 250fn. 55
criticisms of, 60, **67–70**, **90**, 170–1
defences of, **42–3**, **68–9**, 89, 144, 171
and water supplies, 37, **42–3**, 68–9, **89–90**
see also forestry, plantation

African Explosives and Chemical Industry company (AE&CI), 57

African National Congress (ANC), 1, 113, 178

Afrikaner nationalism, 58–61, 70–1, 128, 136, 137, 138, 150, 208–9

agriculture
and desiccation and water supplies, 34, 52, 60, 67, 83, 86, 90, 95
Elsenburg College of Agriculture, 151
expertise, 4–7, 43, 52, 54, 57, 58–9, 71, 83, 98–9, 106, 109, 168
and fire, 4, 52, 54–5, 57–8, 77, 83, 86, 91, 107, 109, 148, 150, 151, 208, 213, 216, 231
history of (colonial and South African), 19, 25, 43, 52, 59–61, 71, 91, 95, 114, 118, 148, 151
research, 8, **57–8**, 64, 91, 104, 106, 109, 148, 231
USA influences, 57, 60, 63, 81, 88
see also farming; pasture science

agriculture, South African Department of, **57**, 61, 68, 77, 83, 91, 114, 148, 208, 213, 216

Agricultural Technical Services and Water Affairs, 90, 95, 174

Department of Agriculture, Forestry and Fisheries, 114

Division of Botany, 49

Division of Plant Industry, 51, 148; Pasture Research and Veld Management, 148

Ministers of, 43, 59, 67–8, 95, 171, 188, 208, 213

policy and recommendations on burning, 58, 79, 86, *see also* agriculture, and fire

alien plants, 169

American Civil War, 152

Anglo Boer War, *see* South African War

animals, 17–19, 30, 64, 78, 87, 89, 96, 125, 135, 153, 168, 235, 242fn. 12
and fire, 9, 18, **23**, 25, 96, 109, 130, 150, 153, 234
see also livestock; grazing; wildlife; *Index of Plants and Animals*

Apartheid, 4, 5, 71, 95, 100, 127, **128**, 133, 176, **184–6**, 187, 190, 195–6, 199, 219
removals (of 'non-whites'), 127, 133, 185–6, 190, 195
see also homelands; science and politics; townships

arson, 4, 20, 85, 150, 151, 167, 190, 191, 199, 205, 213, **239**

atmospheric conditions. *see* weather

Australia, 3, 6, 10, 42, 53, 65–6, 71, 74, 86, 99–102, 107, 108, 113, 168, 169
fire ecology of, 10, 67, 74, 99–101, 107
invasive species from, 2, 39, 64, 108, 132, 157, **168–70**, 180

Australia – *continued*
  see also forestry, Australian; jarrah
    forests; plant introductions,
    from Australia; *Index of Plants
    and Animals*: *Acacia*; *Eucalyptus*;
    Hakea

Barbados (Caribbean), 34
bergies, 191, 219
biocoenoses, 87–8
biodiversity, 7, 8, 58, 61, 102, 107,
    119–20, *122*, 126, 158, 234–5
  and fire, 7, 74, 75, **102–3**, 106–**7**,
    139, 231, 235
biological invasions, 2, 54, 108–**9**,
    133, 142, 152, 162, 168–81, 233,
    265fn. 18
biome, 9, **15**, 87, 100, 106
  see also fynbos, biome
Bokkie the Grysbuck, **153**, *155*, 214, 216
Botanical Society of South Africa, 55,
    57, 139, 141, 146, 151, 156, 176
botanists
  and fire, ix, 2, 6, 24–5, 31, **32–4**,
    47–57, 65, 83, 91, 93, 106, 109,
    139–41, 151, 157–9, 170, 190,
    205, 230–2, 236
  see also Index of People: Bayer, Adolf;
    Bolus, Harriet; Bolus, Harry;
    Bond, William; Compton, R.
    Harold; Cowling, Richard; Hall,
    Anthony; Jordaan, Pieter;
    Levyns, Margaret; MacOwan,
    Peter; Marloth, Rudolf; Moss,
    Charles E.; Pearson, Henry
    Harold; Phillips, E.P.; Specht,
    Raymond
botany, ix, 24, **32–4**, **47–57**, 65, 82,
    91–3, 104, 138, 140–2, 151,
    155–7, 179, 181, 189, 225
bridle paths, 38, 62, 169, 193, 211
British Association for the
    Advancement of Science, 50
British Commonwealth, 3, 64, 70, 105
  Forestry Conferences, *see* British
    Empire, Forestry Conferences
British Ecological Society, 78
British Empire, 3, 30, 53, 60, 61, 65,
    105, 136, 138, 168

Forestry Conferences, **65–70**, 73,
    105, 231
forestry network, 60, 65, 70, 168
British Empire Vegetation
    Committee, 53
browsers (animals), 16, 20, 27, 109
Burma (Myanmar), 66
burning practices
  hunters, 4, 16, 18, 215, 239
  pastoralists, 19, 24–5, 27, 31, 48, 51,
    54–6, 58, 67, 72, 86, 107, 150,
    184, 190
  see also farming methods

cable car, 127, 139, 147, 150, 192,
    196, *203*
Canada, 33, 65, 66, 86, 219
Cape Action for People and the
    Environment (CAPE), 54
Cape Argus, 1, 144, 148
  see also media, newspapers
Cape Colony, 6, 19, 21, 32–44, 56–7,
    60–1, 125, 137, 168, 188, 230–1
Cape Coloureds, 3, 85, 128, 133, 179,
    186, **187–9**, 201, 207, 210
Cape Divisional Council, 129, 149,
    200, 208, 211, 215, 218, 219, 223
  see also, Cape Town, local
    government
Cape Floristic Region, 10, **15–16**, 54
Cape Peninsula, ix, 1–3, 9, *17*, 18, 19,
    24, 26–7, 29, 32, 36, 38, 41–4, 55,
    76, 84, 96, 117–32, *122*, *140*,
    148–83, *160*, 184–6, 190, 195–9,
    203, 208, 210, 216, 218, 224, 227,
    234–8
  afforestation of, 37, 69, **163–7**,
    203–4, 217, 233
  biophysical template, 117–23,
    197–8, 236, *237*, 238
  fires incidence on, 41, 67, 121, 124,
    151, 156, 167, 194–5, 200, 201,
    202, *204*, 206, 209, 210, *212*,
    215, *217*, 224, 226, 233–4, 236,
    238, 272fn. 31
  fire management, 1, 20, 43–4, 139,
    157–9, 161, 182, 297, 199,
    200–29, 233

Fire Protection Organisation, 96,
113, 153, 208, 209, 219
fire regime, 117–18, 124, *198*,
233–4, **236**, 238
Southern Peninsula, 118–19, **121**,
125, 129, **133–4**, 173, 192–3,
215–16, 226–8, 258fn. 11
Cape Peninsula mountains, 117
Constantiaberg, 132, 149, *160*, 173,
175, *183*, 206, 218, 226
Kalk Bay, 133
Muizenberg Mountains, 93, 118,
125, 129, 139, 131, 133, 167,
172–5, 194, 202
Noordhoek Peak, 132, 226
Steenberg, 1, 132, 207
Vlakkenberg, 132, 165, 211, 226
Cape Peninsula National Park, *see*
Table Mountain National Park
Cape Peninsula Protected Natural
Environment, 157
Cape Peninsula Publicity Association,
see tourism
Cape Provincial Administration,
110–11, 113, 142, 153, 159, 205,
216, 221, 228
Cape Road Board, 35
Cape Times, 41, 117, 128, 129, 143,
146, 150, 151, 155–8, 172, 179,
180, 189, 190, 194, 197, 205–8,
211, 212, 215, 218, 221, 223,
224, 226
see also media, newspapers
Cape Town
architecture of, 28, 117, 125, 126,
129, 135, 137, 204
local government, 1, 14, 44, 154,
159, 201–2, 204, 205–8, 211,
218, 227, 233
parks and commons, 2, 124, 130,
141, 156, 164, 183, 178, 180,
181, 188, 191, 199–203, 214,
220–1, 223, 225–7
population of, 126, 158, 163, 184–7,
185, 193, **195–6**, 204, 210,
217–18
urbanisation, 2, 38, 117, **126–34**,
141, 154, 156, 180, 184, **186–7**,
196, 227, 233, 241fn. 14

water supply, 37, 40, **42–4**, 126–7,
145, 164–5, 173, **177**, 182, 190,
198, 204, 213, 221
see also Index of Places: Cape Town,
suburbs and areas of
catchment areas, 3, 64, 67–9, 72, 73,
79, **83–9**, **91**, 103, 106, 110–**13**,
143, **148**, 174–7, 182, 200, 202,
205, 211, 214, 231–2
Cecilia plantation, 2, 36, 132, 164–6,
167, 178, *203*, 211, 216, 228
Cederberg Mountains, 32, 82, 86, 94,
101–2, 110, 112
Centre for Invasion Biology,
Stellenbosch, 175
Centre for Scientific and Industrial
Research (CSIR), 106, 109, 111,
176, 178
Division of Forest science and
Technology, 111
see also Cooperative Scientific
Programs
chaparral, 76, 104
cigarettes, *see* fire, causes of, smokers
Civilian Conservation Corps
(USA), 207
Civilian Protection Service, 209, 210
climate, 7–8, 10, 15, 44, 68, 88, 124,
168, 234, 236
and fire, 58, 74, 100, 117, 118,
234, **236**
and vegetation, 8, 15, 51, 56, 58,
69–70, 83, 88, **99**, 100, 104,
119, 180
see also drought; self-organising
maps; weather; winds
climax vegetation, **51**, 53, 56, 69, 77,
88–9, 96, 101
climbing, *see* mountaineering
colonial botanists (British, at the
Cape), 32–4, 231
see also Index of People: Brown, Rev.
Croumbie; Pappe, Carl Wilhelm
Ludwig
Committee for the Prevention and
Control of Forest and Veld Fires
on the Cape Peninsula, 208

commons (Table Mountain as), 63,
125, 130, 137, **146–7**, 158, 188,
196, 224
community (ecological concept), 105
see also bioconose; ecology,
ecosystems concept
Compton Herbarium, 94
conservation
nature, 4, 5, 8, 32–3, 54, 57, 68, 73,
78, 82, **90–5**, 101, 108, 110–**11**,
113, 136, 139–42, 144, 148–58,
173, 176–7, 182, 186, 188,
216–18, 232
soil, *see* soil conservation
controlled burning, *see* prescribed
burning
Cooperative Scientific Programs,
106, 109
see also Fynbos Biome Project
Cooper's Hill, *see* Royal Indian
Engineering College
crops, 4, 18, 19–20, 23–4, 28
cereals, 20, 24, 28, 37
maize, 50
vines, *see* vineyards
wheat, 20, 24, 30, 242fn. 12
crown lands, 39–40, 63, 67, 71, 93, **95**
see also forests, Crown

defence forces (South African), 200,
204–5, 209, 213, 215, 223
air force, 218, 219, 223, 228, 229
army, 201, 205, 207
navy, 134, 200, 207, 213, 219,
227, 228
see also Royal Navy (British)
deforestation, 33–4, 42, 51, 56, 65, 68,
144, 217, 231
demography (human), 184, 196
of plants, 103
see also plant population biology
Department of Water Affairs and
Forestry, 114, 175
desiccation, 34, 51–2, 67–9, 81
development (South African policy
of), 95, 113, 155–6, 158, 176, 178
Reconstruction and Development
Programme, 176, 178
see also industrial development

disequilibrium, 99, 109
disturbance, 15, 51, 53, 56, 64, 67, 88,
95, 99, 104–6, 118, 199
drift sands, 32, 35–6, 39, 82, 93, 94,
128, 131–3, 163, 169, 180, 227
drought, 33–5, 37, 52, 57, 59, 67, 90,
99, 111, 148, 185, 230, 238
see desiccation; Dust Bowl
(American Midwest)
Drought Investigation Commission
(1923), 60, 52, 81
Dune stabilisation, *see* drift sands
Dust Bowl (American Midwest), 52, 60
Dutch East India Company, *see* VOC

ecological imperialism, 168
ecologists, *see Index of People*: Bews,
John William; Clements, Frederic;
Cody, Martin; Lindeman,
Raymond; Mooney, Harold;
Odum, Eugene; Phillips, John F.V.;
Tansley, Alfred
ecology, 96–106, 232
Clementsian succession theory, 64,
87, 98, 101, 104, 231
ecosystems concept, 97–8, 106
equilibrium (ideas of), 98
disequilibrium, 109
holistic approach to, 87–8
MacArthur school, 104, 106
non-equilibrium, *see* disequilibrium
restoration, 76, 229
and war, 83, 148–9
see also disturbance; fynbos,
ecology; MEDECOS
ecosystems
climate dependent, 58
fire dependent, 58
hybrid, 6, 143, 144, 180, *183*, 234
services, 3, 56, 177, 233
see also Mediterranean type climate
ecosystems
electrification
of Cape Town, 127, 186–7
of railways, 85, 127
endemism/endemic, **119**, 120, 142,
153–4, 202, 258fn. 4

engineers
City of Cape Town, 117, 129, 154–5, 194, 217
civil, 129
French, 42, 68
see also Index of People: Morris, Solomon; Riley, D.G.D.
environmental degradation, 59, 64, 83, 230–1, 275fn. 1 (conclusion)
environmental history, ix, 3, 8, 10, 234
environmentalism, 82, 94, 98–9, 111, 117, 154–7, 193
erosion, *see* soils, erosion
Europe(an), 4–6, 18, 23, 27, 29–30, 62, 77, 80, 136, 139, 168–9, 179, 187, 192
scientific influences from, 30, 36, 41, 43–4, 47, 56, 63, 65, 69, 80, 100, 104–6, 136
evapotranspiration, 90
evolution, convergent, 74, **99**, 101, 103
experiments, 24, 59, 71, 76–7, 80, 86, 90, 104, 111, 168
extinction, 75, 80, 96, 108, 152, 156

farming, 4, 18–19, **23–4**, 27, 31–2, 34–5, 37, 41, **50–9**, 67, **79**, 85, 86, **111**, 125, 149, **188–90**
flowers, 111, 188–90
see also agriculture; crop; farming methods, livestock; pastoralism
farming methods
Dutch settlers and Boers, 19–20, 23, 34, 52, 55, 58–9, 67, 74, 77, 79, 81, 86–7, 190, 230
Khoikhoi, 16, 18–19, 22, 27, 55, 230
see also veld burning
fire, aesthetic responses to, 25, 26, 31, 41, **130**, **143**, **148**, 156
fire, 'natural'
as a natural process, 41, 43, 63, 66, 75
see also biodiversity, and fire; fire causes
fire, and regeneration (plants), 67, 74, 80, 91–3, 101, 102, 232
see also senescence

fire, anthropogenic aspects, 153, 184–96, 208, 234, 239
damage, 89, 176, 205, 214, 224, 228–9, 246fn. 40
management, *see* fire management
people injured or killed by, 110, 142, 213
uses of, 4, 8, 17, 20, 25, 37–8, 50, 55–8, 64–7, 72, 83, 86–8, 91, 95–6, 113, 124, 127, 174, 182, 186–90, 192–3, 196, 199, 230, 239
see also arson; fire causes; media; fire risk
fire awareness, *see* propaganda
fire behaviour, 102–3, 108
fire brigades, 20, 113, 186, 191, 199–202, 204, 209, 214–15, 218, 223, 239
fire, protection measures
areas protected, 4, 44, 66, 72, 73, 81, 121, 132
block (patch) burns, 74, 77, 91, 110, 111, 112, 153
communications, *see* fire fighting, communications
cost/funding of, 73, *112*, 205, 209, 213, 218, 224, 233, 246fn. 40
fire belts/breaks, 63, 72, 77, 86, 89, 110, 166–7, 169–70, 174, 214–15, 216, 219, 244n40
fireplaces (safe), 192, 194, 215
labour, 38, 63, 77, 85, 110–12, 190, 201, 207–9, 211, 221, 223–4
notice boards, 192, 205
patrols/rangers, 63, 206, 220
watching (lookouts), 41, 42, 77, 215
weeding, 42, 173
fire, causes, 65, 191, 199, **239**
anthropogenic, 73, 77, 84, 96, 101, 148, 184, 191–3, 199, 208, 239
arson, *see* arson
burning to clear vegetation, 191, 199, 239
campers/hikers/picnickers, 84, 101, 128, 154, 191–3, 199, 205, 246, 218, 224, 239
categories of (causes), 239
children, 191, 199, 239

fire, causes – *continued*
  falling rocks, 18, 75, 239
  fireworks, 191, 215, 239
  hunters, 4, 18, 215, 239
  lightning, 75, 111, 239
  rubbish burning, 191, 199, 205, 239
  runaway (prescribed / intentional)
      fires, 20, 26, 35, 108, 110–11,
      167, 239
  smokers, 22, 128, 193, 207, 208, 239
  socio-economic, 187–96
  steam locomotives, 84–5, 127, 191,
      199, 239
  vagrants, 21, 27, 147, 191, 199, 218,
      239, *see also bergies*
  woodcutters/gatherers, 38, 39, 63,
      141, 184, 190, 207
fire danger rating systems, 108, 114
fire extent, 38, 72, 74, 89, 110, 112,
      114, 150, 196, 205–6, 209, 211,
      214–17, 220, 224, 226, 227,
      228, 236
fire fighting, 21, 28, 63, 65, 77, 85, 89,
      110, 113, 127, 143, 151, 153, 190,
      192, 197, 198, **199–204**, 206–29
  access/transport, 85, 89, 198, 213,
      215, **222**, 223
  by air, 198, 218, 219, 221, 223–5,
      228–9
  challenges for (Cape Peninsula),
      198–9, 223, 236
  communications, 41, 89, 205, 213,
      215, 223
  equipment, 21, 63, 113, 218, 224
  funding, 77, 209, 213, 221, 224
  strategies/methods, 21, 198, 213,
      214, 218, 220–1, 222, 228
  volunteers, 207, 209, 212–13
  see also fire brigades; transport,
      traffic and motor vehicles
fire frequency, *see* fire regimes
fire incidence
  forestry lands, *85*, 89, 95–6,
      110, 113
fire intensity, *see* fire regimes
fire management, 1, 3, 5–10, 20, 33–5,
      43–4, 57, 64–7, 73, 75–8, 80–1, 84,
      86, 88–9, 91, 95, 101–3, 106–14,

120, 148, 158–9, 182, 197,
      199–200, 203, 226, 232–3
  exclusion and suppression, 57, 65,
      67, 73, 77, 80, 86–7, 88, 91, 92,
      96, 101, 151, 158, 194, 212,
      218, 220, 232–3, 263fn. 58
fire management
  see also prescribed burning; and fire,
      protection measures
Fire Protection Associations, 113,
      114, 153
Fire Protection Committees, 86, 91,
      96, 108, 153, 212–14, 216, 218
Fire Protection Districts (or Areas), 86,
      153, 210
fire regimes, 3, **9**, **10**, **16**, 20, 27, 39,
      94, **100–1**, **103**, 107, 230, **233–8**
  fire frequency (fire return intervals),
      76, 77, 97, 100, 101, 106, 107,
      124, 157, **226–7**, **236**, 253fn. 29
  fire intensity, 76, 80, 101, **102**, 106,
      107, 233, 256fn. 62
  fire seasonality, 10, 16, 18, 20, 23,
      31, 39, 41, 58, 66, 74, 77, 92,
      97, 101, 107, 108, 196, *198*,
      233–4, 238
  see also ignitions
fire risk, 96, 108, 110, 114, 127, 132–3,
      144, 150, 163, 187, 192–3, 195,
      197, 198, 200
fires, forest, 35, 40, 64–5, 102,
      113, 210
fishing industry, 131–2
floods and fire, 86, 88, 97, 159, 176
flower/s
  arranging, 189
  cultivated, 188
  picking/harvesting, 57, 111, 141,
      187–8, 199, 205
  selling/sellers, 187–90, 206, 210
  shows, 141, 170, 189
  reserves, 57, 68, 92–3, 152, 214
foresters, *62*, *92*, and *see Index of
      People*: Banks, C.H.; Boocock, J.J;
      Cajander, Aimo; Hutchins, Lister,
      the Count, Harison, Fourcade,
      Henry; Jarman, Frank; Jolly, N.W.;
      Keet, Johan; Kessell, S.L.; Kotzé,
      P.C.; Kruger, Frederick J.; Legat,

Charles; Poole, Charles Lane;
    Shebbeare, E.O.; Simmons, B.R.;
    Wicht, Christiaan L.
forest reserves, 70, 72, 73, 84, 111,
    131, 133, 165
forestry, by country
    Australia, 3, 6, 65–7, 69, 86, 101–2,
        107–8, 126, 170
    Canada, 65, 66, 86, 117
    India, 3, 6, 33, 36, 40–3, 63, 65,
        66–7, 80, 138
    France, 42, 113, **36–7**, 68; *see also*
        Index of People: de Regné
    Nyasaland (Malawi), 67
    Southern Rhodesia (Zimbabwe), 67
    USA, 9, 61, 63, 64, 86, 102, 108,
        207, 256fn. 62
forestry, conservation, 68, 90, **94–5**,
    100, 101, 110–11, 232
    catchment management, 64, 67, 72,
        77, 79, 84–9, 91, 94, 103, 106,
        110–12, 232
    invasive plant control, **112**, 171–7,
        182, 233
    MAREP (Management, Research,
        Planning), 106
    planners, 86, 106, 232
    see also Our Green Heritage
        campaign; outdoor recreation;
        wilderness
forestry industry, South Africa, 94–5,
    113, 114, 178
forestry, municipal, 36, 163, 165, 167,
    181, 191, 193, 200, 202, *203*,
    215, 217
forestry, plantation, 60, 61, 113, 217,
    246fn. 38
    fire damages, 89, 110, 112, 205, 214,
        224, 228–9
    timber requirements, 84, 164,
        209–10, 231
    see also afforestation; plantations
forestry department (state forestry)
    Conservancies, and Forest Regions,
        33, 40, 60, 61, 84, 86, 96, 110
    Conservators of Forests, 35, 40, 61,
        63, 64, 71, 78, 145, 171, 193
    expenditure, 73, 219, 251fn. 64

funding, 104, 110–12, 174, 176–7,
    192, 219
    history of, 3, 34, 36, 38–41, 57,
        60–1, *62*, 63, 68, 84, 85, 95,
        110–14, 165, 252fn. 10
    Ministers of, 1, **43**, 68, 96, 114, 171,
        176, 177, 188
forestry department (state forestry)
    of the Cape Colony, 38–41, 57, 60–1
    policy on burning, 38, 47, 58, 67,
        72, 74, 77, 79, 81, 83–4, 89, 91,
        93–4, 104, 214
    training, 63, 69, 113
    see also Department of Water Affairs
        and Forestry
forestry research, 61, 64, 70, 71, 72,
    73, 89–90, 94–5, 100, 103,
    168, 176
Forestek, 176
Forestry Research Institute, 94, 111
forests
    Crown, and state, 38, 111, 226
    demarcated and undemarcated, **63**
forests (indigenous), 3, *16*, 33, **47**, 64,
    69, 77, 87, 96, 163
    Afrotemperate, 33, 119
    and climate, 68, 69, 70, 96
    fire protection and management, 3,
        69, 72, 167, 246fn. 40
    see also Forests, Crown; water
        supplies, trees and forests and
France, 24, 29, 36, 37, 42, 68, 113
frontier, closing of, 59
fuel (vegetation, for fire), 100, 121,
    167, 176, 196, 198, 207, 233, 236
fynbos
    and fire, 2, 16, 31, 34, 43, 48, 52, 53,
        55–7, 68, 72–9, 81, 83, 84, 86,
        89–93, 96, 99–111, 113, **120–4**,
        131, 132–4, 141, 152–3, 157,
        176, 193, 196, 223, 226, 231–3,
        236, 253fn.5
    and soils, 54, 100, 119
    Biome, 9, **16–17**, 18, 43, 52, 53, 105
    categorisations of, 54, 103, **120**,
        136, 180
    conservation, 68, 101, 108, 111,
        151, 157

fynbos – *continued*
    diversity, 57, 74–5, 100, 102, 106–7,
        109, 119–20, 231
    ecology, 15, 54, 88, 91, 95, 97, 106,
        107, 99–111, 119–20, 150, 176,
        217, 231–2
    regeneration strategies, 43, 57, **75**,
        96, 107, 232; *see also* senescence
Fynbos Biome Project, **105–6**, 108,
        109, 177
Fynbos Forum, 177, 178
fynbos (threats to), 260fn. 34
    cultivation, 24
    flower picking, 57, 188
    grazing and browsing, 57, 64, 75,
        78, 109
    introduced plants, ix, 64, 76, 108–9,
        112, 132, 168–75, 177, 178,
        180, 181, 227
    urban development, 76, 129–134
    see also veld burning

gardens and gardening, 2, 48, 49, 55,
        82, 125, 126, 129, 132, 135–6,
        138, 146, 154–5, 164, 168–9,
        179–81, 190, 192, 202, 207
    landscape gardening, 125–6, 232
geology, 8, 118–19
geophytes, 16, 75, 77, 93
German immigrants, 39, 163
Germany, 39, 42, 48, 70, 163
government
    local, 113, 146–7, 149, 208, 221,
        224–5
    provincial, 126, 153, 176, 221
grasslands and grasses, 64, 69, 83, 84,
        148, 231
    C_3, 58
    C_4, 58
    and fire, 58, 72, 83, 148, 231,
        252fn. 6
    and rainfall, 58, 84
    see also sweetveld; sourveld; *Index of
        Plants and Animals*; grass species
grass/land science, *see* pasture science
grazing, 18, 64, 65, 67, 69, 72, 81, 83,
        86–7, 109, 199, 203
    see also browsing; livestock
Groote Schuur, *see Index of Places*

heather (referring to fynbos), 39, 82
heritage, natural, 69, 117, 135, 151,
        161, 268fn. 67
homelands (Apatheid), 4, 95
hunter-gatherers, 18–19, 118
hydrology
    and fire, 86–9
    research, 71, 86, 89–90, 231
    see also catchment areas; drought

identity
    colonial, 126
    imperial, 105, 136, 138
    national, 138, 152, 161, 182, 232
    see also nationalism; South
        Africanism; xenophobia
ignitions, 16, 22, 83, 84, 101, 117–18,
        124, 130, 167, 191–3, 196, 199,
        210, 216, 233–4, 236, 239
    see also fire, causes
indicator species, 64, 107
indigenous vegetation types
    savannah, 58, 60, 65, 81, 83
    strandveld, 15
    thicket, 120
    see also forest (indigenous); fynbos;
        grasslands; renosterveld
industrial development, 128–9
    Epping Industria, 129
    Retreat West Industrial Area, 129
    see also fishing industry
informal settlements, 127, 186–7, 228
    Imizamo Yethu, 187, 228
    Masiphumelele, 187
    Red Hill, 187
insurance, 126, 213, 229
Intermountain Fire Sciences
        Laboratory, Missoula,
        Montana, 102
International Biological Programme
        (IBP), 98–9, 106
International Council of Scientific
        Unions (ICSU), 105
International Union of Biological
        Sciences, 98
International Union of Forest
        Research Institutes (IUFRO), 105
introduced plants, 78, 154, 162–83,
        168, *see also* plant introductions

invasion biology, 169, 175
invasive species, ix, 76, 106, 108, 148,
    154, 168–91, 265fn. 18, 266fn.22,
    268fn. 67
and fire, 39, 64, **108**, 178, 227–9
see also biological invasions;
    introduced plants; *Index of*
    Plants and Animals

jarrah forests, 66
Jonkershoek, **71**, *72*, 79, *92*, *102*, 107
forestry research at, 70–2, 76, 81, 86,
    93, *102*, 107–8, 175, 231

Kew, Royal Botanic Gardens at, x, xi,
    6, 30, **32**
directors of, **32**, 48, 138, 141, 170
see also Index of People: Hill, Arthur;
    Prain, David
Khoikhoi, 18–20, 23, 24, 27, 55, 230
Kimberley diamond mines, 124
King's Blockhouse, 127, 145–6,
    147, 201
Kirstenbosch
National Botanic Garden, **55**, 93,
    138, **154–6**, 173, 192–3, 205,
    218, 223
Wild Flower Protection Society
    Committee, 173–4
Kramat (Lion's Head), *225*, 274 n86

labour, 63, 111–12, 176, 185
see also fire protection measures,
    labour
land ownership (including private),
    35, 38, 40, 67, 85, 87, 91, 95, 127,
    148, 152, 159, 163, 167, 176, 200,
    205, 207–8, 211–12, 228
Landsdiens, 95
legislation (fire-related), 63, 65, 73, 78,
    91, 94, 96, 110, 113, 148, 159,
    188, 194, 205, 212, 220
Ordinance 28 of 1846 (for
    'preservation' of the 'Cape
    downs'), 39
Forest and Herbage preservation Act
    No.18 of 1859, 33–5, 230
Madras Forest Act of 1882, 39

Forest Act No.28 of 1888, 38, 63,
    230–1
Forest Act No.16 of 1913, 73
National Parks Act of 1926, 147
Forest and Veld Conservation Act
    No.13 of 1941, 79, 212
Soil Conservation Act No. 45 of
    1946, 60, 86, 153, 210
Forest Act of 1968, 94, 96, 194, 216
Mountain Catchment Areas Act 63
    of 1970, 91, **232**
Forest Amendment Bill of 1972, 94
Environment Conservation Act 73
    of 1989, 158
National Veld and Forest Fire Act
    No.101 of 1998, 113, 197
see also punishments/penalties
livestock, 18–20, 22–7, 31, 35, 37–8,
    40, 53, 67, 76, 124–5, 130–1, 166
see also animals, and fire; grazing

Macchia (used to describe Fynbos), 54
major fires, 110, 260fn. 28, 30, 31
'Great fire' (Cape Town) of 1798, 28
Great fire (Southern Cape) of
    1869, 35
Van Staden's River Valley, 13
    December 1865, 34–35
Lion's Head, January 1909, 191, 204
Devil's Peak, Christmas Day 1935,
    143–7, 170, 188, 190, 193, *206*
Cape Mountains (inland), December
    1942, 148–9, 209
Hottentots Holland mountains,
    February 1947, 211
Hout Bay, February 1948, 210–11
Camp's Bay, April 1973, 157, 191,
    194, **217–18**
Devil's Peak, February 1974,
    168, **219**
Devil's Peak, March 1982, 159, 174,
    191, 194, **220–1**
Platteklip Gorge, December 1986,
    158, **221–2**, *222*
Devil's Peak, February 1991, 159,
    168, 176, **224–5**
Hout Bay, January 1992, 226
Noordhoek, January 1992, 195, 226

major fires – *continued*
  Red Hill and Silvermine, January
    2000, 1, 133, 154, 160, 178–9, ,
    182, 187, 197, **227–9**, 233
Marloth Flower Reserve, 93
MEDECOS (Mediterranean Ecosystem
  Conferences), 8, 97, **99–101**, 104,
  175, 232
media, 1, 179
  books (covering fire in the region),
    1, 34, 53, 94, 104, 153, 154
  films, 82, 209, 216
  newspapers, ix, 73, 144, 150, 155–6,
    179, 190, 197, 205, 208,
    212, 217
  radio, 209
Mediterranean type climate
  ecosystems/regions, 10, 74, 99,
  101, 232
Mexico, 10, 62, 71
motor vehicles, *see* transport
Mountain Club, 136, 141, 152, 157,
  158, 172, 223
mountaineering, 136, 137, 138, 148,
  192, 193, 213, 218
mountain ranges, 84
  Caledon, 145, 151, 188, 253fn. 25
  Cederberg, *see* Cederberg Mountains
  Drakensberg, 94
  Franschhoek, *see Index of Places*
  Groot Winterhoek, 110
  Hex River, 110
  Hottentotsholland, *see Index of
    Places*
  Klein River, 141
  Langeberg, 54, 96
  Matobo Hills, 137
  Olifants Hoek, 151
  Outeniqua, 30, 89, 216
  Swartberg, 54, 102
municipal departments and divisions
  Engineer's Department, 117, 129,
    136, 194, 217–19, 223
  Fire Brigade, 186, 191, 193, 195,
    198, 199, **200–1**, 202, 204,
    209–17, 226, **239**
  Lands and Forests Branch, 181, 193,
    202, 209–11, 213

Parks and Forests Division, 199, 214,
  217–21, 223–5
Parks and Gardens, 180, 191, 200,
  201, 202, 214
municipalities, ix, 38, **113**, 126, **127**,
  136, 137, **143**, 148, 151, 156, 159,
  165–7, 172, 174, 190, 205, 208,
  214–16, 227–8, 232–3, 287
  Cape Town, 44, 164, 165–7, 174,
    180, 181, 186, 188, 191–3,
    199–205, 208–9, 212, 214–20,
    223, 226–7
  Fish Hoek, 208, 212
  Simon's Town, 208, 212, 219, 227
  South Peninsula, 227–8
  Woodstock, 41
  Wynberg, 164
municipal plantations, 166–7, 191,
  202, *203*, 217

Natal Province (KwaZulu-Natal), 3–4,
  49–51, 54, 67, 84, 90, 91, 94
national heritage, *see* heritage, natural
nationalism
  nature and, 126, 232
  see also Afrikaner nationalism;
    identity; xenophobia
national parks, 1, 78, 109, 124, 141,
  147–8, 159, *160*, 162, 178, 181,
  186, 195, 206, 226–8, 235
  Bontebok National Park, 109
  Kruger National Park, 141
  see also Table Mountain National
    Park
National Parks Board, 148
  see also nature conservation
    authorities; South African
    National Parks Board
National Veld Trust, **81–3**, **150**
National Wool Growers
  Association, 90
native locations, *see* townships
natural resources, 3–4, 6, 61, 82,
  129, 150
  minerals , 126, 127, 146, 147, 164,
    219, 221
  timber, 23, 24, 35, 36, 38, 40, 44,
    52, 53, 58, 61–3, 65, 68, 77, 84,

113, 114, 126, 163–71, 181,
200, 210, 215, 219, 230, 231
nature conservation, *see* conservation,
nature
nature conservation authorities, 5, 91,
94–5, 110–11, 113, 153–4, 156–7,
176, 182, 216
CapeNature, 113, 178
Cape Province Department of
Nature Conservation, 153, 154
South African National Parks Board
(SANParks), x–xi, 159, 179,
181, 227
nature conservation NGOs, 159
False Bay Conservation Society, 218
Fynbos Forum, 177
Society for the Protection of the
Environment, 156
South African Nature
Foundation, 158
Table Mountain Preservation
Board, 153
Wildlife Protection Society, 149
WWF South Africa, 178
nature reserves, 2, 57, **78**, 108, **109**,
124, 126, 133, 134, 141, 151–2,
157, 167, 178, 184, 196, 205, 224,
227–8
Cape of Good Hope (Cape Point),
109, 124, 129, 134, 149, 156,
175, 193, 215, 228, 258fn. 11
De Hoop, 109
Helderberg, 109
Silvermine, 133, 154, 193–4,
216, 227
Upper Kirstenbosch Nature Reserve,
151, 192
New Zealand, 65, 66, 144, 168, 221
Nooitgedacht, 204
Northern Forest Fire Laboratory,
Montana, USA, 108

orchards, 2, 24, 114, 126
Ossewabrandwag, 150
Ou Kaapse Weg (road), 129, 133,
194, 227
'Our Green Heritage' campaign, **94–5**,
111, 193, 243fn. 26

outdoor recreation, 187, 191–5,
211, 220
barbecues (braais), 191, 194, 195
in forestry areas, 94, 95, 111, 211
picnics/places, 84, 101, 128, 133,
153, 163, 166, 171, 180, 187,
191–5, 205, 211, 215, 218, **224**,
227, **239**
Oxford, *see* universities; Rhodes
scholar
Forestry Institute, 105

parks and commons, *see* Cape Town,
parks and commons
pastoralism, 18
pasture, 4, 17–18, 20, 23–5, 31–2, 37,
51–2, 57, 59, 60, 64, 148
pasture science, **51–2**, **57**, 59, 60, 76,
148, 248fn. 28
American influences, 51, 52, 57,
58, 76
John William Bews and, 51, 57
Frederic Clements and, 51, 58
Smuts and, *see Index of People*: Smuts
pests, 25, 42, 53, 108–9, 150, 166
Place, sense of, 144–5, 179
plantations
as fire hazards, 40, 42, 44, 132, 144,
167–8, 181, 193, 206, 236
fires in, 35, 38, 62, 80, 84, 89, 110,
112, 132, 142–4, 165, 167,
172–3, 181, 190, 198, 205–6,
211, 214–17, 220, 224, 227,
229, 236, 239
private, 40, 42, 167
protection of, 35, **38**, 41–2, 44,
62–3, 65, **73**, 84, **89**, 97, *112*,
113, 169, 180, 202, *203*, 219,
246fn. 38, 246fn. 40
plantations (individual)
Cecilia, 36, 132, 164, 165, *166*, 167,
178, 216
Constantia, 132, 167, 203
East End, 164
Groot Constantia, 165, 216
Hout Bay (extension of Tokai), 165,
203, 215, 216
Kloof Nek, 166
Newlands Forest, 166

plantations – *continued*
  Nieuwberg, 68
  Peak (or Devil's Peak), 38, 132,
    142–5, 164, 165, *166*, 167, 169,
    205–6, 215, 216, 217
  Roeland Street, 191
  Tokai, 36, 132, 132, 164, 165, *166*,
    175, 178, *183*, 205, 211, 214,
    215, 216, 227
  Uitvlugt, 36, 131, 164, *166*, 170
  Worcester, 36
  Wynberg Rifle Range, 127, 164,
    165, 170
plant collectors / collecting, 24, 47, 48,
    57, 90, 93, 94, 136, 141, 168, 171
  see also Index of People: Bolus, Harry;
    Lückhoff, Carl; Stokoe, T.;
    Kensit, William; le Vaillant,
    François, Sparrman; Anders;
    Thunberg, Carl
plant introductions
  from Australia, ix, 39, 62, 71, 162–3,
    168–9, 179
  from California (USA), 169
  from Europe, 169
  from Mexico, 71
plant population biology, **104–6**
plants, regeneration of, 2, 43, 57, 64,
    66–7, 69, 75, 78, 80, 91, 101, 103,
    107, 143, 217, 236
  see also fire, and regeneration
    (plants); senescence
plant succession, 64, 83, 101, 231
plant transfers, 168, 265fn. 17
  see also introduced plants
population biology, *see* plant
    population biology
population (ecological concept), 105
population of Cape Peninsula, *see*
    Cape Peninsula, population; Cape
    Town, population of
prescribed burning, 72, 73, 75, 77,
    80–1, 85–9, 91, 95, 102, 103, 104,
    107, 110, 111, 152, 167, 216,
    232–3, 246fn. 40
  dangers of, 110, 152, 167
  techniques/tactics, 77–8, 91, **97**,
    213, 246fn. 40

to control invasive plants, 109,
    174–6, 182
to preserve indigenous vegetation,
    91–2, 103, 107, 246fn. 40
to promote flowering, 75
to reduce fire incidence, 65, 77, 81
to reduce soil erosion, 104
to reduce water loss, 77, 90–1, 96,
    104, 111
progress, ideas of, 56, 57, 59, 61, 117,
    129, 130, 138, 156
propaganda (regarding fire), 77, 79,
    82, 96, 154–*5*, 192–3, 208–10,
    212, 214, 216, 233
public opinion
  fires, 2, 40–2, 44, 55, 67, 73–4, 78–9,
    85, 111, 122, 128, 142–4, 150,
    152, 157, 170, 175–6, 192, 193,
    204, 208, 213, 214, 216,
    220, 233
  plantations, 40–2, 44, 69, 70, 144,
    164–5, 170–1, 193
punishments/penalties, 21, 23, 42, 63,
    96–7, 194, 205, 210, 216,
    249fn. 40

railways, *see* transport
rainfall, regions, 15, 79, 83, 84, 88,
    117, 255fn. 57
  see also climate
renosterveld, 15–19, 25, 39, 53–4, 101,
    109, 119–*21*, 125, 134, 163, 180,
    231, 236
  and fire, 25, 27, 53–4, 236
  and soils, 54, 119
  see also Index of Plants and Animals:
    renoster bush
reproduction (plants), *see* plants,
    regeneration
Republic of South Africa, departments
    and divisions, 67, 95, 114, 157,
    159, 177
  see also agriculture, South African
    Department of; Forestry
    Department Research
reservoirs, *see* water supply
resprouters (fynbos regeneration
    strategy), 75, 103, 107
Rhodes' Memorial, 146–*8*

Rhodes Scholar, 69, 73
rivers, 131
  see also Index of Places: Eerste;
  Palmiet
roads, 84, **129**, 150, 155–6, 186, 192–4
Royal family (British), 207, 210–11
Royal Indian Engineering College,
  61, 73
Royal Navy (British), 30, 32, 200, 230
Royal Society of South Africa, 53, 73,
  78, 81, 91, 151–2, 210, 232
Rural Amenities Bill (UK), 188

science and politics, 105, 106, 138,
  149, 177
Scientific Committee on Problems of
  the Environment (SCOPE),
  105, 175
seeders (fynbos regeneration strategy),
  75, 77, 91–3, 97, 101, 103, 107
self-organising maps (SOMs), 124
senescence, 120, 142
serotiny, 107, **108**, 142, **169**
slaves, 4, 21–2, 24, 26–7, 31
smokers, *see* fire, causes of
Society for the Promotion of Nature
  Reserves (UK), 78
soil conservation, 4, 51, 60, 81, 83–4,
  86–7, 91, 153, 210, 212, 213
Soil Conservation Board, 83, 87
soils
  erosion, 60, 67, 70, 73, **81–3**, 88, 97,
    107, 150, 171, 213, 229
  and fire, 27, 30, 34, 37–9, 44, 50, 52,
    56, 58, 67, 72, 76, 80, 81, 83,
    86–8, 91, 149–53, 164, 176,
    192, 210–11, 216, 229, 231
  nutrients, 54, 56, 100, 119, 231
  and vegetation, 54, 56, 100,
    119, 180
soil types
  from granite, 54, 119
  from Malmesbury slates, 54
  from sandstone, 76, 119
sourveld, 58
South Africa
  economy of, 30, 112, 129, 150, 174,
    176, 185, 209, 219, 221, 233,
    244fn. 5

political parties and figures, 59, 61,
  83, 94–5, 110, 113, 129, 137,
  149, 158, 161, 224
politics of, 61, 70, 105, 113–14, 137,
  150, 159, 177, 208–9
South African Association for the
  Advancement of Science, 48,
  50, 138
South African Forestry Association,
  80, 95
South African Forestry Company
  Limited (SAFCOL), 113, 165, 203,
  227, 228
South African Forestry Research
  Institute (SAFRI), 61, 94, 111, 176
South Africanism, 59, 138
South African National Botanical
  Institute (later SANBI), 159, 180
South African National Parks Board
  (SANParks), x–xi, 159, 179,
  181, 227
South African War, 4–5, 48, 57, 59–61,
  126, 138, 150, 163, 185, 192
South African Water Catchments
  Association, 90
sprouters, *see* resprouters
steam locomotives, *see* fire, causes
sweetveld, 58

Table Mountain, ix, 1–2, 4, 15, 22, 24,
  26–7, 32, 35, 37–8, 40–4, 47, 53,
  55, 69, 70, 76, 94, 118–19, 124–7,
  130, 131, 135–61, 164–75, 178,
  180–2, 186–8, 190, 192, 194, 198,
  200, 202, *203*, 204, 206, 208, 213,
  218, 220–4, 226, 234, 236
  cultural attitudes to, 136–7, 144,
    150, 152, 156, 161, 182,
    186, 234
  forests (indigenous), 119, 167, 203
  nature conservation, 135, 149, 156
  see also Index of Places: Table
    Mountain, parts of; Cape
    Peninsula, afforestation
Table Mountain National Park, 1, 78,
  123, 124, **159**, *160*, 162, 178, 181,
  186, 195, 227–8, 235
Table Mountain Preservation Board,
  see Nature conservation NGOs

Tall Timbers Fire Ecology conference, 256fn. 62
timber, *see* natural resources, timber
time, ecological and human, 144
Tokai Arboretum, 36, 162, 163, 179
Tokai Forestry School, 63, 73
tourism, 95, 137, 139, 153, 191–5, 215, 226
    Cape Peninsula Publicity Association, 192
    figures for, 195, 215, 226
townships, 63, 127, 185
    Crossroads, 201
    Khayelitsha, 185
    Langa, 128, 164, 207
trade, 22, 24, 30, 129
transpiration, 67, 69, 88
transport, 63, 124–5, 154–5, 192
    Cape Flats Railway, 127
    Cape Town Harbour, 129
    electric tram, 127
    infrastructure, 84, 126–30, 233
    railways and trains, 84, 191–2
    roads, *see* Ou Kaapse Weg; roads
    traffic and motor vehicles, 125, 128, 205–6, 222
    union and castle steamship line, 192
    see also cable car
treason (veld burning as), 150, 209
trees, 65, 67, 163
    and fynbos, 31, 47, 108, 146, 231
    broadleaved, 65
    introduced, 69, 70, 71, 108–9, 144, 163, 179
    see also Index of Plants and Animals

Union of South Africa, 49, 52, 57, 59, 81, 135, 138
universities (and 'academics'), 2, 50, 64, 152, 157, 178, 189
    Cambridge, 49, 52, 53, 55
    (of the) Cape of Good Hope, 50
    Cape Town, 53, 74, 93, 104, 106, 125, 148, 157, 158, 170, 173, 177, 190
    Edinburgh, 49, 52, 64
    Manchester, 52

Natal (formerly University College), 49, 50, 91
Oxford, 69, 70, 105, 109
Pretoria, 57
Stanford, 100
Stellenbosch, 63, 71, 91
University College of North Wales, 104
UCLA, 104
Witwatersrand, 49
    see also agriculture, Elsenburg College of Agriculture; Tokai Forestry School; *Index of Places*: Nancy; Tharandt
urbanisation, *see* Cape Town, urbanisation; fynbos, threats to; wildland urban interface

value/s, 177, 234
    amenity values, 152, 157, 172, 232
    economic, 95, 161
    see also Aesthetics, of the natural world
Van Riebeeck Park, 139, 173, 178, 181, 214, 223
vegetation
    classification of, 54, 117–18, 180
    see also plants; trees; grasslands; *Index of Plants and Animals*
veld burning, 4, 16, 34, 37, 43, 54–5, 57, 58, 67–8, 70, 75, 83, 141, 231
    and desiccation, 67, 71, 81, 86
    labour, 63, 110–12
    problem of, 20, 23, 30, 34, 37, 47–8, 52, 54–6, 67–8, 70, 83, 87, 148, 151, 173, 190, 209–12, 231
    reasons for, 58, 86–7, 148, 188
    season for, 58, 86, 92, 97, 107–8
vineyards, 20, 24, 27, 114, 126, 131, 132, *183*, 179, 200, 227, 229
veld, romance of, 58–60, 138
VOC, 18–20, 27–8, 82, 163
    opposition to veld burning, 175
    see also Index of People: Van der Stel; Van Riebeeck'

water cycle, 71, 88
see also transpiration
water supplies
conflicts over, 42–4, 90, 231
conservation of, 42, 62, 67, 72, 75, 81, 84, 87–9, 94–6, 106, 166, 176–8, 233
reservoirs, 42–4, 68, 127, 130, 136, 145, 164, 171, 173, 198, 226, 228
threats to, 43, 67–71, 76, 80–1, 84, 86, 151, 172, 177, 182
trees and forests and, 64, 67–70, 145, 163, 164–5, 170, 172, 181, 182, 231
see also Cape Town, water supply; desiccation; rivers; transpiration
weather
and fire, 24, 73, 78, **108**, 110, **124**, 208, 210, 214, 227–9, 233–8
humidity, 198, 214, 227
South Atlantic high pressure system, 124
temperature, 51, 102, 119, 124, 198, 210–11, 214, 227–8
rainfall, 73, 84, 88, 119, *198*, 210, 211, 216, 217, 236–7, 238
see also climate; self-organising maps; winds
wetlands, 131, 133
wilderness, 94

Wild Flower Protection Act, 188
Wild Flowers Protection Society, 141
Wildland urban interface (WUI), 1, 9, 21, 184, 196, 200, 227
wildlife, 9, 17, 19, 25, 30, 64, 89, 96, 109, 135, 153, 234, 242fn. 12
see also Index of Plants and Animals
winds, 21, 35, 38, 40, 51, 80, 119, 124, 129, 131, 132–3, 136, 146, 163, 169, 175, 193, 198–9, 214, 218, 220–3, 227, 228, 236
Berg, 35, 246fn. 40
Gales, 119, 218, 220, 223, 228
Southeaster, 21, 23, 37, 119, 124, 175, 204–8, 219–21, 224, 227, 236
wine, 20, 30, *see also* vineyards
woodcutters, 3, 63, *see also* fire, causes of
Working for Water, 176–8, 227, 233
Working on Fire, 208
World War One, 65, 126, 132, 164, 185, 203, 204
World War Two, 5, 83, 84, 98, 126, 150, 152, 165, 166, 203, 207–9
WWF South Africa, 178

xenophobia, 182

Yale Forestry School, 71
Ysterplaat Aerodrome, 125

Printed and bound by CPI Group (UK) Ltd, Croydon, CR0 4YY